Sound,
Structures,
and Their
Interaction

**Sound,
Structures,
and Their
Interaction**

Miguel C. Junger
and
David Feit

second edition

The MIT Press   Cambridge, Massachusetts, London, England

Reproduction in whole or in part is permitted for any purpose of the United
States Government.

Copyright © 1972, 1986 by the Massachusetts Institute of Technology

This book was set in Baskerville by Asco Trade Typesetting Ltd., Hong Kong,
and was printed and bound by Halliday Lithograph in the United States
of America.

Library of Congress Cataloging-in-Publication Data

Junger, Miguel C.
  Sound, structures, and their interaction.

  Includes bibliographies and index.
  1. Acoustical engineering. 2. Vibration. 3. Sound. I. Feit, David.
II. Title.
TA365.J85   1986   620.2   86-2743
ISBN 0-262-10034-7

Contents

Preface to the Second Edition   xi

1   Statement of the Problem   1
1.1   Introduction   1
1.2   Assumptions   2
1.3   Formulation of the Structural Response   3
1.4   Formulation of the Acoustic Pressure Field   7
1.5   The Integral Equation of the Structure-Fluid Interaction   8
1.6   Historical Development of Structural Acoustics   10

2   The Wave Equation and Its Elementary Solutions   16
2.1   Coupled Space and Time Dependence of Sound   16
2.2   The One-Dimensional Wave Equation for Plane Waves   16
2.3   Harmonic Time Variation   22
2.4   Steady-State Plane Waves   24
2.5   The Three-Dimensional Wave Equation   27
2.6   The Uniformly Pulsating Spherical Source   29
2.7   Specific Acoustic Impedance of Spherical Waves   31
2.8   The Pressure Field inside a Fluid-Filled Pulsating Sphere   34
2.9   An Elementary Interaction Problem: Liquid-Filled Elastic
      Waveguides   37
2.10 Cylindrical Waveguides: An Introduction to
      Two-Dimensional Pressure Fields   40

3   Applications of the Elementary Acoustic Solutions   44
3.1   Introduction   44
3.2   The Point Source   44
3.3   Rigid and "Pressure-Release" Boundaries   46
3.4   The Image Source Simulating a Rigid Boundary; Directivity   48
3.5   The "Pressure-Release" Boundary   52
3.6   Linear Arrays of Point Sources   54
3.7   Far-Field Conditions   60
3.8   Acoustic Power and Intensity   64
3.9   The Pulsating Gas Bubble   67
3.10 Dispersion and Attenuation: Sound Propagation in Bubble
      Swarms   69

4   The Pressure Field of Arbitrary Source Configurations   75
4.1   General Formulation of the Radiation Problem   75
4.2   The Free-Space Green's Function   75

4.3   The Helmholtz Integral Equation   79
4.4   The Sommerfeld Radiation Condition   82
4.5   Physical Interpretation of the Helmholtz Integral   83
4.6   Approaches to the Solution of the Helmholtz Integral
       Equation   85
4.7   Analytical Solution of the Helmholtz Integral Equation   86
4.8   Rayleigh's Formula for Planar Sources   88
4.9   The Scattered Field   89

5   **Planar Sound Radiators**   92
5.1   Source Geometry and Analytical Formulations   92
5.2   Pressure Field of the Circular Piston Using Rayleigh's
       Formulation   95
5.3   The Transform Formulation of Axisymmetric Pressure Fields   100
5.4   The Transform Solution of the Circular Piston   105
5.5   The Far-Field of Rectangular Radiators Evaluated by
       Rayleigh's Formula; the Rigid Piston   109
5.6   The Transform Solution of Rectangular Sound Radiators   112
5.7   Equivalence of Rayleigh's and the Stationary-Phase
       Formulations of the Far-Field   117
5.8   The Far-Field of Rectangular Radiators Displaying a
       Sinusoidal Acceleration Distribution   119
5.9   Physical Interpretation of This Solution; Coincidence   126
5.10   Other Standing-Wave Configurations of Rectangular
       Radiators   128
5.11   The Infinite Planar Radiator with a Sinusoidal Acceleration
       Distribution   131
5.12   The Radiating Strip of Infinite Length and Finite Width   135
5.13   Acoustic Resistance of Circular Pistons and of Surface-
       Radiating Rectangular Source Configurations with
       Sinusoidal Acceleration Distributions   139
5.14   Acoustic Resistance of Edge-Radiating Rectangular Source
       Configurations with Sinusoidal Acceleration Distributions   141
5.15   The Resistance of Corner-Radiating Rectangular Source
       Configurations with Sinusoidal Source Configurations   144

6   **Convex Sound Radiators**   151
6.1   Characteristics of Convex Boundaries   151
6.2   The Green's Function for the Spherical Radiator   154

6.3	The Pressure Field of a Spherical Radiator	157
6.4	Circular Piston and Point Sources on a Spherical Baffle	159
6.5	The Radiation Loading of Spherical Radiators	161
6.6	Concentrated Force Applied to the Acoustic Medium	164
6.7	Cylindrical Radiators with Spatially Periodic Configurations	166
6.8	Radiation Loading of Infinite Cylinders with Standing-Wave Configurations	168
6.9	Transform Formulation of the Pressure Field of Cylindrical Radiators	173
6.10	Stationary-Phase Approximation to the Far-Field of Cylindrical Radiators	175
6.11	Piston in a Cylindrical Baffle	177
6.12	Far-Field of Cylinders with Standing-Wave Configurations of Finite Axial Extent	178
6.13	Comparison of Planar and Cylindrical Standing-Wave Radiators; Specific Acoustic Resistance	181
6.14	Nodal Planes and Acoustic Intensity in a Three-Dimensional Pressure Field	184
6.15	Far-Field of Slender Bodies of Revolution	188
7	**Vibration of Beams, Plates, and Shells**	**195**
7.1	Introduction	195
7.2	Longitudinal Vibrations of an Elastic Bar	195
7.3	Flexural Vibrations in an Elastic Bar	197
7.4	Group Velocity	200
7.5	Rotatory Inertia and Transverse Shear Effects: Timoshenko Beam Equation	201
7.6	Forced Vibrations of an Infinite Elastic Beam	205
7.7	Vibrations of a Finite Elastic Beam	206
7.8	Flexural Vibrations of Thin Elastic Plates	209
7.9	Point Excitation of an Infinite Plate	210
7.10	Flexural Vibrations of Finite Elastic Plates	213
7.11	Thick-Plate Theory; Timoshenko-Mindlin Plate Theory	214
7.12	Introduction to the in Vacuo Vibration of Shells	215
7.13	Equations of Motion for Cylindrical Shells	216
7.14	Planar Vibrations of a Thin Cylindrical Shell	218
7.15	Forced Planar Vibrations of a Thin Cylindrical Shell	219
7.16	Nonplanar Vibrations of a Cylindrical Shell	222
7.17	Spherical Shells; Equations of Motion	228

7.18   Free Axisymmetric Nontorsional Vibrations of a Spherical
        Shell   230
7.19   Forced Vibrations of a Spherical Shell   231

8   Sound Radiation from Submerged Plates   235
8.1   Coincidence Frequency   235
8.2   Phase Velocity of Flexural Waves in a Submerged Plate   236
8.3   Effectively Infinite Locally Excited Plates   239
        8.3.1 The Plate Response   239
        8.3.2 The Pressure Field in Response to a Point Force   245
        8.3.3 Pressure Maximum; Effect of Structural Damping   247
8.4   Infinite Line-Driven Elastic Plate   250
        8.4.1 The Plate Response to a Line Force   252
        8.4.2 Response Green's Function for a Line-Loaded Plate   253
        8.4.3 Scattering of a Flexural Wave by a Plate Discontinuity   255
8.5   Pressure and Power Radiated by an Infinite Plate Driven
        by Distributed Loads   257
        8.5.1 Transform Solutions   257
        8.5.2 Examples of Load Distributions   258
8.6   Power Radiated by Plates   261
8.7   Sound Radiation from Rectangular Plates   264
        8.7.1 Plate Response   264
        8.7.2 Far-Field Sound Pressure   268
8.8   Low-impedance Layers   272

**9   Sound Radiation by Shells at Low and Middle
        Frequencies**   279
9.1   Introduction   279
9.2   Characteristic Equation of the Submerged Spherical Shell   280
9.3   Natural Frequencies, Modal Configurations, and Radiation
        Damping of Submerged Spherical Shells   281
9.4   Response and Pressure Field of Point-Excited Submerged
        Spherical Shells   284
9.5   Normal Modes of Fluid-Filled Spherical Shells   287
9.6   Normal Modes of the Infinite, Submerged Cylindrical Shell   289
9.7   Natural Frequencies of Infinite Submerged Cylindrical Shells   290
9.8   Intermodal Fluid Coupling in the Submerged Finite
        Cylindrical Shell   294

9.9   Approximations to the Radiation Loading of Finite
        Cylindrical Shells   298
9.10   The Far-Field of Point-Excited Cylindrical Shells   300
        9.10.1 The Simply Supported Shell   300
        9.10.2 Interpretation of the Far-Field Results   302
        9.10.3 Low-Frequency Sound Radiation by Free-Free
               Cylindrical Shells   304
        9.10.4 Low-Frequency Sound Field of Freely Floating
               Noncylindrical Shells of Revolution   306
9.11   The Effect of Structural Damping on Sound Radiation   306
9.12   Uncoupled Modes in a Submerged Structure   309

10   Scattering of Sound by Rigid Boundaries   313
10.1   Scattering and Echo Formation   313
10.2   Formulation of the Scattering Problem   314
10.3   The Infinite Plane Reflector   317
10.4   The Spherical Scatterer   318
10.5   The Infinite Cylindrical Scatterer   321
10.6   The Cylindrical Scatterer of Finite Length   323
10.7   Asymptotic Formulation of the Scattered Field of Slender
        Bodies of Revolution   324
10.8   Nature of the Kirchhoff Approximation; Surface Pressure 329
10.9   Kirchhoff Scattering from Cylinders and Spheres; Fresnel
        Zones   330
10.10   Reflection from a Rectangular Baffle   334
10.11   The Helmholtz Reciprocity Principle   338

11   Elastic Scatterers and Waveguides   342
11.1   The Effect of Scatterer Elasticity   342
11.2   Sound Reflection by an Infinite Elastic Plate   344
11.3   Sound Transmission through an Infinite Elastic Plate   346
11.4   Sound Transmission through Finite Plates; Reciprocity   348
11.5   The Spherical Shell as an Acoustic Scatterer   352
11.6   The Scattering Action of the "Pressure-Release" Sphere 358
11.7   Structure-Acoustic Medium Reciprocity Relation Illustrated
        for the Spherical Shell   360
11.8   Rayleigh Scattering by Compressible, Movable Spheres   362
11.9   Extension of the Rayleigh Scattering Formulation to Slender
        Bodies of Revolution   364

11.10 The Cylindrical Shell as a Scatterer 369
11.11 Sound Sources Located on an Elastic Baffle 373
    11.11.1 Sound Source Located on a Planar Elastic Baffle 373
    11.11.2 Sources on Elastic Spherical and Cylindrical Baffles 376
11.12 Sound Propagation in Fluid-Filled Elastic Waveguides 378
    11.12.1 Dispersion Relations for Cylindrical Waveguides 378
    11.12.2 Elastic Cylindrical Hoses and Shells as Waveguides 382
    11.12.3 Modal Amplitudes and Impedance in Waveguides 384

12 High-Frequency Formulation of Acoustic and Structural Vibration Problems 388
12.1 Watson's Creeping Wave Formulation of the Diffracted Field 388
12.2 Point Source on a Rigid Spherical Baffle 388
12.3 High-Frequency Response of a Spherical Shell 397
12.4 The Point-Excited Spherical Shell in Vacuo 398
12.5 The Submerged Spherical Shell 405
12.6 The Point Source on a Cylindrical Surface 408
12.7 Cylindrical Shells 418
    12.7.1 Cylindrical Shell in Vacuo 424
12.8 Pressure Radiated by a Point-Excited Cylindrical Shell 429
12.9 Spherical Shell Radiated Field 433

Glossary 437

Index 445

Preface to the Second Edition

Like the 1972 (first) edition, this text is intended for the applied physicist and engineer acquainted with the mathematical tools found in graduate textbooks. A familiarity with elementary theory of vibrations and strength of materials is desirable. No prior acquaintance with acoustics is expected from the reader.

The primary difference between this book and more familiar texts is the space assigned to the effect of radiation loading exerted by the ambient fluid on the vibrations of elastic structures and the resulting modification of radiated and scattered pressures. Unlike the standard modern acoustic texts, this book returns to the tradition of Rayleigh's *Theory of Sound* by covering the vibrations of elastic shells. The presentation of plate vibrations includes the Timoshenko-Mindlin correction required to generate meaningful high-frequency results. The chapters dealing with acoustics are self-contained. They address primarily sound radiation and scattering, to the exclusion of numerical solutions, statistical techniques, and consequently flow-related phenomena and other broad-band excitations.

Even though the original title has been retained as being still appropriate to the material covered, there are substantial differences from the 1972 edition. To retain the manageable size of the original edition, the theories of plate and shell vibrations have been combined into a single chapter and the chapter dealing with acoustic transients has been dropped. There is an increased emphasis on asymptotic solutions. Acoustics in the first edition was limited to rigorously tractable geometries: the plane, the cylinder, and the sphere. Had we wanted to discuss radiation and scattering by slender bodies of revolution, we would have had to use prolate spheroidal wave harmonics where applicable, and for nonspheroidal geometries we would have referred readers to papers using numerical methods. These configurations are covered in this new edition, but in preference to rigorous formulations, the pressure fields are computed asymptotically by means of simple mathematical models that are solvable in terms of familiar cylinder functions. The chapter on sound radiation by submerged plates has been extensively rewritten to incorporate some of the new results in this area developed over the past decade—in particular, a closer examination of the near-field, the effect of stiffeners and compliant layers, and the relation of load distribution to far-field directivity and acoustic power. The more concise analytical treatment of sound radiation by simply supported cylindrical shells has been supplemented with a study of low-frequency radiation by free-floating, not necessarily cylindrical shells of revolution. Other new subjects covered in this second edition are the acoustics of bubble swarms, the propagation of

sound waves in elastic pipes, and the insertion loss of finite panels. Both Rayleigh and Kirchhoff scattering receive more extensive treatment. Sound radiation by a source placed in a planar elastic baffle is used to illustrate the reciprocity principle, which is then used to analyze the far-field of sources located on elastic spherical and cylindrical baffles. The introductory chapter has been supplemented with a historical review of the development of structural acoustics.

Except for the extensive bibliography associated with that historical section, references listed at the end of each chapter are intended to supplement the material in this text either by providing the point of departure for the analysis presented here or by extending the analysis to areas not covered. Since, with the exception of the mathematical foundation, the development is relatively self-contained (the required knowledge of acoustics and theory of structures being derived or restated in the text), the references cited at the outset of an analysis are primarily mathematical in nature, thus sparing the reader the task of correlating the notations used in different texts on acoustics and plate and shell theory.

While our main goal is to present the underlying theories, we illustrate their application by means of problems selected for their practical interest. We hope to provide readers with the analytical tools for studying practical problems of interest to them. If an apology is needed for not having included those particular problems, we gladly accept the reproach that Shakespeare has Hamlet address to Horatio: "There are more things on heaven and earth, Horatio, than are dreamt of in your philosophy."

We are happy to acknowledge the moral and financial support of individuals and agencies within the U.S. Navy that enabled us to generate much of the material that is not part of the acoustician's stock in trade. Finally, it is with pleasure that we acknowledge the consistent helpfulness and patience displayed by our respective coworkers: J. M. Garrelick, J. E. Cole, III, and Rudolph Martinez at Cambridge Acoustical Associates, Inc., and numerous staff members at the David W. Taylor Naval Ship Research and Development Center.

Miguel C. Junger, Cambridge, Massachusetts
David Feit, Bethesda, Maryland

September 1985

Sound,
Structures,
and Their
Interaction

1.1 Introduction

The physical manifestation of sound is a time-dependent pressure fluctuation around the static pressure in a compressible fluid, such as air or water. These pressure fluctuations can be generated by a vibrating elastic structure—for example, a bell in contact with the atmosphere. The fluctuating pressure on the surface of the structure constitutes the *radiation loading*. To obtain the resultant load, this distributed loading is combined with the prescribed forces that are the prime sources of structural vibrations. Because of the low density of air compared to structural materials, radiation loading exerted by the atmosphere is generally small enough to have a negligible effect on structural vibrations. Radiation loading modifies the motion of a structure vibrating in the atmosphere only under unusual circumstances—for example, when a volume of air in contact with the structure is confined in a small enclosure, or when the structure is exceptionally light, such as a loudspeaker cone. Consequently, in most situations, the dynamic response of a structure in the atmosphere, excited by prescribed driving forces, can be determined as though the structure were vibrating in a vacuum. Subsequently the uncoupled acoustical problem of evaluating the pressure field generated by a velocity distribution prescribed over the boundary of the acoustic medium can be solved independently.

Until World War I, acoustics was almost exclusively concerned with air-borne sound. It was during that period that Rayleigh published the first modern text on acoustics, his *Theory of Sound* (1894). Being concerned primarily with airborne sound, he divided his work into two relatively independent parts, the first dealing with the vibrations of elastic structures in vacuo, and the second with various aspects of acoustics proper. This division of material set a precedent followed in most subsequent theoretical textbooks on acoustics.

A structure vibrating in contact with a fluid of comparable density experiences radiation loading comparable to its inertial and elastic forces. Radiation loading thus modifies the forces acting on the structure and, since these acoustic pressures depend on the velocity, a "feedback coupling" between the fluid and structure exists. Hence the elastic and acoustical problems must be solved simultaneously. This interaction analysis is confined here to fluids not experiencing steady flow, thus precluding instabilities arising from feedback coupling. These situations are dealt with in the extensive hydro- and aeroelastic literature.

So far we have discussed the sound pressures generated by a vibrating boundary. Acoustics is also concerned with the distortion of a sound field

by the presence of an object that reflects or scatters the incident sound waves. A second feature characteristic of a dense acoustic medium is that a sound scatterer or reflector can generally not be considered as rigid—the precise meaning of this word will be defined later in the chapter. The scattering action of a submerged structure is thus modified by the dynamic elastic response of the structure to the incident wave.

Section 1.5 introduces the reader to the coupled equations that govern the dynamic response of a structure vibrating in a dense fluid, for excitations in the form of either driving forces or sound waves incident upon the structure. The acoustical and structural terms of these equations are formulated, respectively, in terms of Green's functions[1] and mechanical influence coefficients.[2] The fundamental concepts of acoustics are presented in chapters 2–6. Chapter 7 deals with *in vacuo* vibrations of plates and shells. Chapters 8 and 9 explore the vibrations and associated sound radiations of, respectively, plates and shells in an acoustic medium. Scattering by rigid boundaries is the subject of chapter 10. Chapter 11 is concerned with the effect of elasticity in scatterers and waveguides, and with the insertion loss of elastic plates. Chapter 12 presents Watson's creeping wave formulation of diffracted fields and extends his technique to high-frequency vibrations of elastic shells.

1.2 Assumptions

To define the range of validity of the theories to be developed, the underlying assumptions will be stated. Both the fluid and the structure are assumed to obey linear constitutive equations. For the structure this means that each component of stress is a linear function of the strain components and, for the fluid, that pressure deviations from the hydrostatic pressure are a linear function of fractional density change (*condensation*) and hence of volume strain (*dilation*). The equations of motion of these media are thus restricted to small signals. Specifically, time-dependent functions represent small fractional deviations from the time-independent equilibrium condition. For example, the oscillatory acoustic pressure $p(t)$ in a liquid is negligible compared with the hydrostatic pressure P_s. This results in dropping terms of order $(p/P_s)^2$ and higher.[3] Linear theory can predict approximately the onset of a nonlinear phenomenon such as cavitation but does not permit an analysis of the phenomenon. Strains are similarly assumed to be negligible compared with unity, in both the structure and the fluid, whose response is thus linearly related to the excitation.

Frictional dissipation of energy is assumed to take place in the solid medium, since this is necessary in dealing with real structures and in arriving

at meaningful solutions even in highly idealized systems excited at resonance. The dissipating mechanism will be simulated by means of a small, structural loss factor associated with a stress component proportional to the strain rate, which is combined with the usual component proportional to strain. The nature of this common approximation, as compared to more sophisticated theories of solid damping, is discussed in textbooks on elastic waves.[4]

Solid and fluid media are assumed to be homogeneous. For the latter, this precludes refraction and volume reverberation. These effects are dealt with extensively in various textbooks on sound propagation in a notoriously inhomogeneous acoustic medium—the ocean.[5] This restriction to a homogeneous medium would make any calculation of the pressure at long ranges from the source unrealistic. Rather, the present theory is intended to provide the pressure at a relatively short range, a result that can then be used, without further consideration of the source, to compute the long-range pressure by means of a realistic mathematical model of the atmosphere or the ocean.

The medium is taken to be initially at rest. The analysis of harmonic, steady-state situations is thus a limiting case of a process initiated sufficiently long ago to ensure that the initial transients have been extinguished, even for vanishingly small dissipation. The acoustic fluid can, for our purposes, be assumed inviscid. In the ocean or the atmosphere this does not restrict the validity of the theoretical results to ranges shorter than those already imposed by the neglect of refraction and volume reverberation. An acoustic medium thus idealized can exert only normal loads on a structure. Conversely, only the normal displacement component of the structural response is directly coupled to a fluid embodying no viscosity.

This chapter presents a formulation of the integral equation representative of structure-fluid interaction phenomena. The purpose of this derivation at this early stage is to clarify the nature of the uncoupled acoustic and elastic solutions that are combined to formulate the interaction problem. The construction of the interaction integral equation illustrates the coupling between the fluid and the structure, but its direct solution will always be circumvented in this book by suitable integral transform or series representations.

1.3 Formulation of the Structural Response
Consider an elastic structure characterized by one dimension that is small compared with its other dimensions—specifically, a plate or shell. This

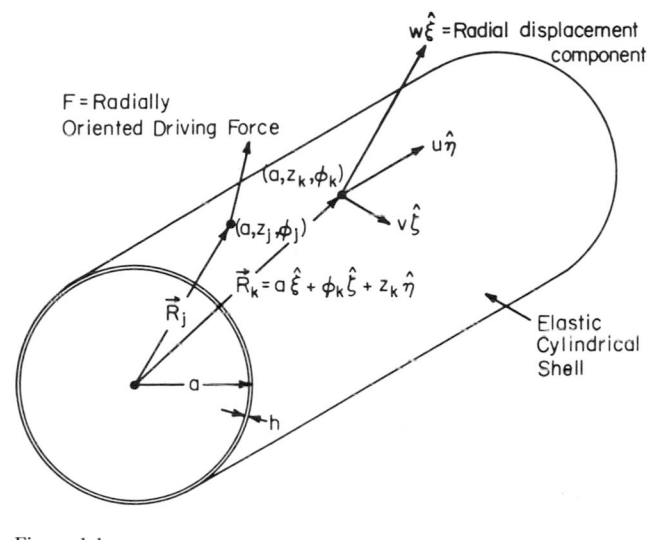

Figure 1.1
Location of drive and field points in the uncoupled structural problem, illustrated for a cylindrical shell ($\vec{R} \equiv \mathbf{R}$).

structure is excited in vacuo by a time-dependent driving force applied in a direction normal to the mean surface, parallel to the small dimension of the structure. In problems analyzed here, the time dependence of this force is harmonic, the exciting force varying periodically with a constant circular frequency ω. In a commonly used notation, which is explored in detail in the next chapter, this harmonic time dependence is expressed in terms of the complex exponential $\exp(-i\omega t)$, where t stands for time. The drive-point location where the force is applied is identified by a position vector $\mathbf{R}_j$, which is a function of the drive-point coordinates. For the cylindrical shell illustrated in figure 1.1, $\mathbf{R}_j$ can be expressed in terms of the cylindrical coordinates $r = a$, $\phi = \phi_j$, and $z = z_j$. In the notation adopted here, the explicit expression for the driving force, in terms of its point of application and time dependence, is

$$F(\mathbf{R}_j, t) = F(\mathbf{R}_j) e^{-i\omega t}.$$

The solution of the equation of motion of the structure vibrating in vacuo predicts the vector displacement of the structure at a field point $\mathbf{R}_k$:

$$\mathbf{d}(\mathbf{R}_k, t) = [w(\mathbf{R}_k)\hat{\xi} + v(\mathbf{R}_k)\hat{\zeta} + u(\mathbf{R}_k)\hat{\eta}]e^{-i\omega t}. \tag{1.1}$$

We shall here retrict our attention to the displacement component $w(\mathbf{R}_k)$ normal to the surface of the structure. The formal solution of the elastic problem can be expressed in terms of an *influence coefficient a* $(\mathbf{R}_k|\mathbf{R}_j)$ defined as the displacement component normal to the plate or shell surface, at field point $\mathbf{R}_k$, in response to a force of unit amplitude applied normally to the plate surface at drive point $\mathbf{R}_j$. The influence coefficient can be computed both for static and time-dependent loads. The response of the structure to a harmonically varying force applied normally to the structure at drive point $\mathbf{R}_j$ can finally be expressed in terms of the influence coefficient as

$$w(\mathbf{R}_k)e^{-i\omega t} = a(\mathbf{R}_k|\mathbf{R}_j)F(\mathbf{R}_j)e^{-i\omega t}. \tag{1.2a}$$

The ω dependence of the influence coefficient is not explicitly displayed. The notation used to denote its dependence on the location of the drive and field points is intended to emphasize the *reciprocity principle*,[2] whereby

$$a(\mathbf{R}_k|\mathbf{R}_j) = a(\mathbf{R}_j|\mathbf{R}_k).$$

For a given structure a force applied at $\mathbf{R}_j$ produces a response at $\mathbf{R}_k$ that equals the response at $\mathbf{R}_j$ when this same force is applied at $\mathbf{R}_k$. A similar reciprocity relation, which will be explored in some detail for acoustical systems, applies to Green's functions.[1] When these reciprocity relations are specialized to homogeneous systems of infinite extent in one, two, or three dimensions, a stronger statement is obtained that makes the Green's function or influence coefficient a function of the single scalar coordinate $|\mathbf{R}_k - \mathbf{R}_j|$, that is, of the distance from the field point to the drive point.

Since the same exponential coefficient multiplies both sides of equation (1.2a), this coefficient can be dropped. Throughout this work, equations relating harmonic functions of time will be thus abbreviated, except when the retention of the exponential adds to the clarity of the development. Using these conventions, the response of the structure in equation (1.2a) is restated more concisely as

$$w(\mathbf{R}_k) = a(\mathbf{R}_k|\mathbf{R}_j)F(\mathbf{R}_j). \tag{1.2b}$$

Because we shall be concerned with the velocity rather than the displacement response of the structure, it is convenient to express this formal

solution in terms of a *mechanical mobility* function $\Upsilon(\mathbf{R}_k|\mathbf{R}_j)$ relating the normal velocity at a field point $\mathbf{R}_k$ to a time-harmonic force applied normally to the structure at drive point $\mathbf{R}_j$:

$$\dot{w}(\mathbf{R}_k) = \Upsilon(\mathbf{R}_k|\mathbf{R}_j) F(\mathbf{R}_j). \tag{1.3}$$

When the field and drive points coincide ($\mathbf{R}_k = \mathbf{R}_j$), Υ is referred to as the *drive-point mobility*. When the field and drive points are restricted not to coincide ($\mathbf{R}_k \neq \mathbf{R}_j$), Υ is the *transfer mobility*. Again because of the reciprocity between the locations of the driving force and of the response,

$$\Upsilon(\mathbf{R}_k|\mathbf{R}_j) = \Upsilon(\mathbf{R}_j|\mathbf{R}_k).$$

The formulation in equation (1.3) can be generalized by considering *vector forces* with components along three orthogonal axes. The mobilities relating a tangential force component to the normal velocity component are, of course, different from, and generally smaller than, the mobilities relating a normal force to a normal displacement. Since the extension to vector exciting forces is straightforward, once the analysis has been performed for forces restricted to a component normal to the surface of the structure, we shall confine our attention to this specialized case.

The structural response to simultaneous normally oriented exciting forces applied at N different points of the structure is obtained by superposition of the responses to the individual forces:

$$\dot{w}(\mathbf{R}_k) = \sum_{j=1}^{N} F(\mathbf{R}_j)\, \Upsilon(\mathbf{R}_k|\mathbf{R}_j). \tag{1.4}$$

If the excitation is in the form of a distributed load, such as the sound pressure, the formal solution becomes a surface integral over the structure. An area element of the structure is, like the drive point of a concentrated force, identified by a position vector, as $dS(\mathbf{R}_j)$. When the distributed force is the radiation loading, the surface of integration coincides with the structure-fluid interface. Taking the pressure as positive in a direction opposite to the response $\dot{w}$, the formal solution of the structure now becomes

$$\dot{w}(\mathbf{R}_k) = -\int_S p(\mathbf{R}_j)\, \Upsilon(\mathbf{R}_k|\mathbf{R}_j)\, dS(\mathbf{R}_j) + \sum_{j=1}^{N} F(\mathbf{R}_j)\, \Upsilon(\mathbf{R}_k|\mathbf{R}_j). \tag{1.5}$$

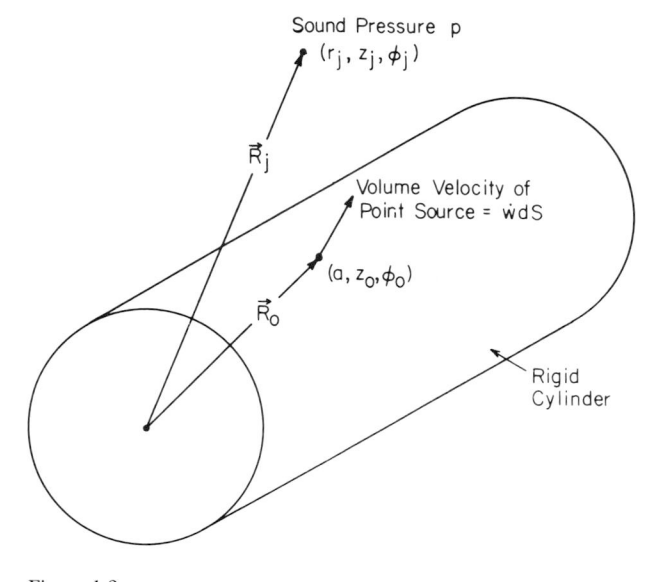

Figure 1.2
Location of source and field points in the uncoupled acoustical problem, illustrated for a cylindrical boundary.

Equation (1.5) represents the formal solution of the in vacuo structural problem, uncoupled from the acoustical problem, specialized to normal velocity components, and restricted to exciting forces oriented normally to the structure.

1.4 Formulation of the Acoustic Pressure Field

The solution of the uncoupled *acoustical problem* consists in finding the pressure p at a field point $\mathbf{R}_j$ in response to an acoustic source at a point $\mathbf{R}_0$ (figure 1.2). For our present purpose, the source locations can be restricted to lie on a *rigid* boundary coinciding with the interface of the structure with the acoustic medium. A rigid surface is defined here as having vanishing mechanical mobilities. The acoustical equivalent of the concentrated driving forces defined in the preceding section is a *source* located on a rigid boundary at a point identified by a position vector $\mathbf{R}_0$. Restricting ourselves once again to harmonic time dependence, a point source is characterized by its *strength*, defined as its volume velocity:

$$\dot{Q}(\mathbf{R}_0, t) \equiv \dot{w}(\mathbf{R}_0)\, dS(\mathbf{R}_0)\, e^{-i\omega t}.$$

The formal solution is conveniently expressed in terms of an acoustical Green's function $G(\mathbf{R}_j|\mathbf{R}_0)$ multiplying the source strength:

$$p(\mathbf{R}_j, t) \equiv p(\mathbf{R}_j)e^{-i\omega t} = G(\mathbf{R}_j|\mathbf{R}_0)\dot{w}(\mathbf{R}_0)\,dS(\mathbf{R}_0)e^{-i\omega t}. \tag{1.6}$$

The resultant pressure produced by N sources is expressible as a sum, in the form of equation (1.4). The limit of a continuous distribution of infinitesimal sources is tantamount to a prescribed velocity distribution over the boundary. Suppressing the harmonic time dependence, the formal solution becomes a surface integral over the boundary

$$p(\mathbf{R}_j) = \int_S \dot{w}(\mathbf{R}_0)\,G(\mathbf{R}_j|\mathbf{R}_0)\,dS(\mathbf{R}_0). \tag{1.7}$$

In the uncoupled acoustical problem, the velocity distribution $\dot{w}$ prescribed over the radiating surface is by definition independent of the radiation loading, just as the driving forces prescribed in the structural problem are independent of the response of the structure.

1.5 The Integral Equation of the Structure-Fluid Interaction
Having stated the formal solutions of the uncoupled structural and acoustical problems, the solution of the interaction problem can now be formulated. Consider an elastic structure excited by N forces F_j, immersed in an acoustic fluid and hence exposed to radiation loading in the form of equation (1.7), with the field points $\mathbf{R}_j$ located on the structure-fluid interface. The acoustic surface pressure now contributes to the oscillatory forces applied to the structure and must therefore be accounted for when computing the dynamic

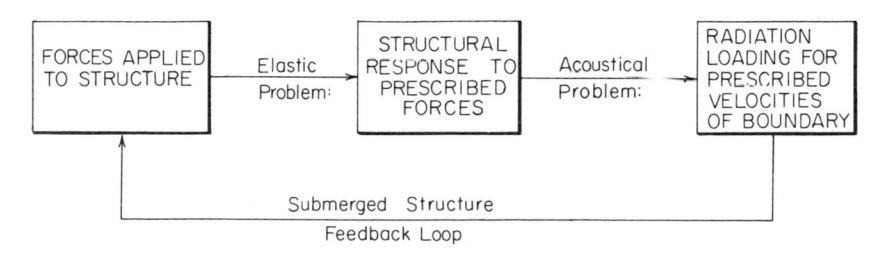

Figure 1.3
Diagram illustrating the dynamic interaction of an elastic structure and a dense ambient fluid of density commensurate with that of the structure. [Reproduced from Greenspon[6]]

response of the structure (figure 1.3).[6] When this acoustic surface pressure is substituted for $p(\mathbf{R}_j)$ in equation (1.5), the dynamic response of the submerged structure becomes

$$
\dot{w}(\mathbf{R}_k) = -\int_S \varUpsilon(\mathbf{R}_k|\mathbf{R}_j) \left[\int_S \dot{w}(\mathbf{R}_0) G(\mathbf{R}_j|\mathbf{R}_0) \, dS(\mathbf{R}_0) \right] dS(\mathbf{R}_j)
$$
$$
+ \sum_{j=1}^{N} F(\mathbf{R}_j) \varUpsilon(\mathbf{R}_k|\mathbf{R}_j).
$$

(1.8)

Since $\dot{w}(\mathbf{R}_0)$ is unknown, the response $\dot{w}(\mathbf{R}_k)$ of the structure is the solution of an integral equation. If the acoustic medium had not been restricted to vanishingly small viscosity, it would also exert shearing loads on the structures, these stresses being proportional to the tangential velocity component of the structure.

The integral equation, equation (1.8), can be written in the standard form of an *inhomogeneous Fredholm equation of the second kind*,[7]

$$
\dot{w}(\mathbf{R}_k) = \int_S K(\mathbf{R}_k|\mathbf{R}_0) \dot{w}(\mathbf{R}_0) \, dS(\mathbf{R}_0) + f(\mathbf{R}_k),
$$

(1.9)

whose kernel is

$$
K(\mathbf{R}_k|\mathbf{R}_0) = -\int_S \varUpsilon(\mathbf{R}_k|\mathbf{R}_j) G(\mathbf{R}_j|\mathbf{R}_0) \, dS(\mathbf{R}_j).
$$

(1.10)

The inhomogeneous term is

$$
f(\mathbf{R}_k) = \sum_{j=1}^{N} F(\mathbf{R}_j) \varUpsilon(\mathbf{R}_k|\mathbf{R}_j).
$$

(1.11)

The discussion up to now has dealt with an elastic structure excited by prescribed concentrated forces. Another situation of interest is a structure exposed to an incident sound wave. The dynamic response of such a structure is the solution of an integral equation of the form of equation (1.9), whose inhomogeneous term embodies the pressure $p(\mathbf{R}_j)$ on the surface of a *rigid* target whose configuration coincides with the outer surface of the actual, elastic target. From equation (1.5)

$$f(\mathbf{R}_k) = -\int_S p(\mathbf{R}_j) \, \Upsilon(\mathbf{R}_k|\mathbf{R}_j) \, dS(\mathbf{R}_j). \tag{1.12}$$

Various approaches to the solution of equation (1.9) are described in the literature.[7] Rather than applying these powerful, general techniques, we shall, as already mentioned, evolve simpler specialized techniques for specific structural configurations, thus circumventing the need for solving the actual interaction integral equation.

Before proceeding with the mathematical development, it appears useful to describe how these problems were initially attacked.

1.6 Historical Development of Structural Acoustics

Even though Rayleigh was primarily interested in situations where radiation loading has a negligible effect on the vibrations of the radiating surface, he formulated the equation of motion of a rigid spring piston radiating into an acoustic fluid. He constructed an equivalent single-degree-of-freedom system vibrating in vacuo, endowed with damping and an increased mass.[8] The earliest interaction solutions formulated for multimodal elastic structures made use of the Laplace approximation, whereby the ambient liquid is deemed incompressible, damping associated with sound radiation being ignored (see section 4.6). In the wake of World War I, in the United Kingdom, Lamb[9] and McLachlan,[10] prompted by the advent of the underwater echo-ranging technique, which came to be called sonar, analyzed the vibrations of submerged circular plates. In the United States, the effect of radiation loading on hull vibrations was studied by Lewis.[11]

The Laplace approximation fails when the acoustic wavelength is comparable to or smaller than the extent of the sound-radiating surface elements that vibrate in phase. The transition from the low- to the high-frequency limit of the radiation loading by an elementary sound source in the form of a reciprocating rigid body was described with unsurpassed clarity over a century ago by Stokes.[12] His statement has been generalized here to include waterborne sound by substituting "fluid" for "gas":

When a body is slowly moved to and fro in any fluid, the fluid behaves almost exactly like an incompressible fluid, and there is merely a local reciprocating motion of the fluid from the anterior to the posterior regions, and back again in the opposite phase of the body's motion, in which the region that had been anterior becomes posterior. If the rate of alternation of the body's motion be taken greater and greater, or, in other words, the periodic time less and less, the condensation and rarefaction of the fluid, which in the first instance was utterly insensible, presently becomes sensible,

and sound waves (or waves of the same nature in case the periodic time be beyond the limits of audibility) are produced, and exist along with the reciprocating flow. As the periodic time is diminished, more and more of the encroachment of the vibrating body on the fluid goes to produce a true sound wave, less and less a mere local reciprocating flow. For a given periodic time, and given size, form, and mode of vibration of the vibrating body, the fluid behaves so much the more nearly like an incompressible fluid as the velocity of propagation of sound in it is greater.

Stokes's description is directly applicable to sound radiation by elastic structures if the expression "anterior" and "posterior regions" is interpreted in terms of adjoining regions of opposite phase on, e.g., a vibrating plate or shell.

Full-fledged interaction problems were first studied during World War II at MIT. Lax[13] refined Lamb's incompressible solution of the clamped circular plate by introducing resistive radiation loading and using an iterative technique to determine the effect of reactive radiation loading on mode shapes. Also at MIT, Fay[14] analyzed the first analytically tractable interaction problem: sound reflection by an effectively infinite elastic plate. The predicted nonspecular reflection was observed by Finney.[15] Also during World War II, in Germany Cremer[16] recognized that the dispersive nature of flexural waves in plates results in a low-frequency region of predominantly reactive radiation loading, and in a high-frequency region of effective sound radiation. The two frequency regions are separated by a coincidence or critical frequency (see chapter 8).

Two other geometries for which the interaction between structural vibrations and the radiation loading exerted by an ambient acoustic fluid is analytically tractable are elastic spheres and infinite cylinders, both solid and in the form of shells. The corresponding scattering and radiation solutions were formulated in Faran's[17] and Junger's[18,19] doctoral theses at Harvard in 1951. The effect of resonances of solid scatterers predicted by Faran's scattering analysis was verified experimentally.[17] Meanwhile, Bleich and Baron at Columbia elucidated the 3-dimensional nature of the vibrations of submerged cylindrical shells.[20] The explicit asymptotic formulation of the far-field pressure in terms of a steepest-descent or stationary-phase evaluation of the integral expression, first introduced for a source on a rigid cylinder,[21] was applied by Bleich to point-excited cylindrical shells.[22,23]

In the Soviet Union, Liamshev[24] analyzed the scattering of sound waves obliquely incident on a cylindrical shell. He verified the importance of coincidence or trace matching between the incident wave and the phase

velocities of free elastic waves in the shell. Warburton.[25] in the United Kingdom, extended the analysis of submerged cylindrical shells by accounting for an acoustic fluid inside as well as outside the shell.

Finally, returning to the planar plates that had been the subject of the earliest interaction studies, Feit[26] formulated the far-field of a point-excited effectively infinite plate in terms of the steepest-descent approximation Bleich had applied to the equivalent cylindrical shell problem (chapter 8).

Room acoustics is a familiar example of the statistical formulation of the response of a multimodal system. Heckl used the statistical properties of the modes of plates[27] and cylindrical shells[28] to evaluate the acoustic power radiated by these structures in response to point excitation. A statistical formulation, which came to be known as "statistical energy analysis," was developed at Bolt Beranek and Newman by Lyon,[29] Maidanik, [29] and Smith.[30,31] It found many applications in interaction problems. For this purpose, modes of vibration of structures were classified according to their effectiveness as sound radiators in terms of the sub- or supersonic nature of the structural standing-wave field characterizing the modal configuration.

The sound field associated with a few source or scatterer geometries, such as a sphere, can be formulated rigorously as a wave-harmonic series (chapters 6 and 10).[6] Unfortunately, if the characteristic dimension of this boundary is large in terms of acoustic wavelengths, the series solution converges slowly. At the beginning of this century, this limitation acquired practical importance in its relevance to radio wave transmission across the Atlantic. In 1919, the British mathematician Watson formulated a technique elaborated by Sommerfeld whereby the cumbersome wave-harmonic series is transformed into a series whose convergence accelerates with decreasing wavelength. In this formulation, the diffracted field is described in terms of waves that are attenuated as they travel over the convex surface at a speed less than the free-space wave velocity. Franz (1954), who extended Watson's formulation, coined the expression "creeping waves" to describe these interface waves. Watson's formulation is readily adapted to the diffraction of sound waves over spherical and cylindrical boundaries (chapter 12), thus establishing a link between wave-harmonic solutions and the description of diffracted fields in terms of rays, an early approach whose most recent and successful embodiment is Keller's Geometric Theory of Diffraction.[32]

The extension of the creeping wave concept to elastic scattering was initiated at the University of Texas by Horton and his coworkers.[33] Starting in the late sixties, Überall and his students at Catholic University systemat-

ically applied this analytical technique to a wide variety of elastic scatterers of increasing complexity,[34] even including layered scatteres.[35] The physical reality of creeping waves was illustrated by means of schlieren photography by Neubauer and his coworkers at the U.S. Naval Research Laboratory.[36] A historical review of more specialized applications of Watson's technique will be found in the introductory section of chapter 12.

This brief historical section has dwelled on early solutions to simple, analytically tractable situations. For lack of space, we cannot describe subsequent developments—in particular, analytical and computer solutions to more specialized problems.[37]

References

1. For a definition, see, for example, A. Sommerfeld, *Partial Differential Equations in Physics* (New York: Academic Press, 1949), pp. 49–51.

2. Y. C. Fung, *Foundations of Solid Mechanics* (Englewood Cliffs, N.J.: Prentice-Hall, 1965), pp. 4–6.

3. A systematic formulation of the approximations inherent in the standard, small-signal equations of acoustics is given by F. V. Hunt, "Propagation of Sound in Fulids," in D. E. Gray, ed., *American Institute of Physics Handbook* (New York: McGraw-Hill, 1957), pp. 3-25–3-56.

4. H. Kolsky, *Stress Waves in Solids* (New York: Dover, 1963), chapter V.

5. See, for example, I. Tolstoy and C. S. Clay, *Ocean Acoustics* (New York: McGraw-Hill, 1966).

6. J. E. Greenspon, ed., *Fluid-Solid Interaction* (New York: American Society of Mechanical Engineers, 1967), p. 82.

7. See, for example, P. M. Morse and H. Feshbach, *Methods of Theoretical Physics* (New York: McGraw-Hill, 1953), pp. 949ff.

8. J. W. Strutt Lord Rayleigh, *The Theory of Sound* (New York: Dover, 1945), 2nd ed., vol. II, p. 169.

9. H. Lamb, "On the Vibrations of an Elastic Plate in Contact with Water," *Proc. Roy. Soc.* A98:205–214 (1920).

10. N. W. McLachlan, "The Accession to Inertia of Flexible Discs Vibrating in a Fluid," *Proc. Phys. Soc. Lond.* 44:546–555 (1932).

11. F. M. Lewis, "The Inertia of Water Surrounding a Vibrating Ship," *Trans. SNAME* 37:1–20 (1929).

12. G. G. Stokes, "On the Communication of Vibrations from a Vibrating Body to a Surrounding Gas," *Phil. Trans. Roy. Soc. (Lond.)* 158:447–463 (1868).

13. M. Lax, "The Effect of Radiation on the Vibration of a Circular Diaphragm," *J. Acoust. Soc. Am.* 16:5–13 (1944).

14. R. D. Fay, "Interaction between a Plate and a Sound Field," *J. Acoust. Soc. Am.* 20:620–625 (1948).

15. M. J. Finney, "Reflection of Sound from Submerged Plates," *J. Acoust. Soc. Am.* 20:626–637 (1948).

16. L. Cremer, *Akustische Z.* 7:81–104 (1942).

17. J. J. Faran, Jr., "Sound Scattering by Solid Cylinders and Spheres," T. M. 22, Harvard Acoust. Res. Lab. (March 1951); also, *J. Acoust. Soc. Am.* 23:405–418 (1951).

18. M. C. Junger, "Sound Scattering by Thin Elastic Shells," T. M. 24, Harvard Acoust. Res. Lab. (July 1951); also *J. Acoust. Soc. Am.* 24:366–373 (1952).

19. M. C. Junger, "Vibrations of Elastic Shells in a Fluid Medium and the Associated Radiation of Sound," *J. Appl. Mech.* 19:439–445 (1952).

20. H. H. Bleich and M. L. Baron, *Free and Forced Vibrations of an Infinitely Long Cylindrical Shell in an Infinite Acoustic Medium*, Contract Nonr-266(08), T. R. 8, Columbia Univ. (Dec. 1952).

21. D. T. Laird and H. Cohen, "The Directionality Pattern for Acoustic Radiation from a Source on a Rigid Cylinder," *J. Acoust. Soc. Am.* 24:46–49 (1952).

22. H. H. Bleich, *Radiation from a Sound Source inside a Cylindrical Shell Submerged in an Infinite Medium*, Contract Nonr-266(08), T. R. 9, Columbia Univ. (Oct. 1953).

23. H. H. Bleich, "Sound Radiation from an Elastic Cylindrical Shell Submerged in an Infinite Medium," *Proc. 2nd U.S. Natl. Cong. Appl. Mech.* (ASME, New York, 1954), p. 213.

24. L. M. Liamshev, "Sound Diffraction by an Unbounded Thin Elastic Cylindrical Shell," *Sov. Phys. Acoust.* 4:161–167 (1958).

25. G. W. Warburton, "Vibration of a Cylindrical Shell in an Acoustic Medium," *J. Mech. Eng. Sci.* 3:69–79 (1961).

26. D. Feit, "Pressure Radiated by a Point-Excited Elastic Plate," *J. Acoust. Soc. Am.* 40:1489 (1966).

27. M. Heckl, "Schallabstrahlung von Platten bei Punktförmiger Anregung," *Acustica* 9:371–380 (1959).

28. M. Heckl, "Schallabstrahlung von Punkförmig Angeregten Hohlzylindern," *Acustica* 9:86–92 (1959).

29. R. M. Lyon and G. Maidanik, "Power Flow between Linearly Coupled Oscillators," *J. Acoust. Soc. Am.* 34:623–639 (1962).

30. P. W. Smith, Jr., "Response and Radiation of Structural Modes Excited by Sound," *J. Acoust. Soc. Am.* 34:640–647 (1962).

31. P. W. Smith, Jr., "Coupling of Sound and Panel Vibration below the Critical Frequency," *J. Acoust. Soc. Am.* 36:1516–1520 (1964).

32. J. B. Keller, "Geometric Theory of Diffraction," *J. Opt. Soc. Am.* 52:110–130 (1962).

33. C. W. Horton, W. R. King, and K. J. Diercks, "Theoretical Analysis of Scattering of Short Acoustic Pulses by a Thin-Walled Metal Cylinder in Water," *J. Acoust. Soc. Am.* 34:1929–1932 (1962).

34. For a review of this vast body of work, see H. Überall, "Surface Waves in Acoustics," in W. P. Mason, ed., *Physical Acoustics*, vol. 10 (New York: Academic Press, 1973), pp. 1–57.

35. G. C. Gaunaurd, "High-Frequency Acoustic Scattering from Submerged Cylindrical Shells Coated with Viscoelastic Absorbing Layers," *J. Acoust. Soc. Am.* 62:503–512 (1977).

36. W. B. Neubauer and L. R. Dragonette, "Observation of Waves Radiated from Circular Cylinders Caused by an Incident Pulse," *J. Acoust. Soc. Am.* 48:1135–1149 (1970).

37. For a more detailed history of the study of elastic structures vibrating in an acoustic fluid see M. C. Junger, "Radiation and Scattering by Submerged Elastic Structures," *J. Acoust. Soc. Am.* 57:1318–1326 (1975).

2 The Wave Equation and Its Elementary Solutions

2.1 Coupled Space and Time Dependence of Sound

In the preceding chapter, sound pressure was defined as the time-dependent deviation p of the local pressure from the ambient static atmospheric or hydrostatic pressure P_s. We assumed these deviations to be sufficiently small compared with the static pressure to permit a linearized approximation to the relation between the pressure fluctuation and the local relative density change—*condensation* or *dilatation* (or *rarefaction*). The local pressure fluctuations are not synchronous in the fluid medium: condensation at one point coincides with rarefaction at another point. We begin by showing that the temporal and spatial pressure fluctuations are correlated by the sound velocity that governs the propagation of a pressure disturbance through an extended fluid. The equation governing the relation between pressure fluctuations in space and time also governs the propagation of certain other disturbances, such as electromagnetic waves. Regardless of the physical nature of the disturbance, equations of this form are called wave equations.

We begin by deriving this equation for one-dimensional waves, that is, for pressure fields that depend on a single spatial coordinate and on time t. The solution is obtained for pressures varying periodically with time. From here, we proceed to study one-dimensional spherical waves radiated by a uniformly pulsating sphere located in an extended fluid medium. Finally, the standing-wave field inside a pulsating fluid-filled sphere is constructed.

2.2 The One-Dimensional Wave Equation for Plane Waves

Consider a volume element of length dz, cross-sectional area S, and mass $\rho S\,dz$ located in an effectively infinite, compressible fluid medium of density ρ. The forces acting on the two faces perpendicular to the z axis are shown in figure 2.1. They are taken positive in the positive z direction. Only the derivative of the fluctuating, acoustic pressure p is included, because the static pressure P_s is, by definintion, independent of location as well as time. Even though this text is intended to present theoretical fundamentals rather than practical applications, occasional reference will be made to the logarithmic notation generally applied to sound pressure. In this notation, the sound pressure level, in decibels (dB), is defined as

$$\mathrm{SPL} \equiv 20 \log\left(\frac{\langle p \rangle}{p_0}\right),$$

where $\langle p \rangle$ is a time-averaged pressure, which for sinusoidally varying signals

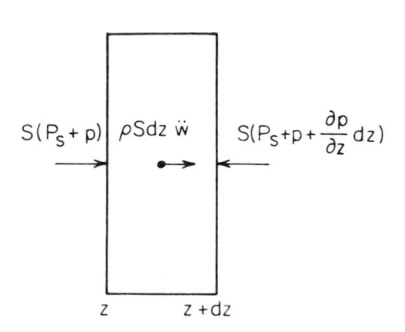

Figure 2.1
Force components acting on a fluid volume element in a one-dimensional pressure field depending only on the z coordinate (S = cross-sectional area of element normal to the z axis).

equals the rms pressure, and where p_0 is a reference sound pressure,[1] which equals either 1 dyne/cm$^2 \equiv 1$ μbar (dB re 1 μbar), 0.0002 dyne/cm^2 (dB re 0.0002 μbar), or 10^{-5} μbar $\equiv 1$ μpascal $\equiv 10^{-6}$ N/m^2 (dB re 1 μPa).

Applying Newton's second law to the fluid element in figure 2.1, we equate the net force to the inertia force of the volume element:

$$(P_s + p)S - \left[P_s + p + \left(\frac{\partial p}{\partial z} \right) dz \right] S = \left(\frac{\partial^2 w}{\partial t^2} \right) \rho S\, dz.$$

Dividing all terms of this equation by the volume $S\,dz$, we obtain

$$\frac{\partial p}{\partial z} = -\rho \frac{\partial^2 w}{\partial t^2}. \tag{2.1}$$

This equation, possibly generalized to include additional body forces such as gravity, is referred to as Euler's equation. Since p is independent of x and y, the other boundaries of the volume element are exposed to identical pressures and are therefore not associated with an unbalanced force.

The compressibility of the medium is now introduced by defining the bulk modulus B that relates the pressure change p applied to a fluid element and the resulting condensation, or fractional density change, $d\rho/\rho$:

$$p = B \left(\frac{d\rho}{\rho} \right). \tag{2.2}$$

From the requirement that the mass of the fluid element of volume V remain invariant,

$$d(\rho V) = \rho\, dV + V\, d\rho = 0,$$

one obtains the relation that allows one to formulate equation (2.2) in terms of the volume strain:

$$p = -B\frac{dV}{V}. \tag{2.3}$$

This is, in effect, the equivalent of Hooke's law extended to fluids. The sound pressure fluctuates too rapidly to permit substantial heat flow from regions of condensation to neighboring regions of rarefaction. In the foregoing equations B is therefore the adiabatic bulk modulus rather than the isothermal modulus measured under static conditions. This distinction is unimportant for liquids, but for gaseous media the difference between the two bulk moduli is considerable. Thus, for isothermal compression of a gas, Boyle's law relates pressure and density:

$$\frac{P_s + p}{P_s} = \frac{\rho + d\rho}{\rho}.$$

When this is subtstituted in equation (2.2), one obtains an isothermal bulk modulus B that equals the static pressure P_s. In the absence of significant heat exchange, the adiabatic gas law relates pressure and density:

$$\frac{P_s + p}{P_s} = \frac{(\rho + d\rho)^\gamma}{\rho^\gamma}$$

$$\approx 1 + \gamma\frac{d\rho}{\rho}, \qquad \frac{d\rho}{\rho} \ll 1,$$

where $\gamma = 1.41$ for the atmosphere. Substituting this relation in equation (2.2), one obtains the adiabatic bulk modulus relevant to airborne sound waves:

$$B = \gamma P_s. \tag{2.4}$$

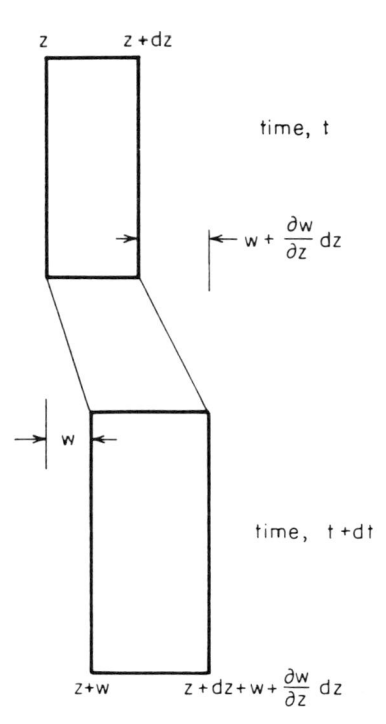

Figure 2.2
Dilatation of a fluid element in a one-dimensional, z-dependent pressure field.

We now apply equation (2.3) to the one-dimensional motion of the mass element in figure 2.2. The differential motion of the two faces perpendicular to the z axis results in the volume change

$$dV = S\left[w + \left(\frac{\partial w}{\partial z}\right)dz\right] - Sw.$$

When one substitutes this expression in equation (2.3), and divides by the volumes $V = S\,dz$, one obtains the relation

$$p = -B\frac{\partial w}{\partial z}. \tag{2.5}$$

To combine equations (2.1) and (2.5), the former is differentiated once with

respect to z, and the latter twice with respect to t. Eliminating the w-dependent term between the resulting two equations, one obtains the desired relation between the temporal and spatial variation of the pressure field,

$$\frac{\partial^2 p}{\partial z^2} - \left(\frac{\rho}{B}\right)\frac{\partial^2 p}{\partial t^2} = 0. \tag{2.6}$$

This partial differential equation in the spatial and temporal coordinate is the *one-dimensional wave equation*. Its solutions are of the form

$$p(z, t) = p\left[z \mp \left(\frac{B}{\rho}\right)^{1/2} t\right], \tag{2.7}$$

where the brackets indicate a functional dependence. Since an observer traveling with velocity

$$\frac{dz}{dt} = \mp \left(\frac{B}{\rho}\right)^{1/2}$$

sees no change in pressure, this is the *sound velocity*

$$c = (B^{-1}\rho)^{-1/2}. \tag{2.8}$$

The reason for formulating this result in terms of the compliance B^{-1} will become apparent in sections 2.9 and 3.10. For an incompressible fluid characterized by an infinite bulk modulus, the sound velocity is infinite and the wave number zero, thus reducing the wave equation, equation (2.6) to the Laplace equation. Since time always increases, the negative sign in equation (2.7) describes waves propagating in the positive direction, and vice versa. Assuming the former situation, the spatial and temporal derivatives are related as

$$\frac{\partial p}{\partial z} = -\left(\frac{\rho}{B}\right)^{1/2}\frac{\partial p}{\partial t}.$$

The acceleration in equation (2.1) can now be expressed in terms of the time derivative of pressure obtained from the equation above. The fluid particle velocity is obtained by integrating the resulting expression:

$$\dot{w} = (\rho B)^{-1/2} \int \frac{\partial p}{\partial t} \, dt$$

$$= p/\rho c.$$

(2.9)

The ratio of pressure to fluid particle velocity in a plane wave is the *characteristic impedance*,

$$\frac{p}{\dot{w}} = \rho c$$

$$= (\rho B)^{1/2}.$$

(2.10)

It equals 42 g/cm^2 sec in air under normal atmospheric conditions, and 1.5×10^5 in the same cgs units in water. These quantities are multiplied by ten in mks units. The fluid particle velocity associated with a given pressure is therefore more than three orders of magnitude larger in air than in water.

The strong pressure dependence of the bulk modulus of gases makes it desirable to formulate an explicit expression for the sound velocity in terms of the static pressure. Substituting the bulk modulus, (2.4), in (2.7), one obtains the sound velocity in gases:

$$c = (\gamma P_s/\rho)^{1/2}.$$

(2.11)

For air at atmospheric pressure ($\gamma = 1.41, \rho = 0.00121$ g/cm^3 = 1.21 kg/m^3) this is 344 m/sec, as compared with a relatively pressure-independent value of 1,460 m/sec for water. Newton, without benefit of the wave equation and, in fact of mathematics, used an intuitive argument to derive the isothermal sound velocity in air (1687):

$$c_i = (P_s/\rho)^{1/2}, \qquad \gamma = 1.$$

This yields a value too small by 19%. It is interesting to recall the curiously gravitational formulation Newton gave his result. He stated the sound velocity as the terminal velocity of an object falling from a height $H/2$, viz.,

$$c_i = (gH)^{1/2},$$

where H is the height of an atmosphere of uniform density for which

$P_s = \rho g H$. The correct adiabatic sound velocity was only derived in 1816 by Laplace.[2]

We now specialize the wave equation to steady-state situations associated with pressures varying periodically with time.

2.3 Harmonic Time Variation

Any function varying periodically with time can be represented as a series of terms each of which varies sinusoidally in time. Therefore, if we consider a function $\Phi(t)$, which is periodic within a *period* T, it can be represented in the form of a Fourier series:

$$\Phi(t) = \frac{A_0}{2} + \sum_{n=1}^{\infty} (A_n \cos n\omega t + B_n \sin n\omega t),$$

where $\omega = 2\pi/T$ is the fundamental frequency. The coefficients A_n and B_n are computed using the relations

$$A_n = \frac{2}{T} \int_0^T \Phi(t) \cos n\omega t \, dt,$$

$$B_n = \frac{2}{T} \int_0^T \Phi(t) \sin n\omega t \, dt,$$

For excitations in the above form, the responses of the linear systems we shall be concerned with can be similarly represented. Consequently, the nth term of the response is associated exclusively with the nth term of the excitation. Therefore no restriction is imposed on the range of applicability of the solution if one considers harmonic excitation that involves a single frequency. Such an excitation can be written in various alternative forms:

$$\Phi(t) = A \cos \omega t + B \sin \omega t$$
$$= C \cos(\omega t + \alpha),$$

where

$$C^2 = A^2 + B^2,$$
$$\alpha = -\tan^{-1}(B/A).$$

This can be expressed in complex notation by introducing *Euler's formula*,

$e^{ix} = \cos x + i \sin x$,

and using Re to indicate "the real part of": $\Phi(t) = \text{Re}[(A + iB)e^{-i\omega t}]$. We shall adopt the extensively used convention whereby the symbol Re is suppressed, it being understood that only the real component of a complex function has a physical meaning. In this shorthand notation,

$$\Phi(t) = C\exp[-i(\omega t + \alpha)].$$

This can be abbreviated further by shifting the origin of the time coordinate:

$$\Phi(t') = Ce^{-i\omega t'}, \qquad t' = t + (\alpha/\omega).$$

The wave equation, equation (2.6), was seen to be a partial differential equation involving spatial coordinates as well as time, whose solutions are of the form $f(z, t)$ or, more generally, $f(\mathbf{R}, t)$. When the time dependence can be factored out, the spatially dependent portion of the solution will be identified as $f(\mathbf{R})$:

$$\begin{aligned} f(\mathbf{R}, t) &\equiv f(\mathbf{R}) \cos(\omega t' + \alpha) \\ &= f(\mathbf{R})e^{-i\omega t}, \qquad t = t' + (\alpha/\omega). \end{aligned} \tag{2.12}$$

As already noted, this time-dependent factor will be suppressed when it does not contribute to the clarity of the development. It is at times convenient to express the spatial dependence in dimensionless form $g(\mathbf{R})$ by factoring out the amplitude F of the function $f(\mathbf{R})$:

$$f(\mathbf{R}) \equiv Fg(\mathbf{R}). \tag{2.13}$$

This notation can be extended to the time derivatives of equation (2.12):

$$\begin{aligned} \dot{f} &= -i\omega f \\ &\equiv \dot{F}g(\mathbf{R})e^{-i\omega t} \end{aligned} \tag{2.14}$$

and

$$\begin{aligned} \ddot{f} &= -i\omega\dot{f} = -\omega^2 f \\ &\equiv \ddot{F}g(\mathbf{R})e^{-i\omega t} \end{aligned} \tag{2.15}$$

In conclusion it is noted, for reference in connection with energy calculations, that the time average of f over a period is zero, but that the average of the function squared is, in the notation of the former of equation (2.12):

$$\langle f^2(\mathbf{R}, t)\rangle = \frac{1}{T}\int_0^T f^2(\mathbf{R})\cos^2 \omega t \, dt$$

$$= \tfrac{1}{2} f^2(\mathbf{R}).$$

$$(2.16)$$

For propagating acoustic waves, the exponential multiplying the pressure will be found to be of the form $\exp(ikR - i\omega t)$. The time average now becomes

$$\langle f^2(\mathbf{R}, t)\rangle = \frac{1}{T}\int_0^T |f(\mathbf{R})|^2 \cos^2 (kR - \omega t) \, dt$$

$$- \tfrac{1}{2}|f(\mathbf{R})|^2.$$

$$(2.17)$$

We now have the tools for solving the wave equation specialized to harmonic time dependence.

2.4 Steady-State Plane Waves

Returning to the wave equation, equation (2.6), we assume a harmonic solution of the form of equation (2.12):

$$p(z, t) = p(z)e^{-i\omega t},$$

which, in accordance with the convention stated earlier, will be written simply as $p(z)$. When this is inserted in equation (2.6), making use of equation (2.15), the wave equation becomes an ordinary differential equation in the spatial coordinate:

$$\frac{d^2 p}{dz^2} + \left(\frac{\omega^2 \rho}{B}\right) p = 0.$$

This equation is referred to as the *one-dimensional Helmholtz equation* or steady-state wave equation. Defining the acoustic wave number

$$k \equiv \omega(\rho B^{-1})^{1/2},$$

$$(2.18)$$

we can write the Helmholtz equation more concisely as

$$\left(\frac{d^2}{dz^2} + k^2\right)p(z) = 0. \tag{2.19}$$

The general solution of this equation is

$$p(z) = A_+ e^{ikz} + A_- e^{-ikz},$$

where A_+ and A_- are as yet undetermined coefficients. Linear combinations of these exponentials form trigonometric functions of argument kz that are also solutions of equation (2.19). To interpret the physical significance of these results we formulate them explicitly in terms of the exponential time dependence $e^{-i\omega t}$:

$$p(z,t) = A_+ e^{i(kz-\omega t)} + A_- e^{-i(kz+\omega t)}. \tag{2.20}$$

If we consider the above solution at a fixed z, we see that is returns to its original value at a later time $t + (2\pi/\omega)$. This time interval is the period T. Conversely, if we now consider the variation of pressure with the spatial coordinate z at a fixed time t, the pressure in the plane $z = z_0$ is the same as in the plane $z = z_0 + (2\pi/k)$. This increment of the z coordinate is the acoustic wavelength

$$\lambda \equiv \frac{2\pi}{k}. \tag{2.21}$$

If z and t are allowed to vary simultaneously, the pressure remains constant provided the exponent in equation (2.20) is maintained invariant:

$$\frac{d}{dt}(\pm kz - \omega t) = 0.$$

This equation defines the speed at which the pressure disturbance propagates, which, by definition, is the *sound velocity* c:

$$\pm\frac{dz}{dt} = \frac{\omega}{k} \equiv c.$$

Substituting equation (2.18) and solving for c, one retrieves the expression for sound velocity in equation (2.8). The Helmholtz equation, equation (2.19), tends to the Laplace equation not only when $B \to \infty$, i.e., when the fluid is incompressible, but also when $\omega \to 0$, i.e., when inertia forces are negligible compared with elastic forces.

To determine the amplitude of the pressure in equation (2.20), a boundary condition must be specified. Consider, for example, the semi-infinite fluid-filled space $z > 0$, whose boundary $z = 0$ vibrates uniformly with velocity

$$\dot{w}(t) = \dot{W}e^{-i\omega t}, \tag{2.22}$$

thus generating sound waves traveling away from the boundary, in the positive z direction. If no other source or boundary is present in the semi-infinite medium, the pressure field is thus represented entirely by the term in equation (2.20) with the positive exponent:

$$p(z, t) = A_+ \exp(-i\omega t + ikz). \tag{2.23}$$

To obtain the pressure amplitude A_+ in terms of $\dot{W}$, we use the fact that the fluid particles in the $z = 0$ plane must vibrate with the same velocity $\dot{w}$ as the boundary, in the absence of cavitation in the fluid or penetration of the boundary by the fluid. The fluid particle acceleration $\ddot{w}$ is related to the pressure gradient by equation (2.1). Specialized to harmonic motion and using the former of equations (2.15) to express the acceleration in terms of the velocity, the pressure gradient becomes

$$\frac{\partial p}{\partial z} = i\omega\rho\dot{w}. \tag{2.24}$$

Differentiating equation (2.23) with respect to z and substituting the results in the above equation, one can solve for the particle velocity at the boundary:

$$\dot{w} = \frac{ikp}{i\omega\rho} = \frac{p}{\rho c}$$

$$= \frac{A_+ e^{-i\omega t}}{\rho c}, \qquad z = 0.$$

When this is set equal to the boundary velocity, equation (2.22), one can solve for A_+ and express the undetermined coefficient in equation (2.23) in terms of $\dot{W}$:

$$p(z, t) = \rho c \dot{W} \exp(ikz - i\omega t)$$
$$= \rho c \dot{v}(z, t). \tag{2.25}$$

We have retrieved the result already obtained without a harmonic-time dependence restriction in equation (2.9). For other than plane waves, the impedance is complex and dependent on both frequency and range and is therefore not, like equation (2.10), a characteristic of the fluid. The ratio of pressure to velocity, defined as the *specific acoustic impedance*, z, can be written in terms of its real and imaginary components, or of a phase angle:

$$\frac{p}{\dot{v}} = z = r + ix$$
$$= |z|e^{i\alpha}. \tag{2.26}$$

The $\cos \alpha$ component of the impedance represents a pressure in phase with the velocity. For $\alpha > 0$, the imaginary component $\sin \alpha$ is associated with a pressure in phase with the particle displacement and thus represents a springlike force. For $\alpha < 0$, the pressure is in phase with the particle acceleration and can therefore be interpreted as an inertia force. Except for plane waves or in the short-wavelength limit, where z tends to ρc, the specific acoustic impedance is a complex quantity. Since both the pressure and velocity have the same time dependence, the real component of z indicates that the fluid reaction on the boundary has a component proportional to the velocity. This component can thus be interpreted as a damping force associated, not with dissipation of energy by friction, but rather with the radiation of acoustic power. Electrical engineers may prefer to view it as a resistive force in analogy with electrical circuit theory. Before going further into the radiation of acoustic power, we shall consider pressure fields with nonplanar wave fronts.

2.5 The Three-Dimensional Wave Equation

A vibrating surface of finite dimensions usually generates a pressure field that is not one-dimensional and hence a function of several coordinates. In general, therefore, the particle displacement is a vector that can be repre-

sented by its components along three orthogonal coordinate axes, equation (1.1). Each of the three components satisfies an equation of the form of equation (2.1). These three equations can be expressed concisely in vector notation by introducing the vector operator $\mathbf{V}$. In rectangular coordinates,[3]

$$\mathbf{V}p = \hat{\eta}\frac{\partial p}{\partial x} + \zeta\frac{\partial p}{\partial y} + \xi\frac{\partial p}{\partial z}.$$

This operator defines the pressure gradient. The generalization of equation (2.1) to three dimensions can now be stated as

$$\mathbf{V}p = -\rho\frac{\partial^2 \mathbf{d}}{\partial t^2}. \tag{2.27}$$

For harmonic time dependence, this reduces to

$$\mathbf{V}p = \rho\omega^2\mathbf{d}$$
$$= i\rho ck\mathbf{d}. \tag{2.28}$$

In the latter expression, ω has been expressed as kc.

In a three-dimensional field, the dilatation in equation (2.5) is the resultant of three differential displacement components $(\partial u/\partial x)\,dx$, $(\partial v/\partial y)\,dy$, and $(\partial w/\partial z)\,dz$. The dilatation can be expressed concisely in terms of the divergence $\mathbf{V}\cdot\mathbf{d}$ of the displacement vector. The three-dimensional form of equation (2.5) thus becomes

$$p = -B\mathbf{V}\cdot\mathbf{d}, \tag{2.29}$$

where, in rectangular coordinates,

$$\mathbf{V}\cdot\mathbf{d} = \frac{\partial u}{\partial x} + \frac{\partial v}{\partial y} + \frac{\partial w}{\partial z}.$$

The divergence $\mathbf{V}\cdot$ of the vector operator $\mathbf{V}$ is the Laplace operator ∇^2. In rectangular coordinates,

$$\nabla^2 \equiv \frac{\partial^2}{\partial x^2} + \frac{\partial^2}{\partial y^2} + \frac{\partial^2}{\partial z^2}. \tag{2.30}$$

To construct the three-dimensional wave equation, one takes the divergence of equation (2.27), and differentiates equation (2.29) twice with respect to time, thus obtaining two equivalent expressions, $\mathbf{V} \cdot (\partial^2 \mathbf{d} / \partial t^2)$ and $\partial^2 (\mathbf{V} \cdot \mathbf{d}) / \partial t^2$. Equating these two results, we construct the three-dimensional wave equation:

$$\nabla^2 p - \left(\frac{\rho}{B}\right) \frac{\partial^2 p}{\partial t^2} = 0. \tag{2.31}$$

For steady-state conditions, this can be written in terms of the wave number in the form of a three-dimensional Helmholtz equation:

$$(\nabla^2 + k^2) p = 0. \tag{2.32}$$

We shall now proceed to solve equation (2.32) for the simplest, physically realizable source configuration, the pulsating sphere.

2.6 The Uniformly Pulsating Spherical Source
In spherical coordinates, the Laplace operator takes the form[4]

$$\nabla^2 \equiv \frac{\partial}{\partial R^2} + \frac{2}{R} \frac{\partial}{\partial R} + \nabla_\sigma^2. \tag{2.33}$$

Here ∇_σ^2 is the two-dimensional or surface Laplace operator to which ∇^2 reduces when $\partial / \partial R, \partial^2 / \partial R^2 = 0$:

$$\nabla_\sigma^2 \equiv \frac{1}{R^2} \left[\cot \theta \frac{\partial}{\partial \theta} + \frac{\partial^2}{\partial \theta^2} + \frac{1}{\sin^2 \theta} \frac{\partial^2}{\partial \phi^2} \right], \tag{2.34}$$

where θ is the polar angle and ϕ the circumferential angle. $\nabla_\sigma^2 p$ vanishes if the field displays spherical symmetry, that is, if it is independent of the spherical angles of latitude θ and longitude ϕ. This condition is satisfied by the pressure radiated by a sphere vibrating harmonically with uniform radial velocity $\dot{W} \exp(-i\omega t)$. This sphere has its center at the origin of coordinates, $R = 0$, and has radius a. On the radiating surface $R = a$, the pressure field must satisfy the boundary condition that the radial component of $\mathbf{d}$ equal the velocity $\dot{W}$ prescribed on the spherical radiator. If one uses

the radial component of the vector equation equation (2.28), this boundary condition becomes

$$\frac{dp}{dR} = i\rho c k \dot{W}, \qquad R = a. \tag{2.35}$$

Furthermore, the pressure must satisfy the Helmholtz equation, equation (2.32) with ∇^2 given in equation (2.33) and specialized to $\nabla_\sigma^2 = 0$:

$$\left(\frac{d^2}{dR^2} + \frac{2}{R}\frac{d}{dR} + k^2\right)p(R) = 0. \tag{2.36}$$

The pressure satisfying this equation is one-dimensional, in that it depends on the single coordinate R, but the wave fronts are not plane. It can be verified that the general solution of this equation is

$$p(R) = \frac{1}{R}(A_+ e^{ikR} + A_- e^{-ikR}). \tag{2.37}$$

Referring back to the discussion of the plane-wave solution, equation (2.20), we conclude that the first term represents outgoing waves, whose phase $(kR - \omega t)$ remains constant with increasing R and t, provided $\dot{R}$ equals the sound velocity, ω/k. Conversely, the $\exp(-ikR)$ term corresponds to converging waves, whose phase remains constant with R decreasing at the sound velocity. It will be shown in chapter 4 that converging waves do not arise in the volume $a < R < \infty$ if the only boundaries are at $R = a$ and ∞, and if there are no sources in this volume. Consequently, the second term in equation (2.37) drops out. The undetermined coefficient A_+ is expressed in terms of $\dot{W}$ by means of the boundary condition in equation (2.35):

$$\frac{dp(R)}{dR}\bigg|_{R=a} = \left(ik - \frac{1}{a}\right)\frac{A_+}{a}e^{ika}$$

$$= i\rho c k \dot{W}.$$

Solving for A_+, and substituting the result in equation (2.37), we obtain an explicit solution in terms of the velocity amplitude of the source:

$$p(R) = \frac{\rho c k a (ka - i) a \dot{W}}{[(ka)^2 + 1] R} \exp[ik(R - a)]$$

$$\simeq -\frac{i \rho c k a^2 \dot{W}}{R} e^{ikR}, \qquad ka \ll 1 \tag{2.38}$$

$$\simeq \frac{\rho c a \dot{W}}{R} \exp[ik(R - a)], \qquad k^2 a^2 \gg 1.$$

The pressure amplitude decays with increasing range as R^{-1}. It will be verified in chapter 3 that this dependence of pressure on range conforms to energy balance requirements. This variation of pressure with range is defined as the *spherical spreading loss*.

2.7 Specific Acoustic Impedance of Spherical Waves

There being no ϕ- and θ-dependent terms in equation (2.38), the pressure gradient in equation (2.28) reduces to dp/dR. The corresponding particle velocity is purely radial:

$$\dot{w}(R) = \frac{1}{i \rho c k} \frac{dp}{dR}$$

$$= \frac{1}{i \rho c k} \left(ik - \frac{1}{R} \right) p(R). \tag{2.39}$$

We are now in a position to compute the specific acoustic impedance defined in equation (2.26). It will be identified with the subscript 0, to distinguish the uniformly expanding wave from more complicated spherical wave fields. Equation (2.39) yields the ratio $p/\dot{w}$:

$$z_0(R) = \frac{\rho c k R (kR - i)}{1 + k^2 R^2}$$

$$\approx -i \rho c k R, \qquad (kR)^2 \ll 1 \tag{2.40}$$

$$\approx \rho c, \qquad (kR)^2 \gg 1.$$

An anticipated in the discussion of equation (2.26), this impedance is complex and dependent on frequency and range. It tends to the characteristic

impedance for large kR. This is to be expected from equation (2.36), since this differential equation reduces to the form of the plane-wave equation, equation (2.19) in the large-R limit, where $d^2p/dR^2 \gg 2(dp/dR)/R$. The impedance is a function of range expressed in dimensionless form as $kR = 2\pi R/\lambda$. It will be found repeatedly that the dependence of a certain characteristic of the pressure field on a linear dimension, such as source dimension or range, is less meaningful than the dependence on the dimensionless ratio obtained by dividing the linear dimension by λ. Other functional relations of the pressure field, such as the variation of pressure amplitude with range, will be found to be more appropriately described by normalizing the range with respect to the source dimension. The independent variable is thus R/a, rather than kR.

The specific acoustic impedance measured on the surface of the source determines the radiation loading defined in the preceding chapter. For the purpose of interpreting the physical meaning of the expression for the surface impedance, $z_0(a)$ is expanded in its real and imaginary components designated, respectively, by r_0 and $-\omega m_0$. The radiation loading can now be written as

$$p(a) \equiv (r_0 - i\omega m_0)\dot{w}$$

$$\equiv r_0\dot{w} + m_0\ddot{w},$$

where it is understood that $\dot{w}$ and the z_0 components are evaluated at $R = a$. The surface pressure is therefore the sum of (1) a term proportional to the surface particle velocity $\dot{w}$, representing a resistive or damping force which embodies energy lost by the pulsating sphere in the form of radiated acoustic energy, and (2) a term proportional to the surface particle acceleration, embodying the inertia force associated with the accession to inertia or entrained mass of fluid set into motion by the pulsating surface of the spherical source. Explicit expressions are obtained by setting $R = a$ in equation (2.40):

$$r_0 = \frac{\rho c (ka)^2}{[(ka)^2 + 1]}$$

$$\approx \rho c (ka)^2, \quad (ka)^2 \ll 1$$

$$\approx \rho c, \quad (ka)^2 \gg 1,$$

(2.41a)

$$m_0 = \frac{\rho a}{[(ka)^2 + 1]}$$

$$\approx \rho a, \qquad (ka)^2 \ll 1$$

$$\approx \rho a (ka)^{-2} \approx 0, \qquad (ka)^2 \gg 1,$$

(2.41b)

In the notation of equation (2.26),

$$|z_0| = \frac{\rho c k a}{[(ka)^2 + 1]^{1/2}},$$

$$\alpha = -\tan^{-1}(ka)^{-1}.$$

(2.42)

For small ka, that is, $a\omega \to 0$ or $c \to \infty$, the impedance tends to

$$z_0 \approx -i\omega m_0 \approx -i\rho c k a, \qquad k^2 a^2 \ll 1.$$

(2.43)

This is the asymptotic "hydrodynamic" solution that ignores sound radiation and is rigorously correct for an incompressible fluid. The pressure is associated with the inertia force of the entrained mass of fluid accelerated by the pulsating sphere. The correspondng mass of entrained fluid, computed from the small-ka limit of (2.41b), equals

$$M_0 = 4\pi a^2 m_0$$

$$= 4\pi a^3 \rho = 3\rho V,$$

(2.44)

where V is the volume displaced by the sphere. In the high-frequency asymptotic limit, the impedance reduces to a resistance whose value coincides with the characteristic plane-wave impedance, equation (2.10). This result is to be expected since the condition $(ka)^2 \gg 1$ implies that the radius of curvature a of the wave front propagating away from the radiating surface is large compared with the wavelength λ, thus approaching the plane-wave situation, which, at lower frequencies, is only obtained for large R [see equation (2.40)]. The frequency dependence of r_0 and m_0 is illustrated in figure 2.3 jointly with the impedance exerted by the fluid contained in a pulsating sphere, derived in the following section.

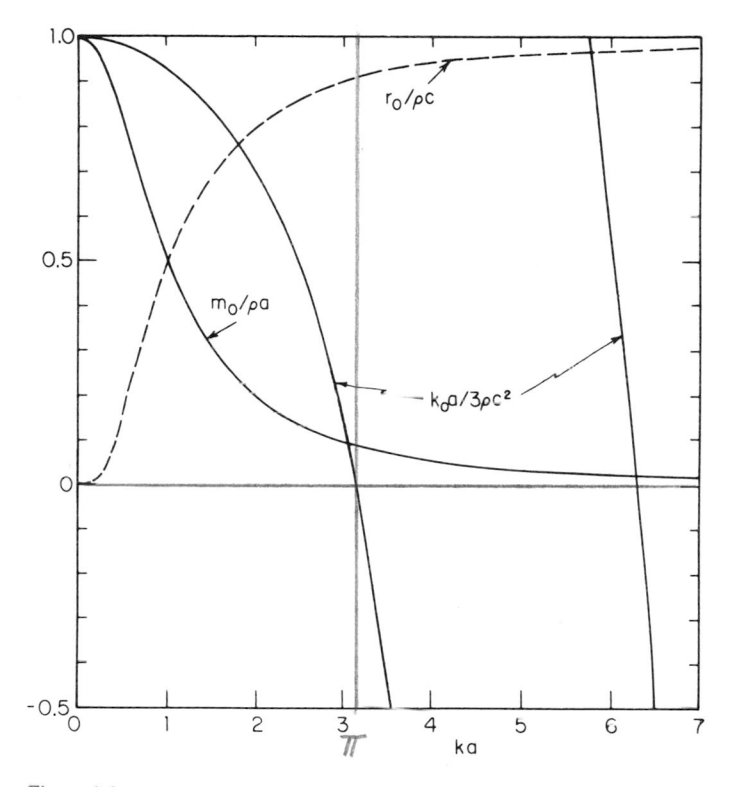

Figure 2.3
Specific acoustic resistance and accession to inertia exerted by an ambient medium on the surface of a uniformly pulsating sphere [equation (2.41)] and effective spring stiffness exerted by a fluid contained in a sphere [equation (2.48)], all per unit area.

2.8 The Pressure Field inside a Fluid-Filled Pulsating Sphere

The pressure in a fluid sphere whose outer boundary vibrates uniformly with velocity amplitude $\dot{W}$ satisfies both the boundary condition in equation (2.35) and the differential equation in equation (2.36). The solution is therefore in the form of equation (2.37). However, the motion inside the boundary is compatible with waves converging on the center of the sphere, as well as with waves expanding from the center, represented, respectively, by the negative and positive exponentials in equation (2.37). The ratio of the corresponding coefficients is determined by the requirement that the pressure remain finite at the center, i.e., that the two exponential terms cancel as $R \to 0$. Since both exponentials tend to unity when their exponents

vanish, the condition at $R = 0$ is satisfied if $A_- = -A_+$. The solution can thus be expressed in terms of a single coefficient A_+:

$$p(R) = \frac{A_+ (e^{ikR} - e^{-ikR})}{R}$$

$$= \left(\frac{2iA_+}{R}\right) \sin kR.$$

The undetermined coefficient is now obtained from the boundary condition in equation (2.35), in the same manner as for the outer pressure field. One thus constructs an explicit expression for the pressure inside the pulsating sphere in terms of the boundary velocity. The result will be expressed, for the sake of brevity, in terms of the spherical Bessel function of order zero j_0, defined as

$$j_0(z) \equiv \frac{\sin z}{z}$$

$$\simeq 1 - \frac{z^2}{6} + \frac{z^4}{120}, \qquad z^6 \ll 1,$$

$$j_0'(z) = -j_1(z) = (\cot z - z^{-1})j_0(z)$$

$$\simeq -\frac{z}{3} + \frac{z^3}{30}, \qquad z^5 \ll 1.$$

(2.45)

The pressure field thus finally becomes

$$p(R) = \frac{i\rho cka j_0(kR)\, \dot{W}}{(ka \cot ka - 1)j_0(ka)}.$$

(2.46)

This is a standing wave field in that there are no propagating constant-phase wave fronts. In chapter 3 it will be shown that such a wave field is characterized by zero time-averaged energy flow. We can thus anticipate that the radiation loading does not contain a resistive term. The surface pressure is obtained by setting $R = a$ in equation (2.46):

$$p(a) = \frac{i\rho cka \dot{W}}{ka \cot ka - 1}.$$

In the low-frequency limit, where

$$ka \cot ka \approx 1 - \frac{(ka)^2}{3},$$

the surface pressure is

$$p(a) \approx -3\frac{i\rho c}{ka}\dot{W}, \qquad k^3 a^3 \ll 1. \tag{2.47}$$

The latter expression indicates that a fluid sphere, small in terms of wavelengths, responds as a distributed spring, in that an outward displacement gives rise to a negative pressure, or inward pull. It is convenient, for the purpose of an application of these results in chapter 3, to express the surface impedance in terms of a dynamic spring stiffness per unit area, k_0:

$$\frac{p(a)}{\dot{W}} = -\frac{ik_0}{\omega}, \tag{2.48}$$

where

$$k_0 = \frac{\rho c^2 k^2 a^2}{a(1 - ka \cot ka)}$$

$$\approx 3\rho c^2/a, \qquad k^4 a^4 \ll 1. \tag{2.49}$$

The dynamic spring stiffness is proportional to the bulk modulus, as seen by solving equation (2.8) for $B = \rho c^2$. The spring constant k_0 has been plotted in figure 2.3. This stiffness becomes an effective mass between the fundamental resonance of the fluid sphere, at $ka = \pi$, and its first antiresonance, at $ka = \tan ka$, i.e., $ka = 4.49$. The practical application of this result in chapter 3 lies in the small-ka region where the fluid sphere acts as a spring.

    In summary, in the low-frequency limit, when the circumference is much smaller than the wavelength, the fluid loading on a pulsating spherical boundary is springlike if the fluid is enclosed, and masslike if the fluid surrounds the sphere. At high frequencies, when the circumference exceeds a wavelength, the ambient fluid acts as a resistance and the enclosed fluid alternates between a masslike and stiffnesslike loading.

    Being restricted to spherical symmetry, the exterior and interior pres-

sure fields constructed above are, though nonplanar, expressible in terms of a single coordinate. This restriction will now be dropped.

2.9 An Elementary Interaction Problem: Liquid-Filled Elastic Waveguides

To conclude this chapter we examine elastic cylindrical waveguides, thereby introducing the reader to two new concepts. This section illustrates a class of elementary interaction problems dealing with a liquid whose wave motion is coupled to an elastic structure or, in the next chapter, to a bubble swarm, the admittance of the composite system being simply the sum of the admittances of its components. This section also introduces 2- or 3-dimensional solutions of the wave equation formulated as products or functions each of which depends on one single coordinate.

In this elementary interaction problem, the elastic pipe wall is assumed to be locally reacting and therefore characterized by a specific acoustic reactance, x_w, assumed finite but large. x_w is positive for a stiffness-controlled wall and negative for a mass like impedance. Matching this reactance to the pressure field in the pipe, one constructs the boundary condition

$$p/\dot{w} = ix_w, \qquad r = a, \tag{2.50}$$

where r is the radial coordinate in the cylindrical coordinate system. The axial coordinate z coincides with the pipe centerline. The pipe radius is a. Paralleling the development in spherical coordinates, equation (2.35), the radial velocity is related to the pressure:

$$\dot{w} = -i(\partial p/\partial r)/\rho ck. \tag{2.51}$$

The boundary condition (2.50), can now be stated explicitly in terms of the pressure and its derivative:

$$p/(\partial p/\partial r) = x_w/\rho ck, \qquad r = a. \tag{2.52}$$

For airborne sound propagating in an unlined pipe, the wall can be approximated as rigid, requiring $\partial p/\partial r$ to vanish at the boundary. An obvious solution can be constructed by noting that, for a plane wave, the vector particle velocity is normal to the wave front. Consequently, for $x_w = \infty$, the plane-wave solution, equation (2.23), satisfies the boundary condition in equation (2.52). For a water-filled pipe, the pipe's elastic

response has a nonnegligible effect on sound propagation. Neglecting longitudinal pipe vibrations, an axisymmetric sound pressure produces a uniform radial deflection w, which results in a hoop strain ε and stress σ:

$$\varepsilon = w/a,$$

$$\sigma = \varepsilon E/(1 - v^2).$$

Interaction solutions are usually more readily interpreted when the structural elastic constants are expressed in terms of concepts related to wave propagation. For this purpose the velocity of compressional waves in plates is introduced:

$$c_{\mathrm{p}} \equiv [E/(1 - v^2)\rho_{\mathrm{s}}]^{1/2}, \tag{2.53}$$

where ρ_{s} is the density of the elastic material. This velocity is 5,400 m/sec in steel and aluminum and two orders of magnitude smaller in rubber. The hoop stress can now be stated explicitly as

$$\sigma = w\rho_{\mathrm{s}}c_{\mathrm{p}}^2/a. \tag{2.54}$$

If h is the wall thickness, this stress is related to the internal pressure by a simple statement of static equilibrium:

$$h\sigma = ap.$$

Substituting (2.54), one obtains the pipe wall stiffness explicitly in terms of the pipe parameters:

$$p/w = \rho_{\mathrm{s}}c_{\mathrm{p}}^2 h/a^2. \tag{2.55}$$

The fractional change in volume resulting from extension of the pipe wall,

$$dV/V = 2\pi a w/\pi a^2 = 2w/a$$

can now be combined with (2.55) to yield a compliance:

$$\begin{aligned} B_{\mathrm{w}}^{-1} &\equiv (dV/V)/p = 2w/ap \\ &= 2a/\rho_{\mathrm{s}}c_{\mathrm{p}}^2 h. \end{aligned} \tag{2.56}$$

This is effectively the reciprocal of a bulk modulus as defined in equation (2.3), the minus sign being eliminated because the pressure causes an increase in the volume encompassed by the pipe. The sound velocity in (2.8) can now be corrected for wall elasticity by noting that the compressibility of the acoustic fluid and of the pipe wall combine so as to enhance the fractional volume change:

$$c_0 = [\rho(B^{-1} + B_w^{-1})]^{-1/2} \tag{2.57}$$

or, in dimensionless form,

$$c_0/c = [1 + (B_w^{-1}/B^{-1})]^{-1/2}.$$

Substituting equation (2.56) normalized to the compliance $B^{-1} = (\rho c^2)^{-1}$ of the acoustic fluid, this becomes

$$c_0/c = [1 + (2a\rho c^2/h\rho_s c_p^2)]^{-1/2}. \tag{2.58a}$$

This is the *Korteweg-Lamb correction*. For water in a steel pipe,

$$c_0/c = [(1 + 2.03 \times 10^{-2}(a/h)]^{-1/2}.$$

Taking, for example, $a/h = 10$, this yields a velocity ratio of 0.91.

Electrically speaking, the fluid bulk modulus and the wall stiffness in equation (2.57) are in parallel. In an air-filled duct, the compressibility of the gas short-circuits the duct stiffness. In a liquid-filled, compliant hose, the hose wall compliance tends to short-circuit the compressibility of the liquid:

$$c_0/c \simeq (c_p/c)(h\rho_s/2a\rho)^{1/2}, \qquad \rho_s c_p^2 \ll \rho c^2. \tag{2.58b}$$

For example, for water in a soft rubber tube ($\rho_s c_p^2 = 2.6 \times 10^7$ μbar $= 2.6 \times 10^6$ N/m^2), with $a/h = 10$, $c_0/c = 0.008$. These numbers are representative of pressure propagation in blood vessels.[5] A similar phenomenon will be described in section 3.10, where low-frequency sound is seen to propagate through a bubble swarm as though the liquid matrix were incompressible, compliance being effectively associated with the gas bubbles. Another example of elastic and inertia forces associated with two different

media in contact are *interface waves* propagating along an ocean bottom whose bulk modulus is reduced, e.g., by gas bubbles generated by decaying organic matter.[6] This special class of Stoneley waves at the interface of an incompressible liquid and of a massless elastic solid, viz. of media individually incapable of supporting wave motion, was first identified by Biot (1952).

2.10 Cylindrical Waveguides: An Introduction to Two-Dimensional Pressure Fields

Further insight into waveguide mechanics can be gained by developing an elementary theory capable of accounting for mass- as well as spring like reactances. We retain the assumption, to be dropped in section 11.12, that $|x_w|/\rho c \gg 1$. The radial fluid particle velocity and hence $\partial p/\partial r$ vanish on the axis but are finite at the boundary. The simplest function satisfying these conditions displays a linear dependence on radius. Selecting a numerical coefficient which simplifies the algebra, we select

$$\partial p/\partial r = -\tfrac{1}{2}P\beta^2 r \exp(i\gamma z). \tag{2.59a}$$

Here γ is the axial wave number ω/c_0, while β is in the nature of a radial wave number, both potentially complex. β and γ are separation constants that enable us to express the pressure as the product of functions each of which depends on only one coordinate, and therefore satisfies an ordinary rather than a partial differential equation, a procedure that will be discussed further in chapter 4. The solution of the wave equation requires the second derivative with respect to r as well as the pressure itself. For this purpose, the constant of integration is selected to make P the pressure on the axis:

$$p = P[1 - \tfrac{1}{4}\beta^2 r^2] \exp(i\gamma z)$$
$$\simeq P \exp(i\gamma z), \qquad |\beta^2 a^2| \ll 1 \tag{2.59b}$$
$$\partial^2 p/\partial r^2 = -\tfrac{1}{2}\beta^2 p \exp(i\gamma z).$$

The boundary conditions equation (2.52) is now stated in terms of (2.59a) and (2.59b):

$$\frac{x_w}{\rho c k a} = \frac{-2}{\beta^2 a^2} + \frac{1}{2}. \tag{2.60}$$

The inequality $|\beta^2 a^2| \ll 1$ is satisfied when the wall impedance is large compared to ρc, as assumed here.

The relation between β and γ is obtained from the Helmholtz equation, equation (2.32), where the Laplace operator is now expressed in cylindrical coordinates:[4]

$$\nabla^2 \equiv \nabla_\sigma^2 + \frac{\partial^2}{\partial z^2},$$

where

$$\nabla_\sigma^2 \equiv \frac{\partial^2}{\partial r^2} + \frac{1}{r}\frac{\partial}{\partial r} + \frac{1}{r^2}\frac{\partial^2}{\partial \phi^2}. \tag{2.61}$$

Combining this operator with equations (2.59), the Helmholtz equation becomes

$$[k^2 - \beta^2 - \gamma^2]P\exp(i\gamma z) = 0, \qquad |\beta^2 a^2| \ll 1.$$

The equation is satisfied when

$$\gamma = (k^2 - \beta^2)^{1/2}. \tag{2.62}$$

This kind of relation between the wave numbers is typical of wave-equation solutions compatible with separation of coordinates. A consequence of separability conditions such as equation (2.62) is the existence of imaginary wave numbers characterizing exponentially decaying rather than propagating waves. These *dead zones* are associated with long-wavelength situations that tend, in the symptotic $k = 0$ limit, to the *incompressible* pressure field.

The final step consists in determining β from the boundary condition. Consider first the stiffness-controlled pipe wall. Dividing equation (2.55) by $\omega = ck$, one obtains an expression for the pipe wall reactance:

$$x_w = \rho_s c_p^2 h / cka^2. \tag{2.63}$$

Substituting this expression in the boundary condition, equation (2.60), one can solve for the radial wave number:

$$\beta^2 \simeq -2\rho c^2 k^2 a/\rho_s c_p^2 h + O(a^{-2}).$$

For a stiff pipe wall, $\beta^2 a^2$ being small and negative, the pressure in equation (2.59b) displays a quasi-planar wave front, the pressure being higher near the wall. Equation (2.62) now becomes

$$\gamma/k \simeq [1 + (2\rho c^2 a/\rho_s c_p^2 h)]^{1/2}. \tag{2.64}$$

This ratio equals c/c_0 and is therefore consistent with the Korteweg-Lamb result in equation (2.58).

Now consider a mass-controlled tube wall. The specific acoustic reactance is

$$x_w = -\omega\rho_s h = -\rho_s ckh. \tag{2.65}$$

This expression holds far above the breathing mode resonance, whose natural frequency is computed by equating the amplitudes of the spring- and mass like reactances—respectively, equations (2.63) and (2.65):

$$f_0 = \omega_0/2\pi = c_p/2\pi a, \tag{2.66a}$$

or, in dimensionless form,

$$k_0 a = c_p/c. \tag{2.66b}$$

This resonance frequency will be shown in chapter 7 to be in the nature of a characteristic frequency of cylindrical shells. At this frequency, the tube circumference measures on compressional wavelength. To visualize the magnitude of this frequency, one computes $f_0 = 11$ kHz for a 15-cm steel pipe, and 700 Hz for a 2.5-cm rubber hose.

The radial wave number in the frequency range $f \gg f_0$ is obtained by substituting equations (2.65) and (2.59) in the boundary condition, equation (2.60):

$$\left. \begin{aligned} \beta^2 &= \frac{4\rho}{2\rho_s ha + \rho a^2} \\ &\simeq \frac{2\rho}{\rho_s ha} + O(a^{-2}) \end{aligned} \right\} \quad \omega^2 \gg \omega_0^2.$$

Referring to equation 2.59b, this result is seen to imply a pressure that decreases near the wall. From Equation (2.62) the corresponding axial wave number is

$$\gamma \simeq k[1 - (2\rho/\rho_s k^2 ha)]^{1/2}, \qquad \rho a/\rho_s h \ll 1. \tag{2.67}$$

This wave number is imaginary, indicating an exponentially decaying near-field pressure when $k < (2\rho/\rho_s ha)^{1/2}$. This inequality defines a *high-pass cutoff frequency* above which the phase velocity is real:

$$c_0/c = k/\gamma = [1 - (2\rho/\rho_s k^2 ha)]^{-1/2}, \qquad \omega > (2\rho/\rho_s ha)^{1/2} c.$$

With increasing frequency, the phase velocity drops from infinity at cutoff to the sound velocity. This high-frequency behavior is to be anticipated since the mass-controlled wall becomes effectively rigid for sufficiently high frequencies.

In chapter 11 these results will be retrieved as the asymptotic limits of the fundamental mode of propagation in a cylindrical waveguide.

References

1. Units of acoustical quantities and their conversion factors are summarized by L. L. Beranek, "Letter Symbols and Conversion Factors for Acoustical Quantities," in D. E. Gray, ed., *American Institute of Physics Handbook* (New York: McGraw-Hill, 1957), pp. 3.18ff.

2. For a discussion of the validity of this assumption, see P. W. Smith, Jr., "On 'Laplace's Fallacy' and Its Inverse," *J. Acoust. Soc. Am.* 30:364–365 (1958).

3. P. M. Morse and H. Feshbach, *Methods of Theoretical Physics* (New York: McGraw-Hill, 1953), p. 31.

4. Ibid; p. 116.

5. J. Lighthill, *Waves in Fluids* (Cambridge, U.K.: University Press, 1978), p. 99.

6. I. Tolstoy and C. S. Clay, *Ocean Acoustics: Theory and Experiment in Underwater Sound* (New York: McGraw-Hill, 1966), p. 27.

3.1 Introduction

The solutions of the elementary source configurations studied in chapter 2 will now be used to construct the pressure fields of relatively complicated sources and to develop some important concepts of acoustics. First, the spherical source of vanishing radius or *point source* will be used to illustrate the *image method* of simulating plane boundaries on which either the pressure or the normal velocity is required to vanish. The distortion of the sound field by these boundaries is used to introduce the concept of directivity. This concept is developed further by constructing the pressure fields of linear arrangements of point sources designed to produce pressure enhancement in selected regions to the detriment of other regions. Next, the criteria for the near and far-field are developed. The acoustic power flow associated with pressure fields is described. The pulsations of a gas bubble in a liquid are analyzed to illustrate the role of radiation loading in formulating an elementary interaction problem. Finally, this result is used to study sound propagation through a bubble swarm, a solution required in chapter 8.

3.2 The Point Source

The pressure field of a point source is obtained by taking the small-ka limit of the solution derived for the pulsating sphere, equation (2.38). The resulting expression is proportional to $-i\rho cka\dot{w}$, which is conveniently expressed in terms of the acceleration:

$$p(R, t) = \left(\frac{\rho a^2}{R}\right)\ddot{w}(a, t)\exp(ikR)$$

$$= m_0\ddot{w}(a, t)\left(\frac{a}{R}\right)\exp(ikR),$$

(3.1a)

where m_0 is the small-ka approximation of the accession to inertia in equation (2.41b). The pressure amplitude can be interpreted as the spreading loss factor (a/R) times the inertia force of entrained fluid, per unit area of the source. Writing $\ddot{w}(a, t)$ or $\ddot{W}\exp(-i\omega t)$ in terms of the specified source acceleration, we can generalize the above result by displacing the point source to a point $\mathbf{R_0}$ not coinciding with the origin of coordinates (figure 3.1), thus making $R = |\mathbf{R} - \mathbf{R_0}|$. The pressure field of the point source becomes

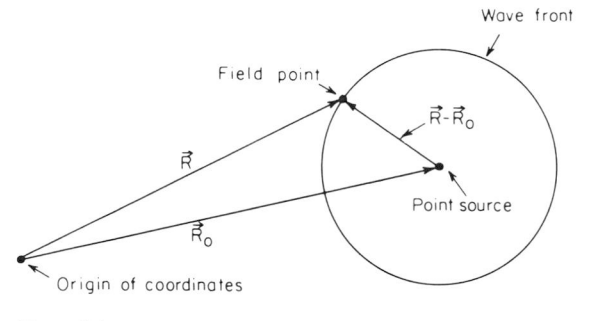

Figure 3.1
Geometry of the pressure field radiated by a point source.

$$p(\mathbf{R}, t) = \frac{\rho a^2 \ddot{W}}{|\mathbf{R} - \mathbf{R}_0|} \exp\left(ik|\mathbf{R} - \mathbf{R}_0| - i\omega t\right). \tag{3.1b}$$

Throughout this book, it will be verified repeatedly that the pressure radiated by an arbitrary source, whose characteristic dimension is small in terms of wavelengths, is determined by its surface-averaged acceleration, or more conveniently its volume acceleration $\ddot{q}(t)$, defined as the surface integral of $\ddot{w}(\mathbf{R}_0, t)$. *A source small in terms of wavelengths can therefore be replaced by a point source embodying the same volume acceleration, regardless of the source geometry or of the local acceleration distribution over the source.* Because the acceleration is uniform for the pulsating sphere, the surface integration for this source configuration reduces to a simple multiplication by the area of the sphere:

$$\ddot{q}(t) \equiv \ddot{Q}e^{-i\omega t}$$

$$= 4\pi a^2 \ddot{W}e^{-i\omega t}.$$

The pressure field of the point source now becomes

$$p(\mathbf{R}, t) = \frac{\rho \ddot{Q}}{4\pi|\mathbf{R} - \mathbf{R}_0|} \exp\left(ik|\mathbf{R} - \mathbf{R}_0| - i\omega t\right). \tag{3.1c}$$

The symbol p_s is used for the spatially dependent factor in the point source solution

$$p_s(R) \equiv \left(\frac{\rho \ddot{Q}}{4\pi R}\right) \exp(ikR). \tag{3.2}$$

This result will be used to illustrate the technique of *image sources* for simulating the effect of plane boundaries on the field of a point source. First however, the more elementary problem of a plane wave incident on a plane boundary is used to introduce the reader to the effect of rigid and "pressure-release" boundaries.

3.3 Rigid and "Pressure-Release" Boundaries

The boundaries considered so far were in motion, thus giving rise to sound waves. We now consider motionless planar boundaries whose presence disturbs an *incident wave* p_i. This disturbance takes the form of a *reflected wave* p_r, which combines with the incident wave to form the resultant field

$$p = p_i + p_r.$$

The reflected wave is determined by the boundary condition prescribed over a planar boundary, which, for the sake of convenience, is taken to coincide with the $z = 0$ plane. We shall consider two types of boundary conditions. The first corresponds to the *rigid boundary* or *baffle* over which the normal velocity component or, in other words, the pressure derivative vanishes:

$$\frac{\partial(p_i + p_r)}{\partial z} = 0, \qquad z = 0.$$

This boundary condition is approximated when airborne sound impinges on a structural wall. We shall consider the simplest situation, a plane harmonic wave traveling in the semiinfinite fluid medium $z < 0$ and impinging normally on the boundary:

$$p_i(z) = P_i \exp(ikz).$$

Other angles of incidence are studied in chapter 10. The reflected wave travels in the negative z direction, away from the boundary:

$$p_r(z) = P_r \exp(-ikz).$$

Noting that

$$\left.\begin{array}{l} \dfrac{\partial p_i}{\partial z} = ikP_i \\[2em] \dfrac{\partial p_r}{\partial z} = -ikP_r \end{array}\right\} \qquad z = 0,$$

we conclude that the boundary condition is satisfied if $P_i = P_r$, the resultant field taking the form

$$p(z) = P_i(e^{ikz} + e^{-ikz})$$
$$= 2P_i \cos kz. \tag{3.3}$$

Thus, *pressure doubling occurs on a rigid plane boundary.*

The second type of boundary condition to be considered requires that the resultant pressure vanish:

$$p_i + p_r = 0, \qquad z = 0.$$

This "*pressure-release*" *boundary* is approximated for waterborne sound by the free surface of the ocean. This can be verified by equating the particle velocities and pressures of the waterborne and airborne sound fields at the interface. Using equation (2.10) and introducing the transmitted airborne pressure $P_t \exp(ikz)$, this velocity continuity condition becomes

$$\frac{P_t}{\rho_a c_a} = \frac{P_i}{\rho_w c_w} - \frac{P_r}{\rho_w c_w},$$

where subscripts a and w refer respectively to air and water. Pressure continuity requires

$$P_t = P_i + P_r.$$

Eliminating P_r between these two equations, we have

$$P_t = \frac{P_i}{[1 + (\rho_w c_w / \rho_a c_a)]}$$
$$\approx \left(\frac{\rho_a c_a}{\rho_w c_w}\right) P_i.$$

The ratio of characteristic impedances of the two media, $42/1.5 \times 10^5 \approx 3 \times 10^{-4}$, is small, making $P_t/P_i \approx 0$. The free surface thus acts effectively as a "pressure-release" boundary.

The ideal "pressure-release" boundary condition is satisfied by setting $P_r = -P_i$. The resultant sound field is

$$p(z) = P_i(e^{ikz} - e^{-ikz})$$
$$= 2iP_i \sin kz. \tag{3.4}$$

Taking the z derivative and referring to equation (2.24), we see that the *particle velocity on the "pressure-release" surface is twice the particle velocity associated with the incident wave.*

When the incident wave is not plane the physical boundary is replaced by an appropriate fictional or *image source* constructed so that the resultant of the original field and the image field matches the desired boundary condition. The resulting sound field no longer depends on the radial coordinate exclusively but is characterized by an angular dependence or *directivity*. Both these concepts are explored in the next section.

3.4 The Image Source Simulating a Rigid Boundary; Directivity
Consider two point sources located respectively by the spherical coordinates, $R_0 = e$, $\theta_0 = 0$, and $\theta_0 = \pi$ (figure 3.2). The distance from either of these two sources to a field point (R, θ) is

$$|\mathbf{R} - \mathbf{R}_0|_{0,\pi} = (R^2 + e^2 \pm 2eR\cos\theta)^{1/2}.$$

The upper and lower sign correspond, respectively, to the source at $\theta_0 = 0$ and $\theta_0 = \pi$. The configuration is axisymmetric about the θ axis, and the sound field consequently independent of ϕ. Combining the contributions of these two sources, obtained from equation (3.2), and reducing the two terms to the same denominator, we have

$$p(R,\theta) = \rho\ddot{Q}\{(R^2 + e^2 + 2eR\cos\theta)^{1/2}\exp[ik(R^2 + e^2 - 2eR\cos\theta)^{1/2}]$$
$$+ (R^2 + e^2 - 2eR\cos\theta)^{1/2}\exp[ik(R^2 + e^2 + 2eR\cos\theta)^{1/2}]\}$$
$$\div 4\pi[(R^2 + e^2)^2 - (2eR\cos\theta)^2]^{1/2}. \tag{3.5}$$

Because of the θ dependence of the pressure field, its gradient has two components:

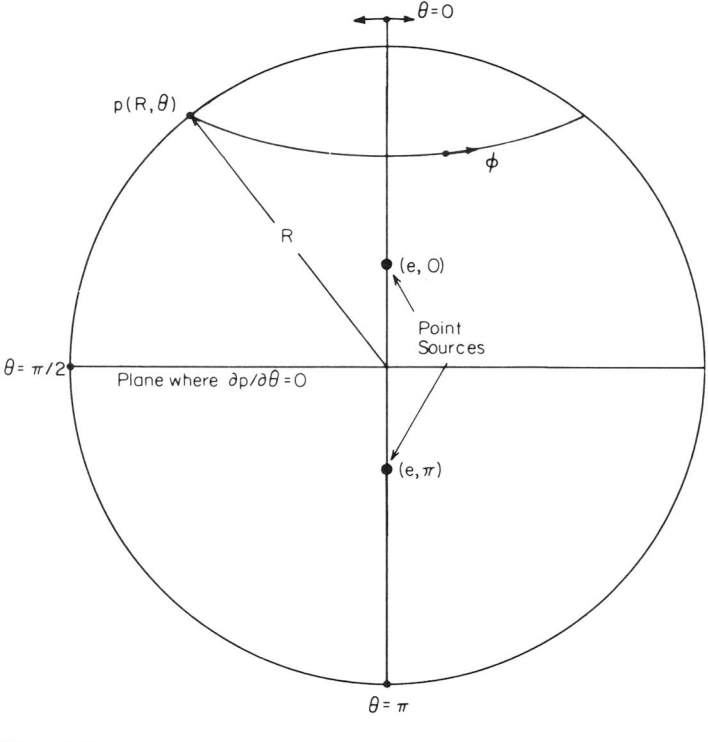

Figure 3.2
Spherical coordinate system used in constructing the pressure field of two point sources vibrating in phase located at $R = e$, $\theta = 0$, and $R = e$, $\theta = \pi$. Configuration is axially symmetrical about $\theta = 0$, thus making the pressure field independent of the circumferential angle.

$$\nabla p = \frac{\partial p}{\partial R}\hat{\xi} + \frac{1}{R}\frac{\partial p}{\partial \theta}\hat{\eta},$$

where the unit vectors $\hat{\xi}$ and $\hat{\eta}$ are associated, respectively, with the radial and tangential directions. By virtue of equation (2.28), the particle velocity is a vector with R and θ components. Differentiating with respect to θ, we verify that $\partial p / \partial \theta$ vanishes as $\theta \to \pi/2$. Consequently the $\hat{\eta}$ component of the particle velocity vector is zero in the ($\theta = \pi/2$) plane. Referring to figure 3.2, we verify that the nonvanishing, $\hat{\xi}$ component is tangential to this ($\theta = \pi/2$) plane. The sound field generated by these two sources is therefore not disturbed if a rigid boundary, coinciding with the ($\theta = \pi/2$) plane is

inserted. The sound field in the half-space $0 < \theta < \pi/2$ bounded by a rigid baffle located in the $(\theta = \pi/2)$ plane is therefore identical with the sound field produced in this half of the unbounded medium, by two sources located symmetrically with respect to this plane. The sound field radiated by an extended sound source located near a baffle can be analyzed in a similar fashion, by replacing the baffle with a symmetrically located source configuration.

The pressure in equation (3.5) and its θ derivative are too complicated to permit ready insight into the θ dependence of the sound field. For this purpose, the solution will be restricted to field points located at ranges that are moderately large compared with the separation between sources. Here,

$$|\mathbf{R} - \mathbf{R}_0|_{0,\pi} \approx R \mp e\cos\theta + O\left(\frac{e^2\cos^2\theta}{R}\right).$$

When this expression is substituted in equation (3.1), and the fields of the two sources are combined, the resultant pressure becomes

$$
\begin{aligned}
& p(R,\theta) \\[4pt]
&\quad \approx \frac{\rho\dot{Q}}{4\pi}\left\{\frac{\exp[ik(R - e\cos\theta)]}{R - e\cos\theta}\right. \\[6pt]
&\qquad \left.+ \frac{\exp[ik(R + e\cos\theta)]}{R + e\cos\theta}\right\} \\[6pt]
&\quad \approx \frac{\rho\ddot{Q}}{2\pi R}\exp(ikR)\,[\cos(ke\cos\theta) \\[6pt]
&\qquad - i\frac{e}{R}\cos\theta\sin(ke\cos\theta)]
\end{aligned}
\qquad \frac{ke^2}{R},\quad \left(\frac{e}{R}\right)^2 \ll 1. \qquad (3.6)
$$

The explicit statement of the inequality involving the wave number is required because dropping the exponent $\frac{1}{2}kR(e/R)^2$ is not justified, however small $(e/R)^2$, unless the exponent is small in terms of radians. For ranges that are so large (or sources that are so close together) that the linear term in (e/R) can be neglected, this reduces to

$$p(R,\theta) \approx \frac{\rho\ddot{Q}}{2\pi R}\exp(ikR)\cos(ke\cos\theta), \qquad \frac{ke^2}{R},\ \frac{e}{R} \ll 1.$$

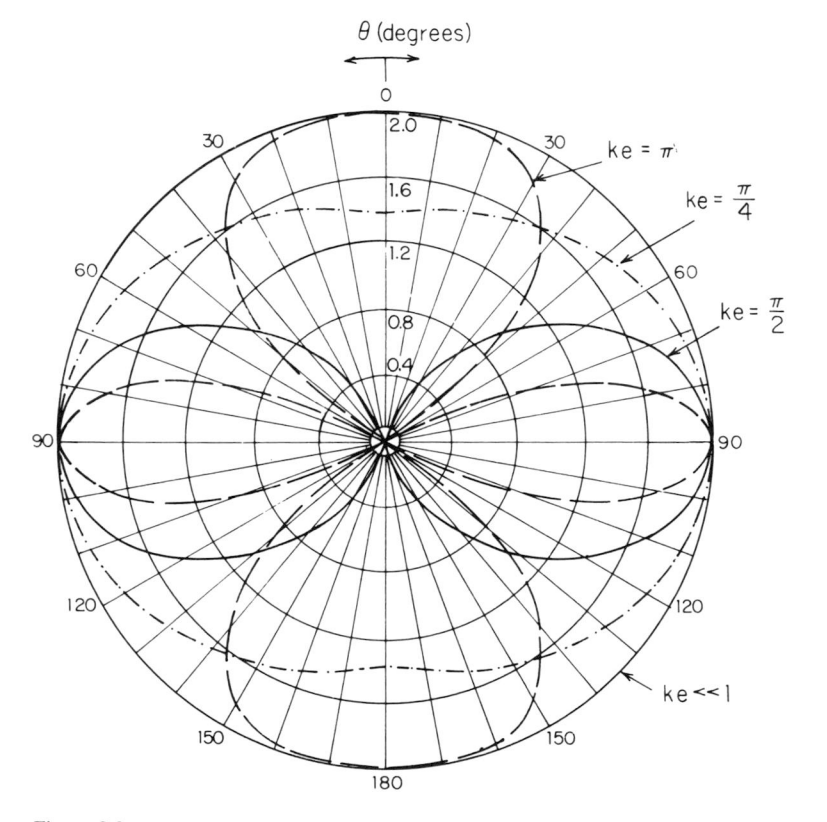

Figure 3.3
Directivity factor of two identical point sources vibrating in phase [equation (3.7)].

The range where these large-R approximations hold is referred to as *the far-field*, a concept discussed further in section 3.7. This result can be expressed in terms of the pressure of a point source, equation (3.2):

$$p(R, \theta) = 2p_s(R) \cos(ke \cos \theta)$$
$$\approx 2p_s(R), \qquad k^2 e^2 \ll 1. \tag{3.7}$$

The ratio $|p(R, \theta)/p_s(R)|$ is the directivity factor plotted in figure 3.3. The increasing complexity of the distribution in angle of the pressure with increasing ke will be found to be characteristic of all sources displaying directivity. When the two sources are close together in terms of wavelengths

$(ke \ll 1)$, this directivity factor tends to a θ-independent value of two. The resulting pressure is twice that of the single point source endowed with the same volume acceleration $\ddot{Q}$. Two point sources can thus be replaced by a single source with volume acceleration $2\ddot{Q}$. This is the first example of the above-mentioned equivalence of a point source and of a more complicated source (in this case, two point sources) with the same resultant volume acceleration, provided the actual source vibrates in phase and is small in terms of wavelengths. The velocity vector corresponding to equation (3.7) is

$$\mathbf{d}(R,\theta) = \dot{w}\hat{\xi} + \dot{u}\hat{\eta}$$

$$\approx \frac{-i\ddot{Q}}{2\pi k R^2 c} \exp(ikR)\left[(ikR - 1)\cos(ke\cos\theta)\hat{\xi}\right. \tag{3.8}$$

$$\left. + ke\sin\theta\sin(ke\cos\theta)\hat{\eta}\right], \qquad \frac{ke^2}{R}, \quad \frac{e}{R} \ll 1.$$

Since the $\hat{\eta}$ component vanishes for $\theta = \pi/2$, the velocity vector is tangential to the plane of symmetry.

3.5 The "Pressure-Release" Boundary
The point sources in figure 3.2 are now assumed to vibrate 180° out of phase, the time-dependent factors of the source at $(e,0)$ and (e,π) being respectively, $\exp[-i(\omega t - \pi)]$, $\exp(-i\omega t)$. The resultant sound field at long range obtained, by superposition, is

$$p(R,\theta)$$

$$\approx \frac{\rho\ddot{Q}}{4\pi}\left(-\frac{\exp[ik(R - e\cos\theta)]}{R - e\cos\theta}\right.$$

$$\left. + \frac{\exp[ik(R + e\cos\theta)]}{R + e\cos\theta}\right), \qquad \frac{ke^2}{R}, \quad \left(\frac{e}{R}\right)^2 \ll 1 \tag{3.9}$$

$$\approx 2ip_s(R)\sin(ke\cos\theta), \qquad \frac{ke^2}{R}, \quad \frac{e}{R} \ll 1.$$

The directivity factor of this pressure field, obtained by dividing equation (3.9) by ip_s, is shown in figure 3.4 for various values of ke. As in the preceding figure, the number of lobes increases with ke. At $\theta = \pi/2$, $\cos\theta = 0$ and hence $\sin(ke\cos\theta) = 0$. The directivity pattern of this source therefore always

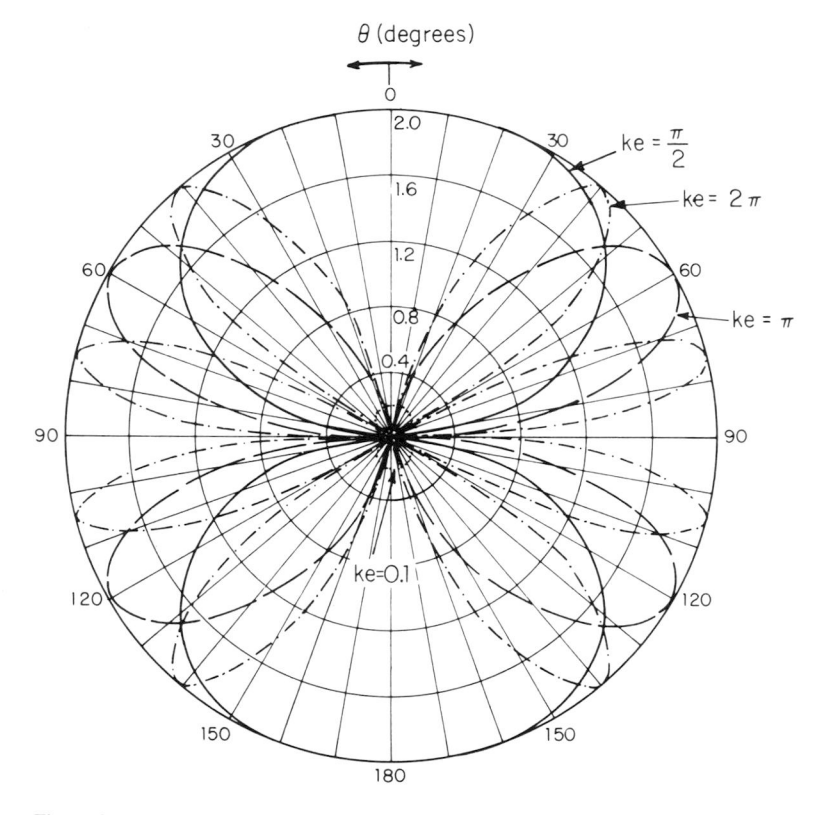

Figure 3.4
Directivity factor of two identical point sources vibrating in phase opposition [equation (3.9)].

displays vanishing pressures in the midplane $(\theta = \pi/2)$. Two point sources at $(e, 0)$ and (e, π) vibrating with equal volume acceleration amplitude but displaying a 180° phase difference in their time-dependent factor are therefore equivalent to a single point source at $(e, 0)$ in the presence of a "pressure-release" plane, that is, of a plane of vanishing pressure, coinciding with the midplane $(\theta = \pi/2)$. It was shown in the preceding section that such a plane is, for practical purposes, the ocean surface. A point source located at a depth e can therefore be analyzed as though it were in an infinite liquid medium, if one introduces a second source identical to the original source, except for a 180° phase shift in the time-dependent factor, located a vertical distance $2e$ above the original source.

The velocity vector $\dot{\mathbf{d}}$ associated with this pressure field is

$$\dot{\mathbf{d}}(R, \theta) = \frac{\ddot{Q}}{2\pi c k R^2} \exp(ikR) \left[(ikR - 1) \sin(ke \cos \theta) \hat{\xi} \right.$$

$$\left. - ke \sin \theta \cos(ke \cos \theta) \hat{\eta} \right], \qquad \frac{ke^2}{R}, \frac{e}{R} \ll 1 \qquad (3.10)$$

In contrast to the velocity in equation (3.8), whose $\hat{\xi}$ (radial) component peaks in the $\theta = \pi/2$ plane, this velocity distribution displays a zero $\hat{\xi}$ component in this plane. It is thus verified that the velocity vector is normal to the pressure-release boundary.

When the sources are close together in terms of wavelengths ($ke \ll 1$), they become a *dipole*, or *doublet*. We can thus replace the sine in equation (3.9) by its argument. Generalizing to an arbitrary origin of coordinates $\mathbf{R}_0$, we describe the pressure field radiated by a dipole in terms of $|\mathbf{R} - \mathbf{R}_0|$:

$$p(\mathbf{R}) = \frac{i\rho \ddot{Q} ke}{2\pi|\mathbf{R} - \mathbf{R}_0|} \cos \theta \exp(ik|\mathbf{R} - \mathbf{R}_0|), \qquad k^2 e^2, \frac{e}{R} \ll 1. \qquad (3.11)$$

The dipole field is negligible compared with the field generated by the two point sources vibrating in phase, in this small-ke range [equation (3.7) with the cosine set equal to unity]. Specifically, the dipole pressure compares with the monopole pressure as ke to unity. This result is not surprising, because the resultant volume acceleration of the dipole is zero, and its characteristic dimension normalized to the acoustic wavelength is equal to ke/π and hence, by definition, small in the $ke \ll 1$ range. When the characteristic dimension $2e$ exceeds a half-wavelength ($ke > \pi/2$), the two out-of-phase sources can no longer be combined into a dipole, and the pressure maximum of the two out-of-phase sources becomes comparable to that of the two in-phase sources. A peculiarity of the dipole is that it is directive in the small-ke limit, where it radiates a figure-eight pattern (see the $ke = 1/10$ curve in figure 3.4).

We shall now illustrate how point sources can be combined to radiate more directive patterns whereby most of the energy is radiated into a small solid angle.

3.6 Linear Arrays of Point Sources
In many practical applications, it is desirable to concentrate the acoustic power radiated by a sound source in a narrow lobe or beam so as to minimize

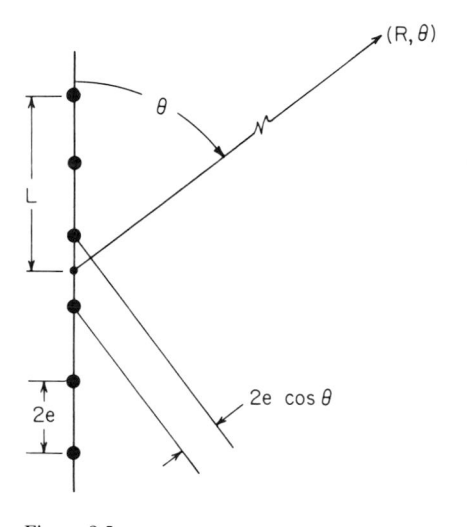

Figure 3.5
Construction of the directivity pattern of a linear array of point sources.

the total power required to achieve a specified pressure at a given range and in a desired direction, or bearing. Furthermore, it is convenient to direct or steer this beam electrically, rather than by rotating the source mechanically. Such a sound-projecting system can be achieved in the form of a linear arrangement or array of point sources whose respective input is individually controlled. A similar incentive exists to construct directive listening systems for the purpose of localizing a sound source in the ocean, or of detecting the presence of a structure by an echo-ranging technique (see section 10.1). In this application, the responsiveness of the listening system must be restricted to acoustic energy incident from a small solid angle. The directive properties of arrays of sound sources and of sound receivers (microphones and, in water, hydrophones) are governed by the same equations.[1]

These equations will be derived here to illustrate applications of the elementary solutions above. Even though we shall refer to sound sources only, it is understood that the results are applicable to receivers as well. These sources are assumed to be small in terms of wavelengths and, consequently, individually omnidirectional. For the sake of simplicity, an even number $2N$ of linearly arranged elements equally spaced a distance $2e$ from each other is assumed. The origin of spherical coordinates is equidistant from the central pair of elements, the elements being located on the $\theta = 0$ axis (see figure 3.5 where $N = 3$). As a result of the difference in path length

from two neighboring elements to a field point, their respective contributions reach the far-field with a phase difference of $(2ke\cos\theta)$. We proceed as in the derivation of the far field radiated by a pair of elements, equation (3.7). The far-field of $2N$ elements is constructed by linearly superposing the pressure contributions radiated by the individual elements and keeping track of the phase shifts between contributions. If each element has a volume acceleration $\ddot{Q}/2N$ and generates a pressure $p_s(R)/2N$, the resultant pressure is

$$
\begin{aligned}
p(R,\theta) &= \frac{p_s(R)}{2N} \sum_{n=-N}^{N-1} \exp[-ik(2n+1)e\cos\theta] \\
&= \frac{p_s(R)}{N} \sum_{m=0}^{N-1} \cos[k(2m+1)e\cos\theta].
\end{aligned}
\tag{3.12}
$$

The sum of cosines in the preceding expression is readily evaluated by means of the trigonometric identity[2]

$$
\cos\alpha + \cos 3\alpha + \cdots + \cos(2N-1)\alpha = \frac{\sin 2N\alpha}{2\sin\alpha}.
$$

Setting $\alpha \equiv ke\cos\theta$ in this relation, we can express equation (3.12) concisely as

$$
p(R,\theta) = \frac{p_s(R)\sin(2Nke\cos\theta)}{2N\sin(ke\cos\theta)}.
\tag{3.13}
$$

The maximum of the directivity factor occurs when $\cos\theta = 0$, in the $\theta = \pi/2$ plane, where both the numerator and denominator vanish. Applying l'Hospital's rule, we find this maximum to be unity. Hence,

$$
p\left(R, \frac{\pi}{2}\right) = p_s(R).
\tag{3.14}
$$

The peak pressure therefore equals the pressure radiated by a point source whose volume acceleration is that of the entire array.

The limiting case of discrete elements is a continuous *source* or *line array* taken here to coincide with the x axis. The total volume acceleration of the array, $\ddot{Q}$ is expressed in terms of its length $2L$ and the volume acceleration

per unit length, $d\ddot{Q}/dx$:

$$\ddot{Q} \equiv 2L \left(\frac{d\ddot{Q}}{dx} \right). \tag{3.15}$$

The summation in equation (3.12) now becomes

$$
\begin{aligned}
p(R, \theta) &= \frac{p_s(R)}{2L} \int_{-L}^{L} \exp(-ikx \cos \theta) \, dx \\
&= \frac{p_s(R) \sin(kL \cos \theta)}{kL \cos \theta}.
\end{aligned}
\tag{3.16}
$$

This is expressed concisely in terms of the spherical Bessel function defined in equation (2.45):

$$p(R, \theta) = p_s(R) j_0(kL \cos \theta). \tag{3.17}$$

The directivity factor (figure 3.6) displays a maximum when the argument of the Bessel function equals zero, in the $\theta = \pi/2$ plane. Pressure nulls occur at $\cos^{-1}(n\pi/kL)$, where $n = 1, 2, \ldots \leq kL/\pi$. Hence, as for the pair of point sources (figure 3.3) increasing frequencies result in an increase in the number of lobes. An important difference is that, for the array, the maximum pressure at $\theta = \pi/2$ or main lobe exceeds the other maxima or side lobes by a ratio of approximately

$$kL \cos \theta_m \approx (m + \tfrac{1}{2}) \pi.$$

The source being cylindrically symmetrical about the θ axis, the main lobe is disk shaped in the three-dimensional pattern. Because the plane of symmetry of this disk is perpendicular to the line array, this source configuration is called a *broadside array*. The physical interpretation of this result is that, for $\theta \neq \pi/2$, $\cos \theta \neq 0$, the pressure contributions from different elements of the line array undergo partial or complete cancellation, because of differences in travel time, which give rise to phase differences $(k \, dx \cos \theta)$ between elements spaced a distance dx (see figure 3.5). For $\theta = \pi/2$, the travel times tend to the same value R/c as $L/R \to 0$, in the far-field, thus excluding phase cancellations.

This interpretation of the broadside characteristics suggests that the main lobe can be steered in any specified direction θ_0 by introducing a

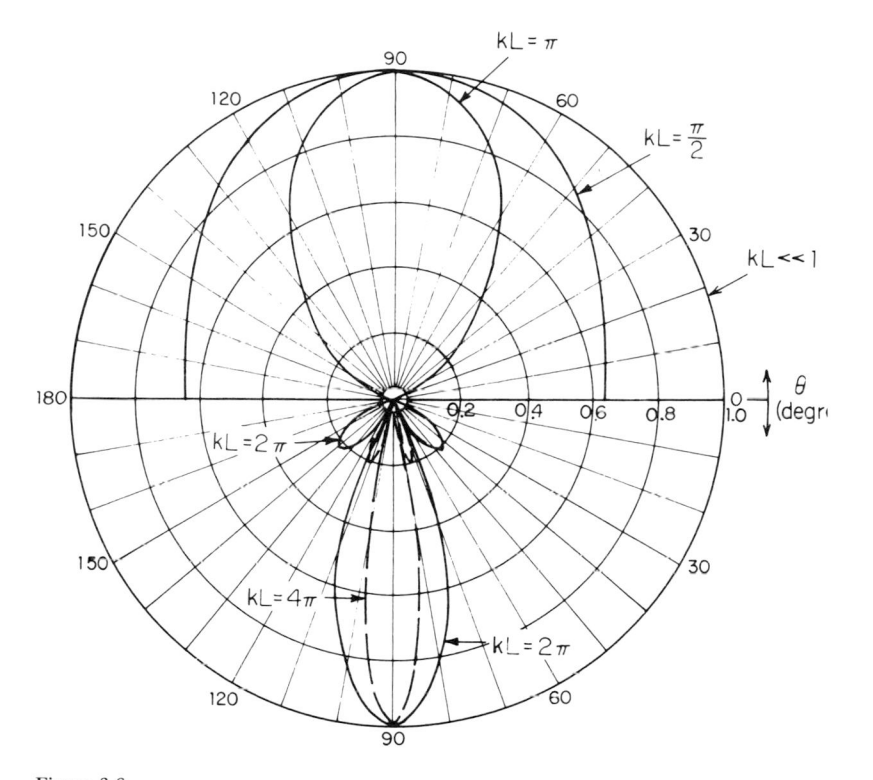

Figure 3.6
Directivity factor of a line source pulsating in phase or "broadside" array [equation (3.18) with $\gamma = 0$]. The directivity factor is defined here as the pressure amplitude normalized to the pressure amplitude radiated by a point source whose volume acceleration equals that of line source. The same normalization constant is used in figures 3.7 and 3.8.

position-dependent phase shift of the signal exciting the elements of the array. This phase shift, which is readily controlled in electrically powered elements, is designed to cancel the phase shift arising from travel time differences. A phase shift in the excitation of the elements results in a volume acceleration per unit length of array whose amplitude is independent of x, but whose phase varies linearly:

$$\frac{d\ddot{Q}}{dx} = \left|\frac{d\ddot{Q}}{dx}\right| \exp(i\gamma x),$$

where γ is a positive or negative constant. In lieu of the phase shift embodied in the integrand of equation (3.16), which is associated exclusively with differences in travel time, the phase angle of the pressure radiated by an element located at x, with respect to the element located at the center of the array, $x = 0$, now becomes $(\gamma - k\cos\theta)x$. The far-field distribution-in-angle of a phased line array is therefore given by the integral

$$p(R, \theta) = \frac{p_s(R)}{2L} \int_{-L}^{L} \exp[i(\gamma - k\cos\theta)x]\,dx$$

$$= p_s(R) j_0 [(\gamma - k\cos\theta)L].$$

$$(3.18)$$

This directivity pattern has two principal maxima or main lobes of amplitude $p_s(R)$ located on the conical surface

$$\theta_0 = \cos^{-1}(\gamma/k).$$

This is illustrated in figure 3.7 for $\gamma = k/2^{1/2}$, that is, $\theta_0 = 45°$ and $135°$.

The two main lobes can be made to sweep through space from $\theta = \pi/2$ to $\theta = 0$ and π by letting γ vary from 0 to k. The latter value of γ identifies an end-fire array that generates a pencil-shaped main lobe or beam whose axis coincides with that of the array (figure 3.8). Physically, this corresponds to a phase shift that cancels the spatial phase shift resulting from differences in travel time along the array axis. The argument of j_0 in equation (3.18) now becomes $k(1 - \cos\theta)L$, which is more conveniently expressed in terms of $\sin(\theta/2)$:

$$p(R, \theta) = p_s(R) j_0\left(2kL \sin^2\frac{\theta}{2}\right).$$

$$(3.19)$$

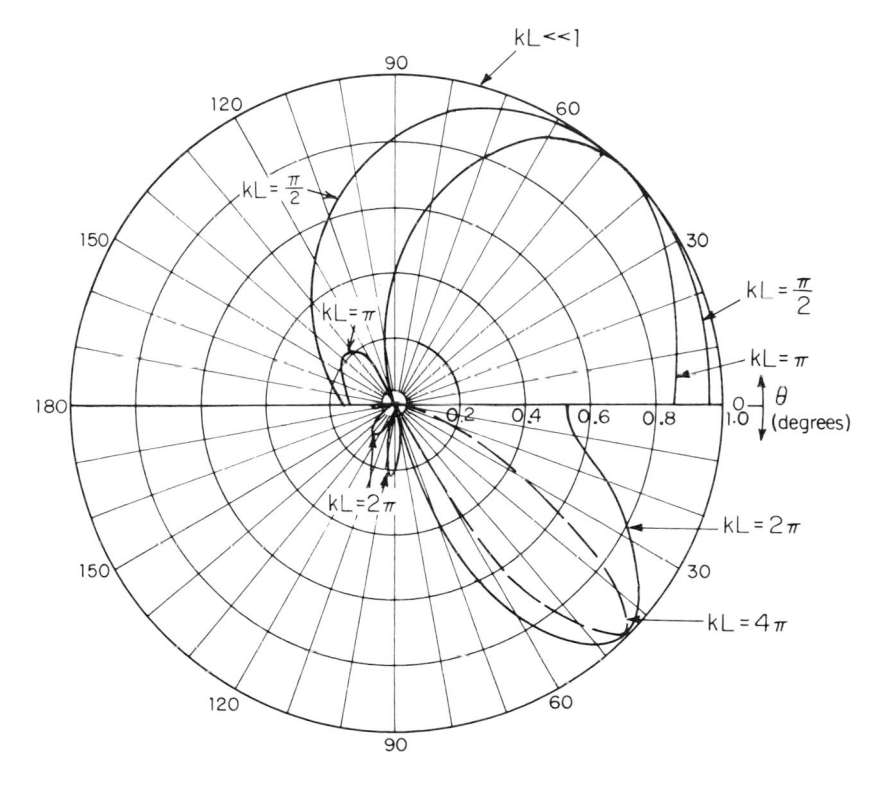

Figure 3.7
Directivity factor of a phased line array [equation (3.18) with $\gamma = k/2^{1/2}$].

In calculating travel time differences in the derivations of section 3.6, it was assumed that the range R is so large that e/R or $L/R \ll 1$, making vectors leading from different portions of the array to the field point effectively parallel. Furthermore, in the step from equation (3.5) to (3.6), it was assumed that ke^2/R (and subsequently kL^2/R) $\ll 1$. This and other aspects of the "long-range" assumption will now be discussed in some detail.

3.7 Far-Field Conditions
The term "long-range" has been used rather loosely throughout this chapter. It will now be defined more precisely as the range required to achieve the following criteria, which characterize the far field:

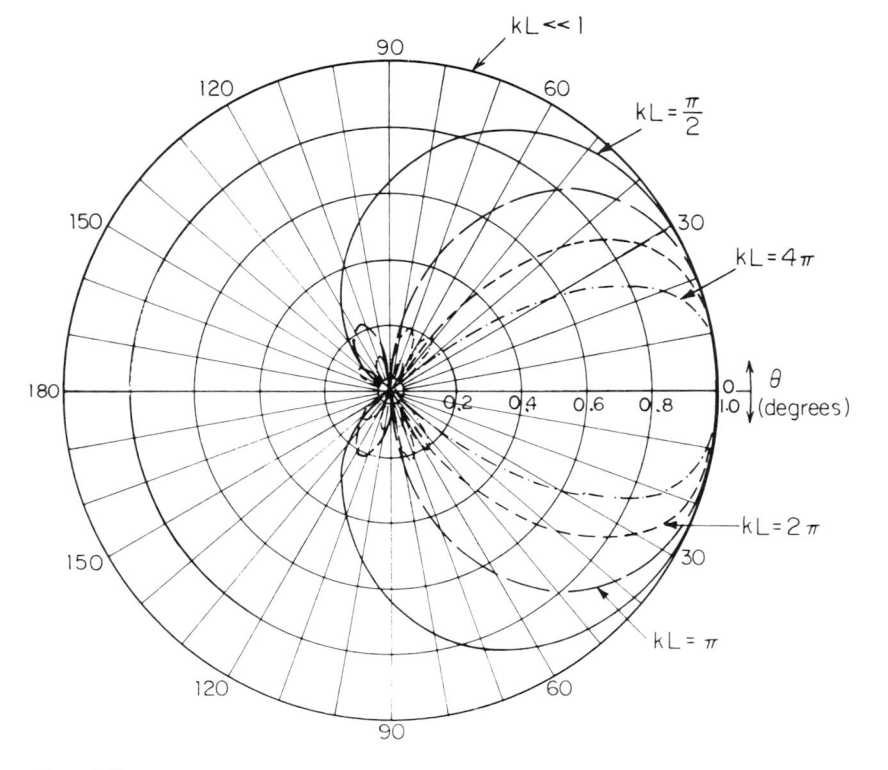

Figure 3.8
Directivity factor of an end-fire array [equation (3.19)].

1. The pressure displays a spherical spreading loss, its amplitude decaying as R^{-1}.

2. The θ and ϕ dependence of the pressure amplitude do not vary with R.

3. The specific acoustic impedance equals the characteristic, plane-wave impedance ρc.

The spatial dependence of the pressure is thus simpler in the far field than in the near-field.

Criteria 1 and 2 define the *Fraunhofer zone*. The region at closer range is associated with *Fresnel's* name. The Fraunhofer criteria are illustrated by the pressure field of the pair of point sources analyzed in section 3.4, at

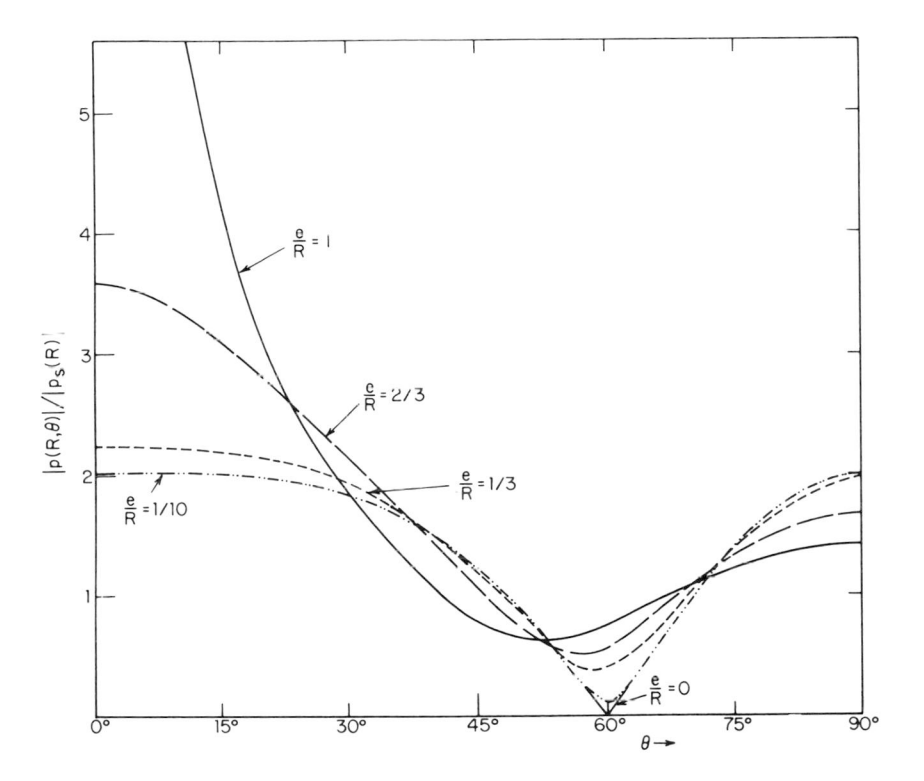

Figure 3.9
Normalized pressure amplitude of two identical point sources vibrating in phase illustrating
transition from near-field [equation (3.6)] to far-field [equation (3.7) and figure 3.3] for $ke = \pi$.

ranges where equation (3.7) applies. The transition from near to far-field
is illustrated in figure 3.9. Under the less restrictive range conditions of
equation (3.6), there is a pressure component that decays as R^{-2}, thus
conflicting with criterion 1. For values of θ that cause the R^{-1} term to vanish,
specifically when $\theta = \cos^{-1}[(n + \frac{1}{2})\pi/ke]$, the first far-field criterion is never
met, however large the range. For this particular source, this anomaly can
arise only if the characteristic source dimension $2e$ exceeds $\lambda/2$. Later on in
this section, we shall see that there is no such limitation for certain other
sources, such as the dipole.

   Excluding bearings for which the R^{-1} term vanishes, it is apparent
that neither the first nor the second far-field criterion is met as long as
the R^{-2} term is not negligible, because its angular dependence differs from

that of the R^{-1} component. Consequently, excluding the zeros of the R^{-1} term, far-field conditions obtain when $e/R \ll 1$ and $ke^2/R \ll 1$. The former condition is purely geometric in that it implies that the source subtends an infinitesimal solid angle as seen from a field point, rays drawn from various source points being effectively parallel. The inequality involving the wave number merely makes the geometric inequality more restrictive if ke is large. In other words, if the source is large in terms of wavelengths, the Fraunhofer region starts at greater ranges as measured in characteristic source dimensions.

The third far-field criterion is met if (1) the $\hat{\boldsymbol{\eta}}$ term in equation (3.8) is negligible compared with the $\hat{\boldsymbol{\xi}}$ term, and (2) if the $\hat{\boldsymbol{\xi}}$ term reduces to its kR component:

$$\mathbf{\dot{d}}(R, \theta) \approx \frac{\ddot{Q}}{2\pi Rc} \exp\left(ikR\right) \cos\left(ke\cos\theta\right)\hat{\boldsymbol{\xi}} \qquad \begin{cases} ke^2/R, & e/R \ll 1, \\ kR \gg 1. \end{cases} \tag{3.20}$$

Referring to equations (3.10) and (3.11), one verifies that for these same restrictions on R, the dipole field also displays a ρc impedance. The third far-field criterion has added a third range restriction $kR \gg 1$. Specifically the range must be large in terms of λ. More generally, the condition $k|\mathbf{R} - \mathbf{R}_0| \gg 1$ ensures that the radius of curvature of the wave front is large in terms of wavelengths. This particular far-field criterion can be achieved on the very surface of the radiator, if its radius of curvature and dimensions are large in terms of λ [see equation (2.40)].

For certain source configurations and boundaries, far-field conditions are never attained. In particular, the formal R^{-1} dependence displayed by the dipole solution, equation (3.11), requires a special interpretation for field points close to the $\theta = \pi/2$, "pressure-release" plane. Thus, at a field point located at long range R, at the same distance from the pressure-release plane as the source, the angular coordinate and the trigonometric functions in equation (3.11) take the form

$$\left. \begin{aligned} \theta &\approx \left(\frac{\pi}{2}\right) - \left(\frac{e}{R}\right) \\ \cos\theta &\approx \sin\left(\frac{e}{R}\right) \approx \frac{e}{R} \\ \sin\left(ke\cos\theta\right) &\approx \frac{ke^2}{R} \end{aligned} \right\} \qquad e \ll R.$$

When this is substituted in equation (3.11), one obtains a solution decaying as R^{-2}. Hence, even though $e \ll R$, far-field conditions are not achieved with respect to spreading loss. This anomalous decay is important in regard to sound propagation in the ocean, where it is referred to as the acoustic *Lloyd mirror effect*,[3] but this anomaly is unimportant in regard to the problems dealt with in this text. For example, the formal solution in equation (3.11) is adequate for computing the acoustic power radiated by a dipole source when using expressions specialized to the far-field, because these expressions are invalid only in the region $\theta \approx \pi/2$, where the pressure is nearly zero and where no significant energy flow takes place. The same applies to the nulls of equation (3.7), which were also shown to conflict with far-field criteria.

Having constructed expressions describing the pressure fields of various sources, we now examine the associated radiation of acoustic energy.

3.8 Acoustic Power and Intensity

In elementary mechanics, work is defined as the time integral of a moving force multiplied by the velocity component parallel to the force. This concept is readily extended to the work performed by an area element dS of the moving boundary of a fluid, the force being the surface pressure multiplied by dS. A surface integration over the boundary yields the resultant work. Work performed by a harmonically reciprocating boundary, typically a sound source, is similarly computed, provided only the real components of pressure and velocity are retained in the integrand. The corresponding average acoustic power radiated by the source is computed by confining the time integration to one period, and dividing the result by the period.

This calculation also yields the acoustic power flowing through a control surface located in the body of the fluid. The pressure in this case is the local sound pressure, and the velocity the fluid particle velocity component normal to the area element of the control surface. The acoustic power radiated by a sound source is computed most conveniently by performing the surface integration over a control surface enclosing the source, at sufficiently large range to satisfy the third far-field criterion. This allows the elimination of either the particle velocity or the pressure in terms of the characteristic impedance defined in equation (2.10). The control surface is typically selected to be a sphere concentric with the source, with $kR \gg 1$ to satisfy the third far-field criterion, pressure and particle velocity being expressed in terms of spherical coordinates.

For a directional source, the power flow through this spherical surface varies with bearing. Specifically, the power flow per unit area, or *acoustic*

intensity, is used to define the directive properties of the source. For a harmonic source, the average of the intensity over one period equals the average over all time, as indicated by equation (2.16). Using equation (2.10), we thus finally obtain the intensity in terms of the pressure amplitude squared,

$$I(R, \theta, \phi) = \frac{1}{2\rho c} |p(R, \theta, \phi)|^2, \qquad kR \gg 1, \tag{3.21}$$

or, alternatively, of the radial particle velocity squared,

$$I(R, \theta, \phi) = \tfrac{1}{2}\rho c |\dot{w}(R, \theta, \phi)|^2, \qquad kR \gg 1. \tag{3.22}$$

The intensity concept can be generalized by defining a vector intensity pointing in the direction of the fluid particle velocity:

$$\mathbf{I} = \langle p\dot{\mathbf{d}} \rangle. \tag{3.23a}$$

For our purpose, we require only the intensity component normal to the control surface.

We shall now specialize these results to specific configurations. For the academic case of infinitely extended plane waves, the intensity is independent of all spatial coordinates. For the uniformly pulsating spherical source, the above equations depend only on R. It is of interest to express the intensity in terms of the velocity amplitude of the source. From equations (2.38) and (2.39),

$$I(R) = \frac{\rho c (ka)^2 a^2 \dot{W}^2}{2[(ka)^2 + 1]R^2}, \qquad kR \gg 1. \tag{3.23b}$$

This intensity decreases with range as R^{-2}, in contrast to the constant plane-wave intensity. Since the pressure field of the uniformly pulsating spherical source is independent of θ and ϕ, the acoustic power is simply the intensity at range R multiplied by the area of the sphere of radius R:

$$\Pi = 4\pi R^2 I(R)$$
$$= \frac{2\pi\rho c (ka)^2 a^2 \dot{W}^2}{[(ka)^2 + 1]}. \tag{3.24}$$

Thus, the power is independent of range, as expected from the requirement that energy be conserved in an inviscid acoustic medium where dissipation cannot occur.

At the surface, or at ranges too short to permit the use of the far-field characteristic impedance ρc, the evaluation of the intensity must account for the nonvanishing phase angle α of the impedance, equations (2.26) and (2.42). The intensity now becomes, in terms of the surface impedance z_0,

$$
\begin{aligned}
I(a) &= \frac{1}{T}\int_0^T \{\mathrm{Re}[\dot{w}(a)]\,\mathrm{Re}[p(a)] + \mathrm{Im}[\dot{w}(a)]\,\mathrm{Im}[p(a)]\}\,dt \\
&= \frac{1}{T}\int_0^T [\mathrm{Re}(\dot{w})\,\mathrm{Re}(z_0\dot{w}) + \mathrm{Im}(\dot{w})\,\mathrm{Im}(z_0\dot{w})]\,dt \\
&= \frac{\dot{W}^2|z_0|\cos\alpha}{2} \equiv \frac{\dot{W}^2 r_0}{2}.
\end{aligned}
\tag{3.25}
$$

If we substitute equation (2.41a) for r_0 and multiplying by the source area $4\pi a^2$, this yields the same expression for power as equation (3.24). Alternatively, the intensity can be computed in terms of the surface pressure amplitude $|p(a)|$:

$$
\begin{aligned}
I(a) &= \frac{1}{T}\int_0^T \left\{\mathrm{Re}\left[\frac{p(a)}{z_0}\right]\mathrm{Re}[p(a)]\right. \\
&\quad \left. + \mathrm{Im}\left[\frac{p(a)}{z_0}\right]\mathrm{Im}[p(a)]\right\}\,dt \\
&= \frac{|p(a)|^2\cos\alpha}{2|z_0|} \equiv \frac{|p(a)|^2 r_0}{2|z_0|^2}.
\end{aligned}
\tag{3.26}
$$

When the sound field is dependent on θ and ϕ, the power is obtained by integrating the radial component of the intensity vector over a large sphere of radius R concentric with the sound source:

$$
\Pi = \frac{R^2}{2\rho c}\int_0^{2\pi}\left[\int_0^{\pi}|p(R,\theta,\phi)|^2\sin\theta\,d\theta\right]d\phi, \qquad kR \gg 1.
\tag{3.27}
$$

If a pulsating boundary encloses a fluid volume, rather than being surrounded by it, it generates a standing wave field characterized by the absence of acoustic power flow. This can be verified for the fluid sphere by

noting that the impedance, acting on the spherical boundary, equation (2.48), is purely imaginary, thus making $\cos \alpha$ and hence the intensity in equation (3.25) vanish. The surface integral of this intensity over the spherical enclosure, in other words the power, is therefore also zero. This could be anticipated from the fact that power flow into a lossless finite system, not containing a power sink, is incompatible with the assumed steady-state conditions.

3.9 The Pulsating Gas Bubble

One of the simplest illustrations of the effect of radiation loading on a vibrating system is the pulsating gas bubble. This analysis is confined to linear pulsations. Deviations from spherical symmetry do not contribute to the monopole strength of the bubble vibrations and can therefore be disregarded. Both the natural frequency and the resonance amplification or quality factor of the fundamental spherically symmetric or "breathing" mode of the bubble will be derived. The procedure of matching the solution of the wave equation in the gas sphere to the solution in the ambient liquid provides another illustration of the coupled solution of an elementary interaction problem between two media. In the next section, these results will be used to illustrate sound dispersion and attenuation in a bubble swarm.

The acoustic pressure field in the sphere, equation (2.46) must be compatible with the field in the ambient liquid, equation (2.38). This requires that there be continuity of pressure and of radial velocity at the gas-liquid interface. Using the subscripts g and l to identify these two media, and taking the bubble diameter to be $2a$, these boundary conditions are

$$p_g(a) = p_l(a), \tag{3.28}$$

$$\dot{w}_g(a) = \dot{w}_l(a). \tag{3.29}$$

These equations will be combined to form the characteristic equation governing the bubble-water interaction. The lowest root of this characteristic equation is the fundamental breathing mode frequency ω_0 of the bubble. Dividing equation (3.28) by (3.29), we have

$$z_{0g}(a) = z_{0l}(a), \qquad \omega = \omega_0. \tag{3.30}$$

It will be assumed, and verified subsequently, that the fundamental natural frequency of the bubble lies in the small-ka region, both for the gas

and for the liquid. The radiation loading on the bubble is therefore primarily in the form of a mass, equation (2.43). The impedance presented by the bubble is in the form of a stiffness, equation (2.48). The boundary condition, equation (3.30), thus becomes

$$-\frac{ik_{0\mathrm{g}}}{\omega_0} = -i\omega_0 m_{0\mathrm{l}}, \qquad k_{0\mathrm{g},1}a \ll 1. \tag{3.31}$$

Solving for the natural frequency, we have

$$\omega_0 = \left(\frac{k_{0\mathrm{g}}}{m_{0\mathrm{l}}}\right)^{1/2}. \tag{3.32}$$

When the small-ka values of k_0, equation (2.49), and of m_0, equation (2.41b), are substituted, an explicit expression for the natural frequency is obtained:

$$\omega_0 = \frac{(3\rho_{\mathrm{g}} c_{\mathrm{g}}^2 / \rho_1)^{1/2}}{a}. \tag{3.33}$$

The bulk modulus of the gas, $\rho_{\mathrm{g}} c_{\mathrm{g}}^2$, is a linear function of the hydrostatic pressure, equation (2.4). The natural frequency thus becomes

$$\omega_0 = \frac{(3\gamma P_{\mathrm{s}} / \rho_1)^{1/2}}{a}. \tag{3.34}$$

For air bubbles in water, the natural frequency, in Hz, is

$$f_0 - \frac{328}{a} P_{\mathrm{s}}^{1/2}, \tag{3.34a}$$

where P_{s} is in atmospheres and a in cm. The corresponding dimensionless frequency normalized to the wave number in water is

$$k_{01}a = \left(\frac{3\gamma P_{\mathrm{s}}}{\rho c^2}\right)^{1/2}$$

$$= 0.014 P_{\mathrm{s}}^{1/2}. \tag{3.35}$$

This natural frequency can be envisioned as pertaining to a spring, represented by the gas bubble, surrounded by a shell of liquid. By virtue of equation (2.41b), specialized to small ka, its radial dimension equals the radius of the sphere. The shell's volume equals three times that of the gas bubble. The shell's outer radius is therefore $4^{1/3}a$.

Radiation damping predominates only for large bubbles whose natural frequency, for air bubbles in water, falls below 1 kHz. For higher natural frequencies, the predominant energy loss is associated with heat transfer across the bubble boundary.[4] For our purposes, we consider the large-bubble situation. The quality factor of a system with natural frequency ω_0 is defined in terms of the ratio of the mass reactance to the resistance of the system as

$$Q_0 \equiv \frac{\omega_0 m_0}{r_0}. \tag{3.36}$$

For the pulsating bubble, the ratio of mass to resistance is merely the ratio of accession to inertia over specific acoustic resistance, equations (2.41a), specialized to small ka:

$$\begin{aligned} Q_0 &= \frac{\omega_0 m_{01}}{r_{01}} \\ &= (k_{01}a)^{-1}. \end{aligned} \tag{3.37}$$

Turning to equation (3.35), the quality factor at 1 atm (atmosphere) is found to be approximately 70.

3.10 Dispersion and Attenuation: Sound Propagation in Bubble Swarms

The solution developed in the preceding section will now be combined with the concept of an equivalent acoustic medium by adding the compliances of the components of a composite medium, as illustrated in connection with elastic waveguides, equation (2.57). This procedure will allow us to study sound propagation through composite media in the form of a liquid matrix through which are dispersed bubbles filled with a gas, or with the same fluid in a different phase, viz., vapor-filled cavities in a boiling liquid. A mathematical model of a bubble swarm as a single equivalent fluid requires that individual cavities be small enough in terms of wavelengths to have a

negligible scattering cross section, i.e., that they not distort the incident sound waves, and that the cavities be sparse to ensure acoustically uncoupled "breathing-mode" responses. Recalling that the entrained mass of liquid can be visualized as a shell of $4^{1/3}a$ outer radius, it is concluded that significant overlap of near-fields of neighboring cavities is avoided if the fractional volume of cavities does not exceed 10%. The example of a composite medium selected here, viz., gas bubble swarms in liquids, is relevant to sound propagation in the ocean, where air bubbles are formed by surface waves or in bodies of water where decaying organic matter generates gas bubbles. This analysis of bubble swarms introduces the concepts of *dispersion*, i.e., of frequency-dependent sound velocity, and of *dead zones*, i.e., frequency bands where the parameters of the medium are incompatible with wave motion.

Only plane waves need be studied to illustrate these phenomena. Because of marked attenuation in certain frequency ranges, the wave number is expressed in terms of its real and imaginary components:

$$k_c = k_r + ik_i. \tag{3.38}$$

Where the subscripts c identifies parameters of the equivalent composite acoustic medium.

The plane wave in equation (2.23) now takes the form

$$p(z) = P \exp(\ k_i z + ik_r z). \tag{3.39}$$

The two parameters describing propagation of an exponentially damped wave, viz., sound velocity and attenuation per unit distance, are stated in terms of the wave-number components as

$$c_c = \omega/k_r,$$
$$A = 8.69 k_i \text{ dB/unit distance.} \tag{3.40}$$

The wave number, even though complex, is formally given by (2.18). The density of the equivalent medium is

$$\rho_c = \alpha \rho_g + (1 - \alpha)\rho_1$$
$$\approx \rho_1, \qquad \alpha \ll 1. \tag{3.41}$$

The subscripts g and l refer respectively to properties of the gaseous inclusion and the liquid matrix; α is the fractional volume of gas. The small-α approximation will be made throughout. The compressibility is the sum of the compressibilities of the two media weighted as to the fractional volume. The gas compressibility is expressed as in equation (3.34) in terms of the hydrostatic pressure. For quasi-static compression, far below the breathing mode resonance of the bubbles, the compressibility is

$$\left. \begin{aligned} B_c^{-1} &= (1 - \alpha) B_l^{-1} + (\alpha/\gamma P_s) \\ &\simeq \alpha/\gamma P_s, \qquad \rho_l c_l^2 \gg \gamma P_s/\alpha \end{aligned} \right\} \qquad \omega^2 \ll \omega_0^2. \tag{3.42}$$

The asymptotic form, applicable when the bubble compliance short-circuits that of the liquid matrix, is compatible with the small-α approximation: for a depth of 100 ft, viz., for $P_s \simeq 4$ atm, the bubble swarm compliance exceeds that of a watery matrix down to the small fractional air volume of $\alpha = 2.5 \times 10^{-4}$. This situation precisely parallels wave propagation in a soft, liquid-filled hose described by equation (2.58b). In that case, compliance is controlled by the hose wall while density is determined by an effectively incompressible liquid. Combining equations (3.39)–(3.42), one obtains the low-frequency velocity of the bubble swarm mixture:

$$c \simeq (\gamma P_s/\alpha \rho_l)^{1/2}, \qquad \alpha \ll 1, \quad \omega^2 \ll \omega_0^2. \tag{3.43}$$

At atmospheric pressure, for a fractional air volume of $\alpha = 10^{-2}$, the low-frequency sound velocity in a bubble swarm in water is only 120 m/sec, a mere 8% of the sound velocity in water. This is consistent with experimental values in figure 3.10. In this frequency range, attenuation is unimportant.

With rising frequencies, the assumption of quasi-static gas compression becomes invalid. As the frequency approaches resonance, the dynamic compliance of the gas bubbles is multiplied by the usual simple-oscillator resonance amplification factor:

$$B_c^{-1} = (1 - \alpha) B_l^{-1} + (\alpha/\gamma P_s) [1 - (\omega/\omega_0)^2 - i(\omega/\omega_0 Q_0)]^{-1}. \tag{3.44}$$

The analogue circuit simulating the dynamic compressibility of the composite medium, (3.44), is illustrated in figure 3.11. Gas bubble stiffness in units of bulk modulus is

$$K_g = \gamma P_s/\alpha. \tag{3.45a}$$

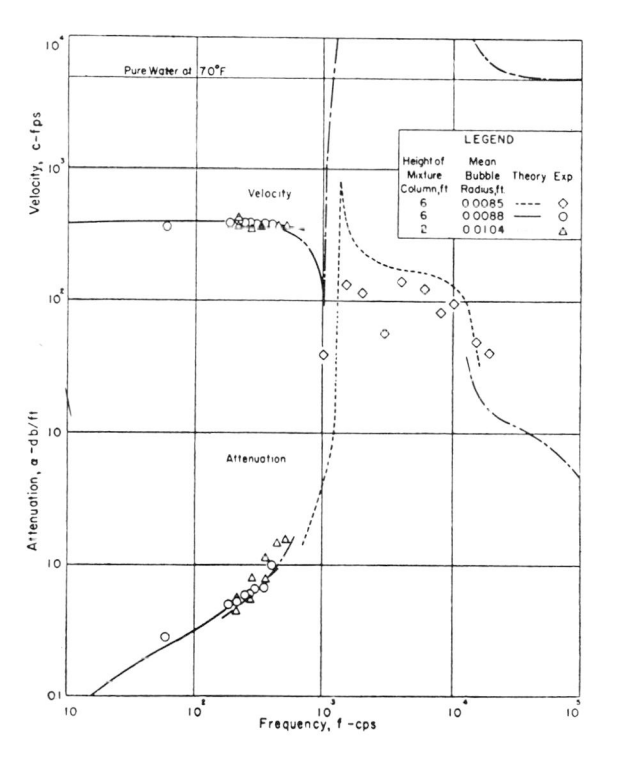

Figure 3.10
Measured sound velocity and attenuation in an air bubble swarm in water: $\alpha = 10^{-2}, P_s \simeq 1$ atm. [Reproduced from Silberman[5]]

The corresponding mass and resistance are

$$M_g = \gamma P_s / \alpha \omega_0^2 = \rho_1 a^2 / 3\alpha,$$

$$R_g = \gamma P_s / \alpha \omega_0 Q_0. \tag{3.45b}$$

At bubble resonance, where the two real terms in (3.44) cancel, the imaginary and real components of the wave number are equal. The sound velocity goes through a minimum and attenuation through a maximum. The interesting feature that distinguishes the effect of bubble resonance from other resonance phenomena is that high attenuation typically extends to higher frequencies over one decade. The reason for this behavior is that the dynamic compliance in (3.44), being mass-controlled, is negative. The compliance of the liquid matrix can again be ignored:

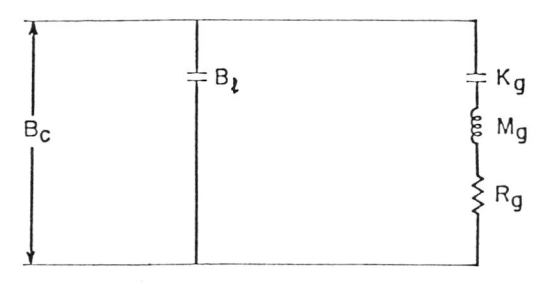

Figure 3.11
Shunt circuit representing the dynamic compliance of the bubble swarm, equation (3.44). For a discussion of the electric impedance analogy used here, see, for example, Olson.[6]

$$\left.\begin{aligned} B_c^{-1} &\simeq -(\omega^2 M_g)^{-1}\\ &\simeq -3\alpha/\rho_1 a^2 \omega^2 \end{aligned}\right\} \quad \omega^2 \gg \omega_0^2. \tag{3.46}$$

Disregarding damping in (3.44), spatial variations of harmonic pressures are governed not by the Helmholtz equation, but by a differential equation of the form

$$d^2 p/dz^2 - \omega^2 \rho_1 |B_c|^{-1} p = 0.$$

The solution describes nonpropagating pressures decaying exponentially with z. It can be verified that when (3.41) and (3.46) are substituted in (2.18),

$$k_i \simeq (3\alpha)^{1/2}/a, \quad \omega^2 \gg \omega_0^2.$$

Referring to (3.40), this is seen to correspond to the attenuation

$$A = 15\alpha^{1/2}/a \quad \text{(dB/unit distance)}. \tag{3.47}$$

Selecting the parameter values and units of figure 3.10, i.e., for $\alpha = 10^{-2}$, $a = 10^{-2}$ ft (0.30 cm), $f_0 = 1.1$ kHz, one computes an astounding attenuation of 150 dB/ft (5 dB/cm). This result is particularly impressive if one considers that the corresponding wavelength in water is 130 cm. As measurements of such large attenuations are limited by flanking path transmission, experimental results in figure 3.10 confirm (3.47) remarkably well. A notable feature of this equation is that it depends on the properties of neither liquid nor gas. It is emphasized that these attenuations are not

the effect of dissipation, but of the absence of elastic restoring forces necessary to support wave motion. In fact, wave motion would be strictly impossible in an ideally lossless medium, i.e., in the absence of the imaginary term in (3.44).

With increasing frequency, the dynamic compliance in (3.46) gradually decreases and becomes comparable to the compliance of the liquid matrix ignored so far. When the two are equal, i.e., when $B_1 = \omega^2 M_g$, the impendance of the shunt circuit in figure 3.11 is infinite, thus giving rise to an *antiresonance* at frequency

$$\omega_a - (B_1/M_g)^{1/2}$$

$$= \left(\frac{3\alpha}{1-\alpha}\right)^{1/2} \frac{c_1}{a}.$$

The resultant compliance is zero, the sound velocity being effectively infinite in a lossless medium. For $\omega > \omega_a$, the compliance tends asymptotically to B_1^{-1}. The high-frequency limit of the sound velocity is therefore that of the liquid matrix. In this frequency range, as in the low-frequency stiffness-controlled range, attenuation is small, being associated primarily with the scattering action of individual bubbles. This phenomenon was not accounted for because it is relatively unimportant in the resonance range and in the dead zone where attenuation is large.

In this and the preceding chapter, we solved the wave equation for relatively simple situations. A powerful, generally applicable formulation of the sound field will now be developed.

References

1. V. M. Albers, *Underwater Acoustics Handbook—II* (University Park, PA: Pennsylvania State University Press, 1965), pp. 155–156.

2. B. W. Jolley, *Summation of Series*, 2nd ed. (New York: Dover, 1961), p. 78, Eq. 420.

3. See, for example, Albers, *Underwater Acoustics*, pp. 47–50.

4. C. Devin, Jr., "Survey of Thermal, Radiation, and Viscous Damping of Pulsating Air Bubbles in Water," *J. Acoust. Soc. Am.* 31:1654–1667 (1959).

5. E. Silberman, "Sound Velocity and Attenuation in Bubbly Mixtures Measured in Standing Wave Tubes," *J. Acoust. Soc. Am.* 29:925–933 (1957).

6. M. F. Olson, "Classical Electro-Dynamical Analogy," in D. E. Gray, ed., *American Institute of Physics Handbook* (New York: McGraw-Hill, 1957), pp. 3-134–3-139.

**The Pressure Field of
Arbitrary Source
Configurations**

4.1 General Formulation of the Radiation Problem

The sound fields radiated by two elementary boundary configurations—the infinite rigid plane and the uniformly pulsating sphere—were constructed in chapter 2. Both these sources are characterized by an acceleration amplitude that is invariant over the radiating surface. The approach developed in this chapter permits the construction of the pressure field in an exterior, effectively infinite region containing harmonically vibrating radiators of unrestricted geometry over whose surface an arbitrary acceleration distribution has been specified. Combining the homogeneous Helmholtz equation governing the pressure with a theorem relating a surface and a volume integral, we shall formulate the pressure field at an arbitrary field point as the surface integral of a linear combination of the weighted pressure and acceleration over the radiating boundary. Since the only prescribed quantity is the surface acceleration, the surface pressure being unknown, the pressure in the field can only be determined after solving an inhomogeneous integral equation in the form of equation (1.9). This equation is constructed by allowing the field point to approach the radiating surface. Like the differential equation from which it arises, this integral equation is associated with the name Helmholtz.

For certain source geometries, the components of the surface integral involving the surface pressure can be made to vanish. The remaining integral thus constitutes an explicit integral expression for the pressure in terms of the prescribed surface acceleration. For other source geometries, not tractable in this fashion, the integral equation can be solved asymptotically in certain frequency ranges. The numerical techniques developed to obtain a direct solution of the integral equation, without these restrictions, will not be described here. The first step in formulating the integral equation is to construct a suitable Green's function.

4.2 The Free-Space Green's Function

The integral equation described in the preceding section is formulated in terms of the free-space Green's function, which, to serve its purpose, must satisfy two mathematical conditions whose physical meaning will become apparent later. In this section, we shall construct the free-space Green's function from these requirements. Unlike the Green's function G, introduced in equation (1.7), which is a function of two independent variables, the source and field point locations, the free-space Green's function depends on a single variable, the scalar distance between the source point $\mathbf{R}_0$ and field point $\mathbf{R}$. To emphasize this difference between the two types of func-

tion, the free-space Green's function is designated by $g(|\mathbf{R} - \mathbf{R}_0|)$, rather than in the form $G(\mathbf{R}|\mathbf{R}_0)$ introduced in chapter 1.

The primary requirement imposed on the free-space Green's function is that it be a solution of the inhomogeneous Helmholtz equation, which was formulated in homogeneous form in equation (2.32):

$$(\nabla^2 + k^2)g(|\mathbf{R} - \mathbf{R}_0|) = \delta(\mathbf{R} - \mathbf{R}_0). \tag{4.1}$$

Here $\delta(\mathbf{R} - \mathbf{R}_0)$ is the three-dimensional Dirac delta function defined by the value of its integral when integrated over a volume V:

$$\int_V \Phi(\mathbf{R}_0)\,\delta(\mathbf{R} - \mathbf{R}_0)\,dV(\mathbf{R}_0) = \Phi(\mathbf{R}), \qquad \mathbf{R} \text{ in } V$$

$$= \frac{\Phi(\mathbf{R})}{2}, \qquad \mathbf{R} \text{ on boundary of } V \tag{4.2}$$

$$= 0, \qquad \mathbf{R} \text{ outside } V.$$

This relation holds if $\mathbf{R}$ and $\mathbf{R}_0$ are interchanged in the delta function. This symmetry, combined with the even, viz. second order of the derivatives in the Laplace operator ∇^2 in equation (4.1) [see equations (2.30), (2.34), and (2.61)], ensures that the free-space Green's function is a solution of the inhomogeneous Helmholtz equation, whether the Laplace operator is formulated in terms of the coordinates of the source point $\mathbf{R}_0$ as ∇_0^2, or of the field point $\mathbf{R}$ as ∇^2.

The second requirement the free-space Green's function must satisfy is the *Sommerfeld radiation condition*,[1] which will be found to ensure that the integral expression for the pressure represents outward traveling waves:

$$\lim_{|\mathbf{R}-\mathbf{R}_0|\to\infty} |\mathbf{R} - \mathbf{R}_0|\left(\frac{\partial g}{\partial|\mathbf{R} - \mathbf{R}_0|} - ikg\right) = 0. \tag{4.3}$$

Unlike the pressure fields constructed in chapters 2 and 3, the solution satisfying equations (4.1) and (4.3) is not subject to any boundary condition at finite range, being adequately defined by the strength of the inhomogenous term in equation (4.1). It is this absence of a boundary condition that associates this Green's function with "free-space," in contrast to the more specialized Green's functions introduced in chapter 1 and discussed further in this chapter.

To construct the solution to equations (4.1)–(4.3), a spherical system of coordinates is selected whose origin coincides with the singularity in equation (4.1). Thus specialized, the distance $|\mathbf{R} - \mathbf{R}_0|$ now reduces to the radial spherical coordinate R. In this spherical coordinate system, equations (4.1), (4.2), and (4.3) and their solution are a function of R only, and not of the angular coordinates. The Laplace operator in equation (4.1) thus simplifies to equation (2.33), with $\nabla_\sigma^2 = 0$. Except at the origin, the Green's function satisfies the homogeneous Helmholtz equation

$$\left(\frac{\partial^2}{\partial R^2} + \frac{2}{R} \frac{\partial}{\partial R} + k^2 \right) g(R) = 0, \qquad R > 0,$$

whose general solution is

$$g(R) = \frac{1}{R} (A_+ e^{likR} + A_- e^{-ikR}). \tag{4.4}$$

The three-dimensional delta function is seen from equation (4.2) to have dimensions of length to the power minus 3. Since the operator multiplying g in equation (4.1) has dimensions of the reciprocal of length squared, g has dimensions of the reciprocal of length, thus indicating that the undetermind coefficients in equation (4.4) are dimensionless.

To satisfy the boundary condition at infinity, equation (4.3), A_- must be set equal to zero. The other coefficient is determined from the strength of the singularity of the inhomogeneous term in equation (4.1). For this purpose, this equation is integrated over a spherical volume element of radius a concentric with the singular point $R = 0$. The definition of δ, equation (4.2), indicates that this integral must equal unity:

$$\lim_{a \to 0} 4\pi \int_0^a (\nabla^2 + k^2) \, gR^2 \, dR = 1.$$

Substituting equation (4.4), with $A_- = 0$, one obtains

$$\lim_{a \to 0} 4\pi A_+ \left(\int_0^a \nabla^2 \left(\frac{e^{ikR}}{R} \right) R^2 \, dR + k^2 \int_0^a e^{ikR} R \, dR \right) = 1. \tag{4.5}$$

The second integral is integrated without difficulty:

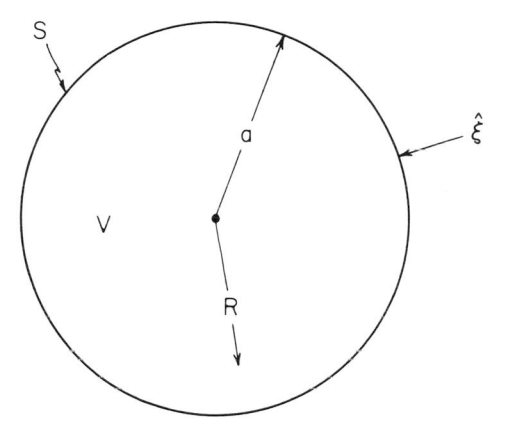

Figure 4.1
Volume and surface integrals used in the construction of the free-space Green's function.

$$k^2 \int_0^a e^{ikR} R \, dR = e^{ikR} (1 - ikR)|_0^a$$
$$= e^{ika}(1 - ika) - 1$$
$$= 0, \quad ka \to 0.$$

Integration of the ∇^2 term in equation (4.5) is most conveniently performed by means of *Gauss's integral theorem,*[2] which will be used repeatedly in this chapter. This theorem relates (1) the volume integral of the divergence of a vector field **F** defined within a volume V bounded by the closed surface S, and (2) the surface integral of the **F** component normal to the boundary (figure 4.1):

$$\int_V \nabla \cdot \mathbf{F} \, dV = - \int_s \mathbf{F} \cdot \hat{\xi} \, dS. \tag{4.6}$$

As elsewhere in this book, the unit vector $\hat{\xi}$ points *into* the volume V.

The volume integral in equation (4.6) can be cast in the form of the ∇^2 integral in equation (4.5) by defining a vector in the form of the gradient of the novanishing A_+-term in equation (4.4):

$$\mathbf{F} \equiv \nabla \left(\frac{A + e^{ikR}}{R} \right).$$

The ∇^2 integral in equation (4.5) is seen to be in the form of the volume integral in equation (4.6) if one notes that $\nabla^2 \equiv \nabla \cdot \nabla$ and $4\pi R^2 \, dR = dV$. In computing the normal component of the above gradient for use in the surface integral in equation (4.6), we note that the unit vector $\hat{\xi}$ points in the direction of decreasing R (figure 4.1):

$$
\left. \begin{aligned}
\mathbf{F} \cdot \hat{\xi} &= \frac{-\partial (A_+ e^{ikR}/R)}{\partial R} \\[2mm]
&= \frac{A_+ e^{ikR}(1 - ikR)}{R^2}
\end{aligned} \right\} \quad R = a
\tag{4.7}
$$

$$
= \frac{A_+}{a^2}, \qquad a \to 0.
$$

Since this formulation is independent of the angular coordinates, the surface integral is obtained simply by multiplying this result by the area of the sphere, $4\pi a^2$. The surface integral in equation (4.6), which is multiplied by a minus sign, thus becomes $(-4\pi A_+)$. Substituting this result in place of the ∇^2 integral in equation (4.5) and remembering that the k^2 integral vanishes as $a \to 0$, one finds the undetermined coefficient to be

$$
A_+ = -(4\pi)^{-1}.
$$

When this is substituted in equation (4.4), with $A_- = 0$, as required by the radiation condition, the solution generalized back to an arbitrary coordinate system $\mathbf{R}_0 \neq 0$, thus replacing the spherical coordinate R by the distance $|\mathbf{R} - \mathbf{R}_0|$, the Green's function is obtained:

$$
g(|\mathbf{R} - \mathbf{R}_0|) = -\frac{e^{ik|\mathbf{R}-\mathbf{R}_0|}}{4\pi|\mathbf{R} - \mathbf{R}_0|}.
\tag{4.8}
$$

We are now in a position to construct the Helmholtz integral equation.

4.3 The Helmholtz Integral Equation
We wish to obtain an integral representation of the pressure-field $p(\mathbf{R})$ satisfying the Helmholtz equation in volume $V(\mathbf{R})$ bounded by the surfaces S_0 and S_1 illustrated in figure 4.2:

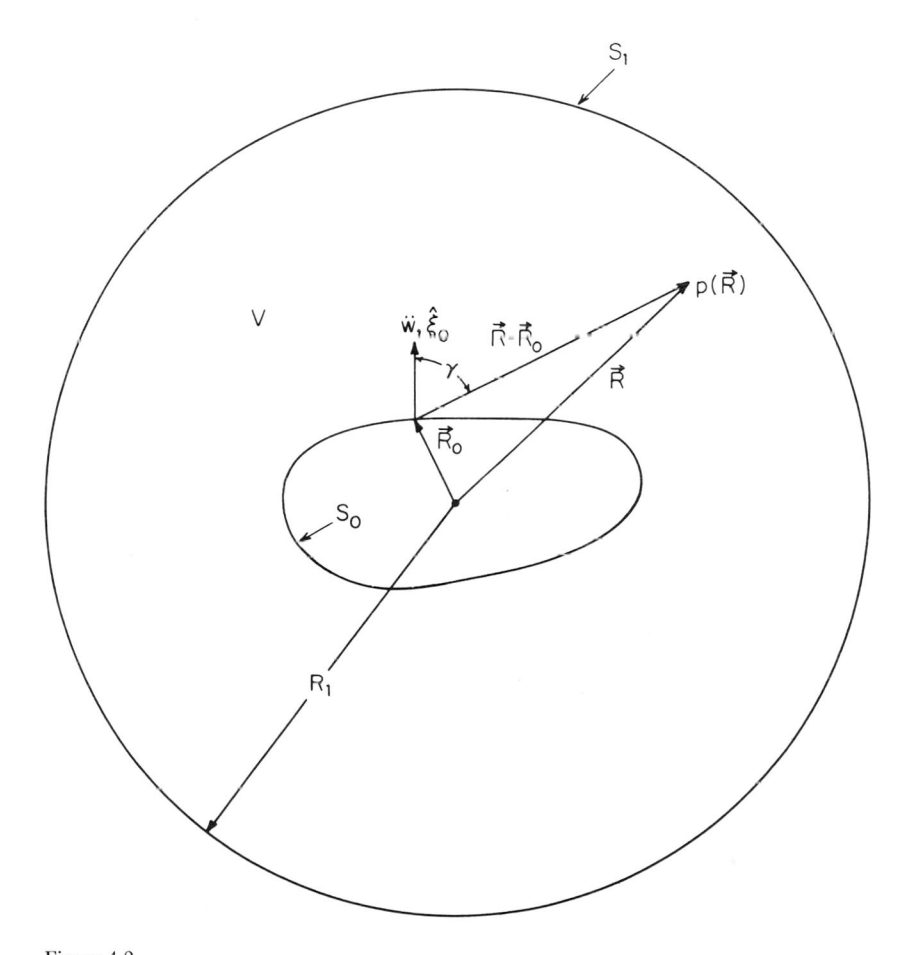

Figure 4.2
Volume and surface integrals used in the construction of the Helmholtz integral equation.

$$(\nabla^2 + k^2)p = 0 \qquad \text{in } V(\mathbf{R}). \tag{4.9a}$$

The solution is subject to the boundary condition prescribed over the radiating surface:

$$\frac{\partial p}{\partial \xi_0} = -\rho \ddot{w} \qquad \text{on } S_0(\mathbf{R}_0). \tag{4.9b}$$

Furthermore, the solution must be compatible with the condition that there are no sources except on this surface.

The desired integral representation is obtained in the form of Gauss's integral theorem, equation (4.6), formulated in terms of the $\mathbf{R}_0$ coordinate system, with

$$\mathbf{F} \equiv p(\mathbf{R}_0)\,\nabla_0 g(|\mathbf{R} - \mathbf{R}_0|) - g(|\mathbf{R} - \mathbf{R}_0|)\,\nabla_0 p(\mathbf{R}_0). \tag{4.10}$$

The integrand of the volume integral, equation (4.6), now becomes

$$\nabla_0 \cdot (p\,\nabla_0 g - g\,\nabla_0 p) = \nabla_0 p \cdot \nabla_0 g + p\,\nabla_0^2 g - \nabla_0 g \cdot \nabla_0 p - g\,\nabla_0^2 p$$
$$= p\,\nabla_0^2 g - g\,\nabla_0^2 p,$$

and that of the surface integral

$$\hat{\xi}_0 \cdot (p\,\nabla_0 g - g\,\nabla_0 p) = p\,\frac{\partial g}{\partial \xi_0} - g\,\frac{\partial p}{\partial \xi_0},$$

where use has been made of the directional derivative in the $\hat{\xi}$ direction, $\nabla_0 \cdot \hat{\xi}_0 = \partial/\partial \xi_0$. Substituting these results in equation (4.6), one obtains Green's identity:

$$\int_v (p\,\nabla_0^2 g - g\,\nabla_0^2 p)\,dV(\mathbf{R}_0)$$

$$= -\int_S \left(p\,\frac{\partial g}{\partial \xi_0} - g\,\frac{\partial p}{\partial \xi_0} \right) dS(\mathbf{R}_0), \qquad S = S_0 + S_1. \tag{4.11}$$

To evaluate the volume integral one solves the homogeneous Helmholtz equation (4.9a), formulated in the $\mathbf{R}_0$ coordinate system, for $\nabla_0^2 p$, and the inhomogeneous Helmholtz equation (4.1) is similarly solved for $\nabla_0^2 g$:

$$\nabla_0^2 p = -k^2 p,$$
$$\nabla_0^2 g = -k^2 g + \delta(\mathbf{R} - \mathbf{R}_0).$$

When this substituted in the volume integral in equation (4.11), the integrand reduces to $p(\mathbf{R}_0)\,\delta(\mathbf{R} - \mathbf{R}_0)$. Referring to equation (4.2), it is seen that when this product is integrated over $V(\mathbf{R}_0)$, the integral yields[3] $p(\mathbf{R})$, if $\mathbf{R}$ defines a point in V, zero, for points not in V, and $p(\mathbf{R})/2$ for points on the boundary. Equation (4.11) thus becomes

$$p(\mathbf{R}) = -\varepsilon \int_{S_0 + S_1} \left(p\frac{\partial g}{\partial \xi_0} - g\frac{\partial p}{\partial \xi_0} \right) dS, \tag{4.12}$$

where $\varepsilon = 0$ for $\mathbf{R}$ not in V, $\varepsilon = 1$ for $\mathbf{R}$ in V, and $\varepsilon = 2$ for $\mathbf{R}$ on S_0 or S_1. On the surface S_0 the integral is interpreted as the Cauchy principal integral value.[3] The integral over the surface S_1 at infinity requires further discussion

4.4 The Sommerfeld Radiation Condition

Unless the surface integral at infinity in equation (4.12) vanishes, one reaches the paradoxical conclusion that the pressure field is not uniquely determined by the boundary condition over the radiating surface S_0, even though no other sources exist. Furthermore, since the contribution of this surface integral to the sound field would be in the form of waves converging from infinity, the source S_0 would be required to act as an energy sink. We shall now show that this paradoxical result is precluded by imposing on the solution the Sommerfeld radiation condition[1] already utilized in constructing the free-space Green's function. In formulating this condition, we can ignore variations of pressure over the outer boundary S_1, and assign a spherical shape to it. The S_1 area integral thus reduces to the product of the integrand of the surface integral in equation (4.12) by the area $4\pi R_1^2$ of the outer boundary, where R_1 is the radius of the sphere. The requirement that this integral vanish for finite R takes the form

$$-\int_{S_1} \left(p\frac{\partial g}{\partial \xi_0} - g\frac{\partial p}{\partial \xi_0} \right) dS = \lim_{R_1 \to \infty} 4\pi R_1^2 \left[g(|\mathbf{R} - \mathbf{R}_1|)\frac{\partial p(\mathbf{R}_1)}{\partial R_1} \right.$$
$$\left. - p(\mathbf{R}_1)\frac{\partial g(|\mathbf{R} - \mathbf{R}_1|)}{\partial R_1} \right] = 0.$$

In the limit of infinite R_1, $|\mathbf{R} - \mathbf{R}_1| \approx R_1$, and from equation (4.8),

$$g(|\mathbf{R} - \mathbf{R}_1|) \approx \frac{-\exp(ik|\mathbf{R} - \mathbf{R}_1|)}{4\pi R_1},$$

$$\frac{\partial g}{\partial R_1} \approx ikg(|\mathbf{R} - \mathbf{R}_1|).$$

When this is substituted in the preceding S_1 integral, the radiation condition becomes

$$\lim_{R_1 \to \infty} R_1 \left[ikp(\mathbf{R}_1) - \frac{\partial p(\mathbf{R}_1)}{\partial R_1} \right] = 0. \qquad (4.13)$$

To satisfy equation (4.13), the pressure must therefore decrease with increasing R_1 as $|\mathbf{R}_1 - \mathbf{R}_0|^{-1}$ or faster. This will be found to be the case for all sources of finite extent. Having thus reduced the surface integral over the boundary at infinity to zero, the surface integral in equation (4.12) is thus restricted to the radiating surface S_0. Introducing the boundary condition, equation (4.9b), and restricting $\mathbf{R}$ to field points in the volume V, equation (4.12) becomes

$$p(\mathbf{R}) = -\int_{S_0} \left(p \frac{\partial g}{\partial \xi_0} + \rho \ddot{w} g \right) dS(\mathbf{R}_0). \qquad (4.14)$$

This powerful formula, obtained by Helmholtz in 1859, allows one to construct a pressure field in space from information on the pressure and pressure gradient (or fluid particle acceleration) over a surface. If one selects an acceleration distribution, one may not also arbitrarily specify the surface pressure. Before this integral representation of the pressure field can be used, the surface pressure on the boundary S_0 must be determined, either analytically or, of course, experimentally.

4.5 Physical Interpretation of the Helmholtz Integral

The surface pressure on the boundary is the solution of the Fredholm integral equation obtained by specializing the field point location to the boundary S_0. Referring of equation (4.12), this is seen to require the multiplication of the surface integral in equation (4.14) by two. In the notation introduced in equation (1.9), this integral equation is

$$p(\mathbf{R}) = \int_{S_0} K(|\mathbf{R} - \mathbf{R}_0|) p(\mathbf{R}_0) \, dS(\mathbf{R}_0) + f(\mathbf{R}), \qquad R \text{ on } S_0(\mathbf{R}_0), \quad (4.15)$$

where $K \equiv -2\partial g/\partial \xi_0$. Before describing the inhomogeneous term, we shall express the kernel explicitly. We refer to figure 4.2 in constructing the directional derivative of the Green's function. This figure indicates that when the angle $\gamma < \pi/2$, that is, when the field point is visible from the source point, the derivative $\partial(|\mathbf{R} - \mathbf{R}_0|)/\partial \xi_0$ is negative, and vice versa. Hence

$$-\frac{\partial g(|\mathbf{R} - \mathbf{R}_0|)}{\partial \xi_0} = \frac{\partial g(|\mathbf{R} - \mathbf{R}_0|)}{\partial(|\mathbf{R} - \mathbf{R}_0|)} \frac{\partial(|\mathbf{R} - \mathbf{R}_0|)}{\partial \xi_0}$$

$$= -g(|\mathbf{R} - \mathbf{R}_0|) [|\mathbf{R} - \mathbf{R}_0|^{-1} - ik] \cos \gamma \qquad (4.16)$$

$$\approx ikg(|\mathbf{R} - \mathbf{R}_0|) \cos \gamma, \qquad k^2|\mathbf{R} - \mathbf{R}_0|^2 \gg 1.$$

The kernel in equation (4.15) can now be expressed in terms of equations (4.8) and (4.16):

$$K(|\mathbf{R} - \mathbf{R}_0|) = -\frac{2g(|\mathbf{R} - \mathbf{R}_0|)}{|\mathbf{R} - \mathbf{R}_0|} (1 - ik|\mathbf{R} - \mathbf{R}_0|) \cos \gamma \qquad (4.17)$$

$$\approx 2ikg(|\mathbf{R} - \mathbf{R}_0|) \cos \gamma, \qquad k^2|\mathbf{R} - \mathbf{R}_0|^2 \gg 1.$$

The inhomogeneous term is

$$f(\mathbf{R}) = -2\rho \int_{S_0} g(|\mathbf{R} - \mathbf{R}_0|) \ddot{w}(\mathbf{R}_0) \, dS(\mathbf{R}_0). \qquad (4.18)$$

The inhomogeneous term is seen to be in the form of a surface integral of a distribution of point sources, equation (3.1).

To interpret the meaning of the kernel integral, we anticipate a result derived in chapter 6, whereby $\partial g/\partial \xi_0$ is the sound pressure generated at $\mathbf{R}$ by a point force of unit amplitude applied at $\mathbf{R}_0$ and acting in the direction $\hat{\xi}_0$. The kernel integral in equation (4.15) is therefore equivalent to a distribution of point forces of amplitude $p(\mathbf{R}_0) \, dS$. The Helmholtz integral thus effectively replaces the radiating boundary by a distribution of point sources and of forces weighted, respectively, according to the surface acceleration and surface pressure. This concept is already inherent in Fresnel's theory of diffraction (1818).

4.6 Approaches to the Solution of the Helmholtz Integral Equation

For arbitrary source configurations with the solution not restricted to a particular frequency range, the surface pressure distribution must be obtained by solving equation (4.15) numerically. A description of alternative techniques and of their respective advantages and limitations can be found in a recent review paper.[4] The surface pressure obtained, equation (4.15), becomes a simple integral representation of the pressure at field points not located on the radiating surface.

The solution of the Helmholtz integral equation can also be circumvented asymptotically in the short- and long-wavelength limits. The former is associated with the *plane-wave* approximation whereby by specific acoustic impedance is approximated by the characteristic impedance, ρc. This situation is achieved when the wavelength is small compared with (1) the radius of curvature of the radiating surface and (2) the dimensions of regions of this surface vibrating in phase. When these two requirements are met, the components of the pressure gradients that are tangential to the boundary become negligible compared with the component normal to the boundary. The surface pressure therefore approaches locally the plane-wave characteristics, even though the pressure is not uniform over the entire boundary. This situation was verified in the large-ka limit for the pulsating sphere, equation (2.41), and will be encountered repeatedly in chapter 5 and 6. The circular rigid piston, which is analyzed in the next chapter, is an exception to this statement, in that its surface pressure does not tend toward $\rho c \dot{W}$ when the piston diameter becomes large in terms of wavelengths. However, if surface averaged, its specific acoustic impedance tends to ρc when the piston diameter exceeds a wavelength. Consequently, the integral representation developed below for the asymptotic short-wavelength limit applies. In this high-frequency range, the surface pressure in equation (4.15) can be replaced by $\rho c \dot{w}(\mathbf{R}_0)$. The pressure is now given explicitly by the integral representation in equation (4.14). Writing $ik\rho c \dot{w}$ as $-\rho \ddot{w}$,

$$p(\mathbf{R}) \approx -\rho \int_{S_0} g(|\mathbf{R} - \mathbf{R}_0|)\,\ddot{w}(\mathbf{R}_0)$$

$$\cdot\,(1 + \cos\gamma)\,dS(\mathbf{R}_0), \qquad k^2|\mathbf{R} - \mathbf{R}_0|^2 \gg 1. \tag{4.19a}$$

Alternatively, this same plane-wave approximation can be used to express the acceleration in (4.14) in terms of the surface pressure:

$$p(\mathbf{R}) = -ik \int_{S_0} g(|\mathbf{R} - \mathbf{R}_0|) p(\mathbf{R}_0)$$

$$\cdot (1 + \cos\gamma)\, dS(\mathbf{R}_0), \qquad k^2|\mathbf{R} - \mathbf{R}_0|^2 \gg 1. \tag{4.19b}$$

The long-wavelength asymptotic solution utilizes the incompressible or Laplace approximation to the surface pressure. In this situation, the term $ik|\mathbf{R} - \mathbf{R}_0|$ can be neglected compared with unity in the expression for the kernel, equation (4.17). In the far-field, specifically when the range is large compared with the characteristic dimension of the radiating surface, the remaining term in equation (4.17) becomes negligible, thus entirely eliminating the surface pressure component of the Helmholtz integral, equation (4.14). The far-field pressure can now be concisely expressed in terms of the source acceleration:

$$\ddot{w}(\mathbf{R}_0) = ?$$

$$p(\mathbf{R}) = -\rho \int_{S_0} g(|\mathbf{R} - \mathbf{R}_0|)$$

$$\cdot \ddot{w}(\mathbf{R}_0)\, dS(\mathbf{R}_0), \qquad |\mathbf{R} - \mathbf{R}_0| \gg L, \quad kL \ll 1. \tag{4.20}$$

This asymptotic formulation was used to compute the far-field of line sources, equation (3.16) and (3.18). For radiators of nonvanishing area, its range of validity can be extended to higher frequencies and shorter ranges if the $\cos\gamma$ factor in the expression for the kernel, equation (4.17), is small over most of the range of the surface integration.[5] The Green's function, and hence f, can be expanded in a power series in kL. The asymptotic solution in equation (4.20) is invalid for acceleration distributions that cause the k-independent term in this series to vanish, thus reducing f in equation (4.18), like the pressure term in equation (4.14), to order kL.

4.7 Analytical Solution of the Helmholtz Integral Equation
The integral equation is circumvented if we can construct a Green's function $G(\mathbf{R}|\mathbf{R}_0)$ that satisfies Neumann boundary conditions:

$$\frac{\partial G(\mathbf{R}|\mathbf{R}_0)}{\partial \xi_0} = 0 \qquad \text{on } S_0(\mathbf{R}_0). \tag{4.21}$$

In this case, the unknown $p(\mathbf{R}_0)$ term in equation (4.14) vanishes, thus reducing the integral equation to a simple integral representation, with no

restriction as to range or frequency:

$$p(\mathbf{R}) = -\rho \int_{S_0} G(\mathbf{R}|\mathbf{R}_0)\ddot{w}(\mathbf{R}_0)\, dS(\mathbf{R}_0). \tag{4.22}$$

A Green's function satisfying equation (4.21) is of the form

$$G(\mathbf{R}|\mathbf{R}_0) = g(|\mathbf{R} - \mathbf{R}_0|) + \Gamma(\mathbf{R}|\mathbf{R}_0), \tag{4.23}$$

For spherical coordinates, see Eq. 6.15

where $\Gamma(\mathbf{R}|\mathbf{R}_0)$ is a solution of the homogeneous Helmholtz equation (4.9a), whose derivative $\partial\Gamma/\partial\xi_0$ equals $-\partial g/\partial\xi_0$ on the surface S_0. A Green's function satisfying equation (4.21) can be constructed if

1. The boundary S_0 is completely defined by specifying the value of a single coordinate $\xi = \xi_0$ of a system of three orthogonal coordinates.

2. The wave equation is *separable* in this coordinate system, a property to be defined in the next paragraph.

Condition 1 is met by boundaries such as an infinite plane, a cylinder of infinite length, and a sphere. In contrast, the exterior far-field of a cylinder of finite length cannot be constructed analytically.

Condition 2 implies that the pressure can be represented by linear combinations of wave harmonics, each harmonic being the product of a function $R_{mn}(\xi)$ of the coordinate normal to the boundary, and of a function $S_{mn}(\eta, \zeta)$:

$$p(\xi, \eta, \zeta) = \sum P_{mn} R_{mn}(\xi) S_{mn}(\eta, \zeta). \tag{4.24}$$

The function $S_{mn}(\eta, \zeta)$ is itself the product of two functions, or surface harmonics, each of which depends on a single coordinate η or ζ. If ∇_σ^2 is the surface Laplace operator stated for spherical coordinates in equation (2.34), then

$$\nabla_\sigma^2 R_{mn}(\xi) \equiv 0, \qquad (\nabla^2 - \nabla_\sigma^2) S_{mn}(\eta, \zeta) \equiv 0.$$

The Helmholtz equation (2.32) can now be stated for each term in the series in equation (4.24) as

$$R_{mn}\nabla_\sigma^2 S_{mn} + S_{mn}(\nabla^2 - \nabla_\sigma^2) R_{mn} + k^2 R_{mn} S_{mn} = 0.$$

It can verified that this equation is satisfied if

$$\nabla_\sigma^2 S_{mn} = -k_\sigma^2 S_{mn},$$
$$(\nabla^2 - \nabla_\sigma^2) R_{mn} = -(k^2 - k_\sigma^2) R_{mn}. \tag{4.25}$$

Source configurations admitting solutions of the form of equation (4.24) include the sphere, the infinite cylinder, the spheroid and certain other shapes of minor importance.[6] The construction of the Green's function in spherical coordinates is illustrated in section 6.2. Such techniques exist for all separable coordinate systems.[7]

When the desired Green's function has been constructed, the pressure field is obtained from equation (4.22). The evaluation of this integral is straightforward if the acceleration distribution can be presented as a series of surface harmonics:

$$\ddot{w}(\eta, \zeta) = \sum \dot{W}_{mn} S_{mn}(\eta, \zeta). \tag{4.26}$$

This is necessarily the case when the boundary forms a closed finite surface, such as a sphere. If the source, for example, a cylinder, has an infinite dimension, but its acceleration distribution is nonzero over a finite region of the boundary, analytical evaluation of the integral in equation (4.22) is possible only in the far-field. This difficulty will be extensively illustrated for planar sources in the next chapter. First, however, we shall derive the Green's function for this source geometry.

4.8 Rayleigh's Formula for Planar Sources

The Green's function satisfying Neumann boundary conditions on the infinite plane $z = z_0$ will now be constructed. It was shown in section 3.4 that the pressure field generated by two identical point sources displays a zero derivative in the direction normal to the plane of symmetry, which is precisely the condition required of $G(\mathbf{R}|\mathbf{R}_0)$ on the plane z. The desired Green's function is, therefore, in rectangular coordinates,

$$G(\mathbf{R}|\mathbf{R}_0) = \lim_{e \to 0} [g(x, y, z|x_0, y_0, z_0 - e) + g(x, y, z|x_0, y_0, z_0 + e)]$$
$$= 2g(x, y, z|x_0, y_0, z_0) \tag{4.27}$$
$$= 2g(|\mathbf{R} - \mathbf{R}_0|),$$

where

$$|\mathbf{R} - \mathbf{R}_0| = [(x - x_0)^2 + (y - y_0)^2 + (z - z_0)^2]^{1/2}.$$

When this is substituted in equation (4.22), one obtains the pressure in terms of the function defined in equation (4.18):

$$p(\mathbf{R}) = -2\rho \int_{S_0} g(|\mathbf{R} - \mathbf{R}_0|)\ddot{w}(\mathbf{R}_0)\, dS(\mathbf{R}_0). \tag{4.28}$$

If the origin of coordinates is placed near the center of the source, the integral $f(\mathbf{R})$ can be simplified by approximating the denominator $|\mathbf{R} - \mathbf{R}_0|$ of $g(|\mathbf{R} - \mathbf{R}_0|)$ by R, thus removing it from under the integral. Equation (4.28) thus reduces to *Rayleigh's formula*,   *see p.96 , Eq.10.39*

$$p(\mathbf{R}) = \frac{\rho}{2\pi R} \int_{S_0} \exp(ik|\mathbf{R} - \mathbf{R}_0|)\ddot{w}(\mathbf{R}_0)\, dS(\mathbf{R}_0), \qquad \mathbf{R}_0 \ll \mathbf{R}. \tag{4.29}$$

Noting that $\ddot{w}\, dS$ is the volume acceleration of an area element of the source S_0, and referring to the solution for the point source, equation (3.1), we conclude that a planar source located in an infinite baffle is equivalent to a distribution of point sources. This interpretation is consistent with Huygens' principle (1690).[8] Equation (4.29) is, like equation (4.14), a remarkable result in that it uses two-dimensional information to construct a field that depends on three spatial variables.

4.9 The Scattered Field

Chapter 10 deals with the scattered field, defined as the distortion of an incident sound field $p_i(R)$ by the introduction of a boundary $S_0(R_0)$. To conclude this chapter we shall adapt the Helmholtz integral in equation (4.14) to the evaluation of the pressure $p_{s\infty}(R)$ scattered by a boundary of   *$\dot{w} = 0$* infinite impedance. The construction of the field scattered by an elastic   *$\ddot{w} \neq 0$* boundary will be formulated in chapter 11.

For the purpose of evaluating the scattered pressure, it is assumed that two, rather than one, closed boundaries exist in the fluid medium. One boundary is responsible for generating the incident wave field; the other is the scatterer surface. The range from source to scatterer is sufficiently large to ensure that the waves are effectively plane when they reach the latter. The volume integral in equation (4.11) yields the total pressure:

$$p(R) = p_i(R) + p_{s\infty}(R). \tag{4.30}$$

The surface integration in equation (4.12) now involves two surface integrals, one over the distant source, the other over the scatterer. The former yields the incident plane waves. The integral over the scatterer surface, $S_0(R_0)$, produces an expression similar to equation (4.14) and (4.15), but restricted to the pressure term if the scatterer is rigid. Specifically, $\ddot{w}$ in equation (4.14) and $f(R)$ in equation (4.15) both vanish. Setting the integral over the distant source equal to the incident pressure, the integral representation of the pressure field becomes

$$p(\mathbf{R}) = p_i(\mathbf{R}) - \int_{S_0} p(\mathbf{R_0}) \frac{\partial g(|\mathbf{R} - \mathbf{R_0}|)}{\partial \xi_0} dS(\mathbf{R_0}). \qquad \text{rigid scatterer}$$

The derivative $\partial g/\partial \xi_0$ is still given by equation (4.16). Transferring p_i to the left side of this equation and using the definition in equation (4.30), the integral representation of the scattered pressure becomes

$$p_{s\infty}(\mathbf{R}) = p(\mathbf{R}) - p_i(\mathbf{R})$$
$$= -\int_{S_0} p(\mathbf{R_0}) \frac{\partial g(|\mathbf{R} - \mathbf{R_0}|)}{\partial \xi_0} dS(\mathbf{R_0}). \tag{4.31}$$

Before evaluating this integral, one must solve the integral equation arising from equation (4.31), for field points $\mathbf{R}$ placed on the scatterer surface. The resulting Fredholm equation of the second kind is similar to the one in equation (4.15), where the nonhomogeneous term $p_i(\mathbf{R})$ takes the place of $f(\mathbf{R})$, equation (4.18). Numerical and asymptotic techniques for solving the integral equation are similar to those developed for the equivalent radiation problem.

References

1. A. Sommerfeld, *Partial Differential Equations in Physics* (New York: Academic Press, 1949), p. 189.

2. P. M. Morse and H. Feshbach, *Methods of Theoretical Physics* (New York: McGraw-Hill, 1953), p. 37.

3. For a discussion of the evaluation of the integral when the field point is moved to the radiating surface, see, for example, G. Chertock, "Sound Radiation from Vibrating Surfaces," *J. Acoust. Soc. Am.* 36: 1305–1313 (1964).

4. H. A. Schenk, "Improved Integral Formulation for Acoustic Radiation Problems," *J. Acoust. Soc. Am.* 44:41–58 (1968).

5. M. Strasberg, "Sound Radiation for Slender Bodies in Axisymmetric Vibration," Paper 0–28, 4th International Congress on Acoustics, Copenhagen, August 1962. This procedure will be illustrated in section 6.15.

6. P. M. Morse and H. Feshbach, op cit. pp. 494–517.

7. Ibid., pp. 828–832.

8. B. B. Baker and E. T. Copson, *The Mathematical Theory of Huygens' Principle*, 2nd ed. (Oxford: Clarendon Press, 1950). Chapter I presents a detailed discussion of the relation between Huygens' principle, Fresnel's theory of diffraction, and Helmholtz's integral formula, as well as the latter's equivalent for arbitrary time dependence (Kirchhoff's formula), which is not covered here.

5.1 Source Geometry and Analytical Formulations

The sound field radiated by planar sources will be constructed in this chapter in terms of the function describing the acceleration distribution over the radiating surfaces. The two analytical formulations developed here apply to spatially nonperiodic acceleration distributions extending out to infinity, such as embodied in the effectively infinite point-excited plate analyzed in chapter 8. In this chapter, these two formulations will be used to construct general solutions that will be specialized to finite radiators surrounded by effectively infinite, motionless baffles. Two types of acceleration distributions are analyzed: (1) accelerations uniform over the vibrating surface; (2) standing waves. The former dynamic configurations characterize rigid pistons, while the latter simulate structural modes of plates in vacuo. Both circular and rectangular pistons are considered. Standing-wave configurations are analyzed only for rectangular boundaries, as these are more representative of structural elements than circular boundaries. The pressure fields will be constructed for standing-wave distributions of acceleration simulating in vacuo modes of simply supported plates. All of these sources are assumed to radiate out of an effectively infinite plane baffle.

In practice, a rigid baffle surrounding the radiator is approximated by a source located in an extended boundary whose impedance is high compared with that of the fluid medium, for example, by a loudspeaker mounted on a large wall, or in the open, on the ground. Alternatively, the baffle can be simulated by the plane of symmetry of two distributions of point sources vibrating symmetrically. This is illustrated for two point sources in figure 3.2. The more complicated situation for extended planar radiators is illustrated in figure 5.1. The finite distance separating the plane of the radiator from the plane of symmetry can be disregarded in the construction of the sound field, as long as this distance is small in terms of wavelengths. Consequently, the distinction between the radiating plane and the plane of symmetry will be disregarded.

A finite, planar radiator located in a baffle is characterized by an acceleration distribution that has a specified finite value on the radiating surface and vanishes elsewhere on the surface of the baffle. Thus the dynamic configuration of the circular piston of radius a is formulated in cylindrical coordinates, as

$$\ddot{w}(r) = \ddot{W}H(a - r), \tag{5.1}$$

where H is the Heaviside function defined by

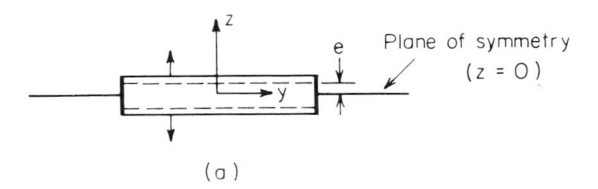

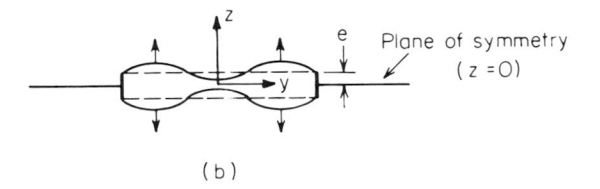

Figure 5.1
Simulation of an infinite rigid baffle surrounding a planar radiator in the form of (a) a rigid piston and (b) a simply supported plate, by the plane of symmetry of two symmetrically vibrating pistons or plates. Distance e can be ignored if $ke \ll 1$.

$$H(x) = 0, \qquad x < 0$$
$$= 1, \qquad x \geq 0. \qquad\qquad (5.2)$$

Before constructing the pressure field of representative planar radiators, we shall describe the two alternative analytical techniques that will be used in this and subsequent chapters.

The pressure field of these planar geometries can be evaluated in the far-field by means of Rayleigh's formula, equation (4.29). In a few cases, the less restrictive integral representation in equation (4.28) can be used to evaluate near-field pressures. There is an alternative formulation that utilizes a *transform* technique. This latter appraoch is suggested by the fact that all of these source geometries have in common a spatially nonperiodic acceleration distribution. For the simple configurations considered here, these distributions can be expressed analytically as a transform adapted to the particular source geometry. Thus the axisymmetric distribution of a rigid circular piston, equation (5.1), lends itself to the use of the *Hankel transform*, defined by the relation[1]

$$\tilde{f}(\gamma) \equiv \int_0^\infty f(r)\, \mathcal{J}_0(\gamma r)\, r\, dr, \qquad\qquad (5.3a)$$

$$f(r) = \int_0^\infty \tilde{f}(\gamma) \mathcal{J}_0(\gamma r) \gamma \, d\gamma. \tag{5.3b}$$

Substituting the acceleration distribution for the piston, equation (5.1), one obtains the transform

$$\tilde{\ddot{w}}(\gamma) = \int_0^\infty \ddot{w}(r) \mathcal{J}_0(\gamma r) r \, dr$$

$$= \frac{\ddot{W} a \mathcal{J}_1(\gamma a)}{\gamma}, \tag{5.4}$$

where use has been made of the integral[2,3]

$$\int_0^a \mathcal{J}_0(\gamma r) r \, dr = \frac{a \mathcal{J}_1(\gamma a)}{\gamma}. \tag{5.5}$$

To construct the pressure field, one takes the transform of the Helmholtz equation formulated in the appropriate coordinate system, which for this configuration is cylindrical. The resulting equation can be solved for the pressure transform $\tilde{p}(\gamma; z)$, expressed in terms of $\tilde{\ddot{w}}(\gamma)$. The actual pressure distribution is finally constructed as the inverse transform

$$p(r, z) = \int_0^\infty \tilde{p}(\gamma; z) \mathcal{J}_0(\gamma r) \gamma \, d\gamma. \tag{5.6}$$

This integral will be evaluated in the far field using a *stationary-phase* approximation.

   Rayleigh's formulation and the transform techniques each have their advantages. Both techniques will be illustrated, beginning with the application of Rayleigh's formulation to the circular piston, in section 5.2. In section 5.3, the transform formulation is used to construct the pressure fields radiated by acceleration distributions that are axisymmetric but unrestricted as to their radial configuration. This result is then (section 5.4) specialized to the pressure field and radiation loading of the circular piston. In section 5.5, Rayleigh's formulation is applied to the rectangular piston. The transform formulation is developed for rectangular radiators (section 5.6). The corresponding stationary-phase approximation is shown to embody the same physical restrictions as Rayleigh's formula (section 5.7). The far-field pres-

sure is constructed for acceleration distributions simulating the in vacuo modes of rectangular plates, the results being interpreted in terms of coincidence effects (sections 5.8–5.10). The pressure field and radiation loading associated with a periodic acceleration distribution, such as a train of waves extending over the plane boundary out to infinity, is analyzed in sections 5.11 and 5.12. Finally, acoustic power calculations are illustrated for some finite radiator configurations (sections 5.13–5.15).

5.2 Pressure Field of the Circular Piston Using Rayleigh's Formulation

The circular piston is located in the $z = 0$ plane, its center coinciding with the origin of coordinates (figure 5.2). The distance between the source point $\mathbf{R}_0$, identified by the cylindrical coordinates $(r_0, \phi_0, z_0 = 0)$, and the field point $\mathbf{R}$, identified in spherical and cylindrical coordinates, respectively, as (R, θ, ϕ) and (r, ϕ, z) is obtained from the trigonometric relations

$$
\begin{aligned}
|\mathbf{R} - \mathbf{R}_0| &= [r^2 + r_0^2 - 2rr_0 \cos(\phi - \phi_0) + z^2]^{1/2} \\
&= [R^2 + r_0^2 - 2Rr_0 \sin\theta \cos(\phi - \phi_0)]^{1/2} \\
&= R - r_0 \sin\theta \cos(\phi - \phi_0) + O\left(\frac{r_0^2}{R}\right).
\end{aligned}
\tag{5.7}
$$

At long range, where $R \gg r_0$, the linear r_0 term must be retained in the exponent of the phase factor in equation (4.29), but the r_0^2 term can be dropped, thus implying that (kr_0^2/R) is negligible. The reason for using different approximations in the denominator and in the exponential is that in the former it is only required that the terms dropped be small compared with R, while in the latter, they must be small compared with unity. Referring to figure 5.2, it is seen that

$$
\begin{aligned}
\overline{OQ} &= r_0 \cos(\phi - \phi_0), \\
\overline{OP} &= \overline{OQ} \sin\theta = r_0 \cos(\phi - \phi_0) \sin\theta.
\end{aligned}
\tag{5.8}
$$

Thus, the power series expression in equation (5.7) approximates the distance from field point to source point by the distance from P to the field point.

When one substitutes equation (5.8) in Rayleigh's formula, factors out the terms that are dependent on source point location, and rotates the ϕ axis to the plane containing both the z axis and the field point, the integral representation of the far-field pressure becomes

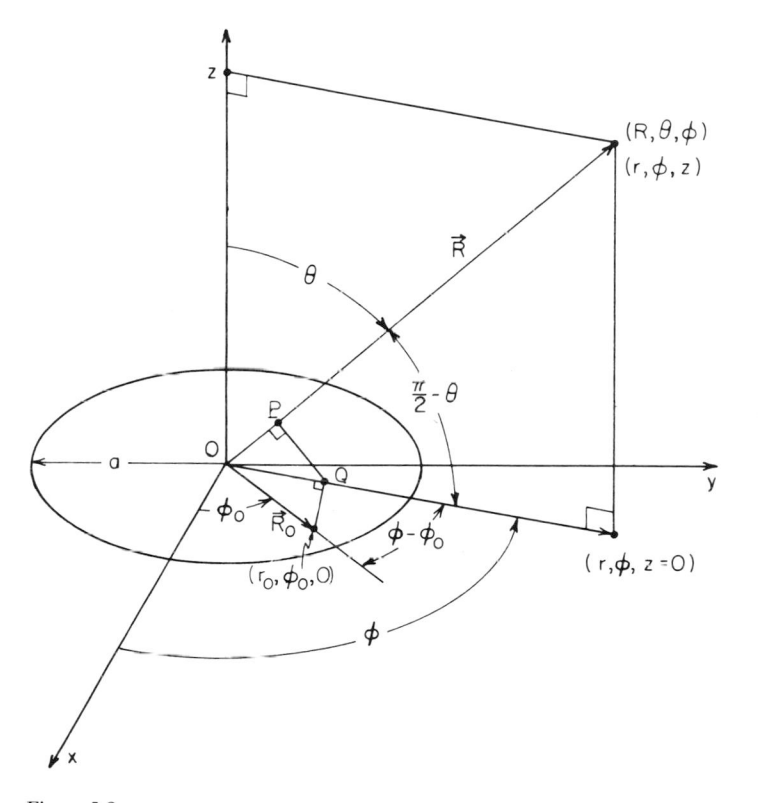

Figure 5.2
Construction of the far-field of a circular piston radiator.

$$p(\mathbf{R}) = \frac{\rho\ddot{W}}{2\pi R} \int_0^a \int_0^{2\pi} \exp(ikR - ikr_0 \sin\theta \cos\phi_0)\, r_0\, dr_0\, d\phi_0$$
$$= \frac{\rho\ddot{W}}{\pi R} e^{ikR} \int_0^a \int_0^{\pi} \cos(kr_0 \sin\theta \cos\phi_0)\, r_0\, dr_0\, d\phi_0$$

$$\left.\begin{array}{c} \\ \\ \end{array}\right\} \quad \begin{array}{l} kr_0^2 \ll R, \\ r_0 \ll R. \end{array}$$

The ϕ_0 integration can be performed by utilizing an integral representation to the Bessel function[4]

$$J_0(x) = \frac{1}{\pi} \int_0^{\pi} \cos(x \cos\phi_0)\, d\phi_0. \tag{5.9}$$

The remaining r_0 integral in the integral representation of the pressure is

now of the form of equation (5.5), with $\gamma = k \sin \theta$. One thus arrives at the far-field expression for the pressure:

$$p(R, \theta) = \frac{\rho \ddot{W} a e^{ikR}}{R} \frac{\mathcal{J}_1 (ka \sin \theta)}{k \sin \theta}, \qquad ka^2, a \ll R. \tag{5.10a}$$

This can be expressed more compactly in terms of the volume acceleration $\ddot{Q} = \pi a^2 \ddot{W}$, as

$$p(R, \theta) = \frac{\rho \ddot{Q} e^{ikR}}{\pi R} \frac{\mathcal{J}_1 (ka \sin \theta)}{ka \sin \theta}. \tag{5.10b}$$

Referring to equation (3.2), we can express this concisely in terms of the pressure $p_s(R)$ of the point source endowed with the same volume acceleration:

$$p(R, \theta) = \frac{4 p_s (R) \mathcal{J}_1 (ka \sin \theta)}{ka \sin \theta}. \tag{5.10c}$$

For a small argument, $\mathcal{J}_1 (x) \approx x/2$. Hence, near the piston axis or in the small-ka limit, the pressure becomes

$$p(R, \theta) \approx 2 p_s (R), \qquad \theta \approx 0 \text{ or } (ka)^2 \ll 1. \tag{5.10d}$$

The directivity factor is plotted in figure 5.3. As in the case of the line arrays studied in chapter 3, the pattern becomes more directive and the number of side lobes increases as the characteristic dimension of the source measured in wavelengths, in this case ka, increases. Analytical evaluation of the directivity factor has been carried out for a wide variety of circular radiators endowed with a nonuniform acceleration distribution.

The pressure can also be evaluated at field points on the piston axis, where $R = z$, without restricting the field point location to the far-field (figure 5.4). Substituting $|\mathbf{R} - \mathbf{R}_0| = (z^2 + r_0^2)^{1/2}$ in equation (4.28) and noting that the ϕ_0 integration becomes unnecessary because of the axi-symmetric location of the field point, one sees that the on-axis pressure, in cylindrical coordinates, reduces to

$$p(r = 0, z) = \rho \ddot{W} \int_0^a \frac{\exp[ik(z^2 + r_0^2)^{1/2}]}{(z^2 + r_0^2)^{1/2}} r_0 \, dr_0.$$

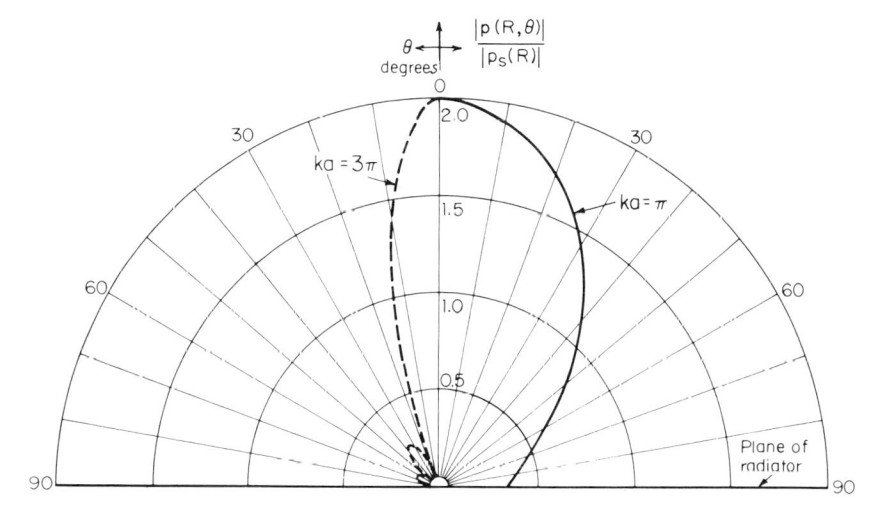

Figure 5.3
Directivity of a baffled circular piston, for two values of ka [equation (5.10)].

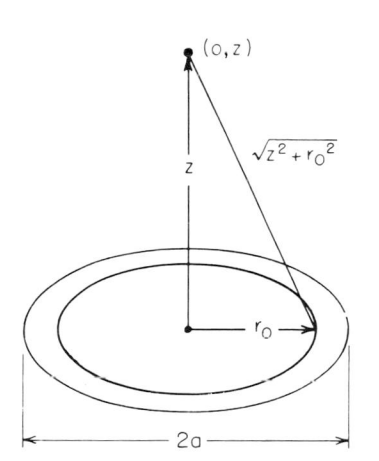

Figure 5.4
Construction of the near-field pressure on the piston axis.

When one sets

$$u = (z^2 + r_0^2)^{1/2},$$

the pressure integral becomes

$$p(0, z) = \rho \ddot{W} \int_z^{(z^2+a^2)^{1/2}} e^{iku} \, du$$

$$= -\frac{i\rho \ddot{W}}{k} \{\exp[ik(z^2 + a^2)^{1/2}] - \exp(ikz)\}. \tag{5.11a}$$

The absolute value of the coefficient multiplying the exponentials equals $\rho c \ddot{W}$. The exponentials can be combined into a sine by using the trigonometric identities

$$|e^{i\alpha} - e^{i\zeta}| = [(\cos\alpha - \cos\zeta)^2 + (\sin\alpha - \sin\zeta)^2]^{1/2}$$

$$= [2 - 2\cos(\alpha - \zeta)]^{1/2}$$

$$= 2|\sin\tfrac{1}{2}(\alpha - \zeta)|, \qquad \alpha \equiv k(z^2 + a^2)^{1/2}, \quad \zeta \equiv kz.$$

The absolute value of the pressure thus becomes

$$|p(0, z)| = 2\rho c \ddot{W} \left| \sin\frac{k}{2}[(z^2 + a^2)^{1/2} - z] \right|. \tag{5.11b}$$

This is an example of a solution that does not appear to tend to a far-field form as $(a/z) \to 0$ unless an asymptotic expansion is used. In this large-z limit the argument of the sine becomes

$$\frac{k}{2}\left[z\left(1 + \frac{1}{2}\frac{a^2}{z^2}\right) - z\right] \approx \frac{ka^2}{4z} + O\left(\frac{ka^4}{z^3}\right).$$

Setting the sine equal to its argument, equation (5.11b) takes on a form consistent with equation (5.10b):

$$|p(0, z)| \approx \frac{\rho a^2 \ddot{W}}{2z} = \frac{\rho \ddot{Q}}{2\pi z}, \qquad \left(\frac{ka^2}{z}\right) \ll 1. \tag{5.11c}$$

At the center of the piston, equation (5.11b) reduces to

$$|p(0,0)| - 2\rho c \dot{W} \left| \sin\left(\frac{ka}{2}\right) \right|, \qquad z = 0. \tag{5.11d}$$

The pressure distribution along the axis is plotted in figure 5.5.

5.3 The Transform Formulation of Axisymmetric Pressure Fields

We shall now demonstrate a transform technique applicable to planar sources displaying cylindrical symmetry, such as the circular piston. This technique will be used in chapter 8 to construct the field radiated by the point-excited, effectively infinite plate. The pressure satisfies the boundary condition, equation (2.28), and the Helmholtz equation, equation (2.32), with the Laplace operator expressed in cylindrical coordinates:

$$\left(\nabla_\sigma^2 + \frac{\partial^2}{\partial z^2} + k^2 \right) p(r, z) = 0, \tag{5.12}$$

where ∇_σ^2 stands for the three r-dependent terms in the definition in equation (2.61). The ϕ derivative drops out for the ϕ-independent configurations of concern here. Applying the transform formula, equation (5.3a), to the Helmholtz equation, one obtains

$$\int_0^\infty \nabla_\sigma^2 p(r, z) \mathcal{J}_0(\gamma r) r \, dr + \left(\frac{\partial^2}{\partial z^2} + k^2 \right) \int_0^\infty p(r, z) \mathcal{J}_0(\gamma r) r \, dr = 0. \tag{5.13}$$

The second integral is identical to the definition of $\tilde{p}$. The first integral yields[5] $(-\gamma^2 \tilde{p})$. One thus obtains a homogeneous differential equation in z that is formally similar to the plane-wave Helmholtz equation, equation (2.19):

$$\left[\frac{\partial^2}{\partial z^2} + (k^2 - \gamma^2) \right] \tilde{p}(\gamma; z) = 0. \tag{5.14a}$$

The general solution of this equation is in the form of equation (2.23), with $(k^2 - \gamma^2)^{1/2}$ taking the place of k. The boundary condition is also similar to equation (2.24):

$$\rho \tilde{\ddot{w}}(\gamma) = \frac{-\partial \tilde{p}(\gamma; z)}{\partial z}, \qquad z = 0. \tag{5.14b}$$

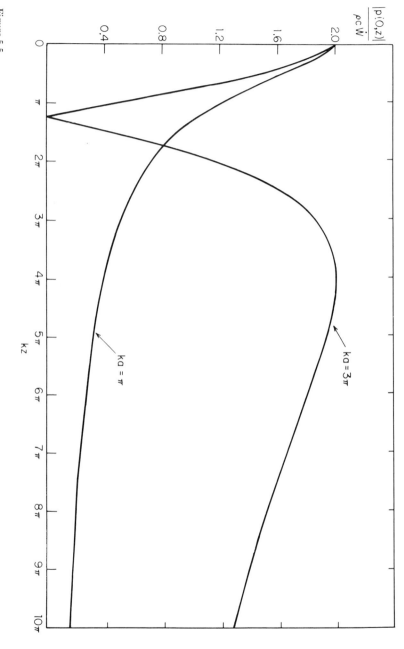

Figure 5.5
Near-field pressure on the piston axis, for two values of ka [equation (5.11)].

Applying Eq. 2.24 to this and taking the limit z=0, yields Eq. 5.1. O.K. ✓
Although the pressure at the surface of the baffle is a function of r, the functions w, ẃ and ẅ do not depend on r.

The transform that satisfies equation (5.14) is

$$\bar{p}(\gamma; z) = \frac{i\tilde{w}(\gamma)\rho \exp[i(k^2 - \gamma^2)^{1/2}z]}{(k^2 - \gamma^2)^{1/2}}. \tag{5.15}$$

Substituting this result in equation (5.6), we obtain an integral representation for the pressure field:

$$p(r, z) - i\rho \int_0^\infty \frac{\tilde{w}(\gamma)}{(k^2 - \gamma^2)^{1/2}} J_0(\gamma r) \exp[i(k^2 - \gamma^2)^{1/2}z]\gamma \, d\gamma. \tag{5.16}$$

When z is set equal to zero, the surface pressure is obtained:

$$p(r, 0) = i\rho \int_0^\infty \frac{\tilde{w}(\gamma)}{(k^2 - \gamma^2)^{1/2}} J_0(\gamma r)\gamma \, d\gamma. \tag{5.17}$$

In the near field, this integral can be evaluated analytically only for a few combinations of a field-point location and acceleration distribution, but it can always be approximated in the far-field by the *method of stationary phase*. For this purpose, the integrand in equation (5.16) is expressed in terms of an exponential as

$$I = \int_{-\infty}^\infty \Phi(\gamma) \exp[i\Psi(\gamma)] \, d\gamma. \tag{5.18}$$

Equation (5.16) can be cast in this form by expressing the Bessel function in terms of Hankel functions of the first and second kind:[6]

$$\begin{aligned} J_0(x) &= \tfrac{1}{2}H_0^{(1)}(x) + \tfrac{1}{2}H_0^{(2)}(x) \\ &= \tfrac{1}{2}H_0^{(1)}(x) - \tfrac{1}{2}H_0^{(1)}(-x) \end{aligned} \tag{5.19a}$$

The integral representation of the pressure equation (5.16), now becomes

$$p(r, z) = \frac{i\rho}{2} \int_0^\infty \frac{\tilde{w}(\gamma) \exp[i(k^2 - \gamma^2)^{1/2}z}{(k^2 - \gamma^2)^{1/2}} [H_0^{(1)}(\gamma r) - H_0^{(1)}(-\gamma r)]\gamma \, d\gamma.$$

The integral of $-H_0^{(1)}(-\gamma r)$ from 0 to ∞ can be written as an integral of $-H_0^{(1)}(\gamma r)$ from 0 to $-\infty$. Reversing the limits of integration, this finally

becomes an integral $+H_0^{(1)}(\gamma r)$ from $-\infty$ to 0. We now have an integral from $-\infty$ to ∞ whose integrand is proportional to $+H_0^{(1)}(\gamma r)$. Switching to spherical coordinates $(r = R \sin \theta, z = R \cos \theta)$ and using the large-argument asymptotic expression for the Hankel function,[7]

$$H_n^{(1)}(x) = \left(\frac{2}{\pi x}\right)^{1/2} (-i)^n \exp\left(ix - i\frac{\pi}{4}\right), \qquad x \gg n^2 + 1, \qquad (5.19b)$$

we see that the integral representation of the far-field pressure takes on the form of the integral in equation (5.18):

$$p(R, \theta) \doteq \frac{\rho e^{-i\pi/4} I}{(2\pi R \sin \theta)^{1/2}}, \qquad (5.20a)$$

where the modulus and phase of the integral in equation (5.18) are, respectively,

$$\Phi(\gamma) = \frac{\tilde{\tilde{w}}(\gamma) \gamma^{1/2}}{(k^2 - \gamma^2)^{1/2}}, \qquad (5.20b)$$

$$\Psi(\gamma; R, \theta) = R[\gamma \sin \theta + (k^2 - \gamma^2)^{1/2} \cos \theta]. \qquad (5.20c)$$

The far-field pressure can now be evaluated by the method of stationary phase. This technique is based on the fact that the resultant contribution of ranges of integration where the modulus varies slowly with γ while the phase fluctuates rapidly is relatively small, because of cancellation between neighboring regions of opposite phase and nearly equal amplitude. The main contribution to the integral arises from the region of integration where the phase of the integrand changes slowly with γ, thus minimizing the cancellation. The corresponding value of γ, designated here as $\bar{\gamma}$, is the point of stationary phase defined by the condition

$$\partial \Psi / \partial \gamma = 0, \qquad \gamma = \bar{\gamma}. \qquad (5.21a)$$

The stationary-phase approximation to the integral in equation (5.18) is[8]

$$I = \frac{(2\pi)^{1/2} \Phi(\gamma) \exp[\pm i(\pi/4) + i\Psi(\gamma)]}{|\partial^2 \Psi / \partial \gamma^2|^{1/2}}, \qquad \gamma = \bar{\gamma}. \qquad (5.21b)$$

The alternative positive and negative signs in the exponential are associated with the corresponding values of $\partial^2 \Psi / \partial \gamma^2$. For the phase angle in equation (5.20c), the point of stationary phase is given by

$$\frac{\partial \Psi}{\partial \gamma} = R\left[\sin\theta - \frac{\gamma}{(k^2 - \gamma^2)^{1/2}} \cos\theta\right]$$

$$= 0 \quad \text{for } \bar{\gamma} = k\sin\theta = \frac{2\pi \sin\theta}{\lambda}.$$

(5.21c)

The second derivative of the phase is

$$\left.\begin{aligned}\frac{\partial^2 \Psi}{\partial \gamma^2} &= -R\cos\theta\left[\frac{1}{(k^2 - \gamma^2)^{1/2}} + \frac{\gamma^2}{(k^2 - \gamma^2)^{3/2}}\right] \\[2mm] &= -R\cos\theta\left(\frac{1}{k\cos\theta} + \frac{k^2 \sin^2\theta}{k^3 \cos^3\theta}\right) \\[2mm] &= -\frac{R}{k\cos^2\theta}\end{aligned}\right\} \quad \gamma = \bar{\gamma}.$$

This quantity being negative, the negative value of $\pi/4$ is used in equation (5.21b). The value of the integral in equation (5.18) thus becomes

$$I = \frac{(2\pi \sin\theta)^{1/2}}{R^{1/2}} \exp\left(ikR - \frac{i\pi}{4}\right) \tilde{w}(k\sin\theta).$$

When this is substituted in place of the integral in the inverse transform for the pressure, equation (5.20), the latter becomes

$$p(R,\theta) = \rho\tilde{w}(k\sin\theta)\frac{e^{ikR}}{R}.$$

(5.22)

The remarkable feature of this result is that it was obtained without specifying the acceleration distribution. Thus, from the continuous wave number spectrum of the acceleration distribution, the solution selects a single, θ-dependent value of γ, which alone determines the far-field pressure to the exclusion of all other wave numbers. The physical meaning of this spatial filtering action will be explored later in this chapter. Equation (5.22)

yields with little effort the far field of any radiator whose acceleration distribution is expressible as a Hankel transform.

Another interesting conclusion can be drawn from the above result: For field points located on the axis of symmetry, $\theta = 0$, the stationary-phase value of $\bar{\gamma}$, equation (5.21c), equals zero, and $\mathcal{J}_0$ equals unity in equation (5.4). Consequently, that integral reduces to $\ddot{Q}/2\pi$ no matter what the acceleration distribution $\ddot{w}(r)$. The pressure in equation (5.22) therefore equals $2p_s(R)$, as already noted in equation (5.10d) for the rigid piston, whatever the dynamic configuration of the source.

5.4 The Transform Solution of the Circular Piston
Setting $\gamma = k \sin \theta$ in equation (5.4), one obtains the stationary-phase value of the transform of the acceleration distribution $\ddot{w}$:

$$\tilde{\ddot{w}}(k \sin \theta) = a \ddot{W} \frac{\mathcal{J}_1 (ka \sin \theta)}{k \sin \theta}. \tag{5.23}$$

When this is substituted in equation (5.22), one obtains the result computed earlier from Rayleigh's formula, equation (5.10a), thus verifying the equivalence of the two formulations. The pressure on the piston axis can be evaluated by setting $r = 0$, $\mathcal{J}_0 = 1$ in equation (5.16), with the acceleration transform given in equation (5.4). This yields the result obtained in equation (5.11) from Rayleigh's formula.

To conclude the discussion of axisymmetric radiators, we compute the radiation loading on a circular piston. Substituting the acceleration transform, equation (5.4), in equation (5.17), the surface pressure becomes

$$p(r, 0) = i\rho \ddot{W} a \int_0^\infty \frac{\mathcal{J}_0 (\gamma r) \mathcal{J}_1 (\gamma a)}{(k^2 - \gamma^2)^{1/2}} \, d\gamma. \tag{5.24}$$

This integral can be evaluated analytically only at $r = 0$ and $r = a$, but if we seek the resultant force, which is frequently of greater practical interest than the local pressure, we obtain a double integral, in r and γ, which can be evaluated analytically. This fluid reaction will be expressed in terms of an accession to inertia M_p and a resistance R_p. To keep track of the phase relations both sides of equation (5.24) are multiplied by $\exp(-i\omega t)$. If we use the notation of equations (2.15), the coefficient multiplying the integral thus becomes:

$$i\rho a \ddot{W} \exp(-i\omega t) = i\rho a \ddot{w} = \rho a \omega \dot{w}.$$

The fluid reaction is defined in terms of a resistance, in units of force over velocity, and of a mass:

$$2\pi \int_0^a p(r,0)\, r\, dr \equiv (R_p - i\omega M_p)\dot{w}. \qquad (5.25a)$$

Comparing the real and imaginary components of, respectively, the right side of the above identity and the integrand in equation (5.24), one sees that the range of integration $k > \gamma$, where the integrand is real, corresponds to $R_p\dot{w}$. The range $k < \gamma$ corresponds to $-i\omega M_p\dot{w}$. Consequently,

$$M_p = 2\pi\rho a \int_0^a \left[\int_k^\infty \frac{\mathcal{J}_1(\gamma a)\,\mathcal{J}_0(\gamma r)}{(\gamma^2 - k^2)^{1/2}}\, d\gamma \right] r\, dr,$$

$$R_p = 2\pi\rho c k a \int_0^a \left[\int_0^k \frac{\mathcal{J}_1(\gamma a)\,\mathcal{J}_0(\gamma r)}{(k^2 - \gamma^2)^{1/2}}\, d\gamma \right] r\, dr.$$

Both r integrations can be performed by means of equation (5.5):

$$M_p = 2\pi\rho a^2 \int_k^\infty \frac{\mathcal{J}_1^2(\gamma a)}{\gamma(\gamma^2 - k^2)^{1/2}}\, d\gamma, \qquad (5.25b)$$

$$R_p = 2\pi\rho c k a^2 \int_0^k \frac{\mathcal{J}_1^2(\gamma a)}{\gamma(k^2 - \gamma^2)^{1/2}}\, d\gamma. \qquad (5.25c)$$

The integral in equation (5.25b) is available in integral tables,[9] in terms of the Struve function[10,11] of order unity $\mathbf{H}_1$:

$$M_p = \frac{\pi\rho a^3 \mathbf{H}_1(2ka)}{k^2 a^2}. \qquad (5.26)$$

To obtain asymptotic values, we note that

$$\mathbf{H}_1(x) = \frac{2x^2}{3\pi} + O(x^4), \qquad x \ll 1$$

$$= \frac{2}{\pi} + O(x^{-1/2}), \qquad x \gg 1.$$

When these relations are substituted in equation (5.26), the low- and high-frequency limits of the accession to inertia are obtained:

$$M_p \approx \frac{8\rho a^3}{3}, \qquad k^2 a^2 \ll 1 \tag{5.26a}$$

$$\approx \frac{2\rho a^3}{k^2 a^2} \approx 0, \qquad k^2 a^2 \gg 1. \tag{5.26b}$$

Characteristically, the accession to inertia tends to a constant value at low frequency, and to zero at high frequency (figure 5.6). The asymptotic expression for the entrained mass can be envisioned as a fluid column whose cross-sectional area A coincides with the vibrating piston,

$$M_p \simeq \rho A H, \qquad k^2 A \ll 1, \tag{5.26c}$$

and whose height is

$$H = (8/3\pi)\, a = 0.85a. \tag{5.26d}$$

Replacing the radius by $(A/\pi)^{1/2}$, this expression can be used to approximate the entrained mass of baffled pistons that even though not circular have an aspect ratio of approximately unity:

$$H = (8/3\pi^{3/2})\, A^{1/2} = 0.48A^{1/2}. \tag{5.26e}$$

The resistance integral in equation (5.25c) can be rewritten in terms of the new variable θ as

$$\gamma = k \sin\theta, \quad d\gamma = k\cos\theta\, d\theta, \quad (k^2 - \gamma^2)^{1/2} = k\cos\theta.$$

The integral thus becomes

$$R_p = 2\pi\rho c a^2 \int_0^{\pi/2} \frac{\mathcal{J}_1^2\,(ka\sin\theta)}{\sin\theta}\, d\theta. \tag{5.27a}$$

Alternatively, this result could have been constructed by substituting the far-field pressure, (5.10a), in the integral expression for power, (3.27), divided by $\omega^2/\ddot{W}^2$. This integral (5.27a) is available in texts on Bessel

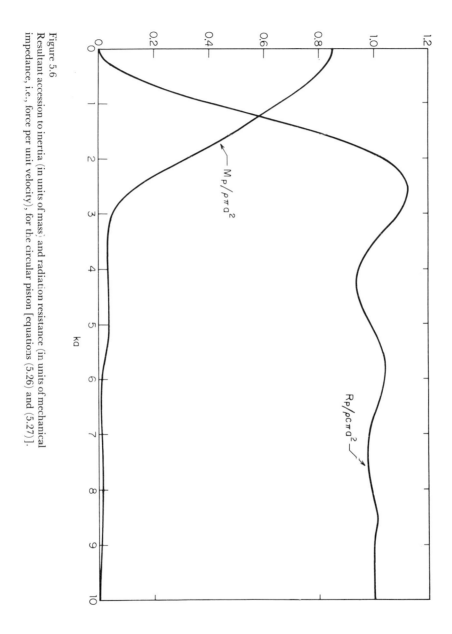

Figure 5.6
Resultant accession to inertia (in units of mass) and radiation resistance (in units of mechanical impedance, i.e., force per unit velocity), for the circular piston [equations (5.26) and (5.27)].

functions.[12] The resistance thus becomes

$$R_p = \rho c \pi a^2 [1 - (ka)^{-1} \mathcal{J}_1 (2ka)]. \tag{5.27b}$$

Asymptotic expressions for the resistance are obtained from the relations[13]

$$\mathcal{J}_1 (x) \approx \frac{x}{2} - \frac{x^2}{16}, \qquad x^5 \ll 1$$

$$\approx 0, \qquad x^{-1/2} \ll 1.$$

The low- and high-frequency limits of the resistance now become

$$R_p \approx \frac{\rho c \pi a^2 (ka)^2}{2}, \qquad k^4 a^4 \ll 1 \tag{5.27c}$$

$$\approx \rho c \pi a^2, \qquad k^2 a^2 \gg 1. \tag{5.27d}$$

Like the entrained mass, the low-frequency resistance can be expressed in more general form in terms of the piston area:

$$R_p \simeq 2\pi \rho c \frac{A^2}{\lambda^2}, \qquad \frac{A^2}{\lambda^4} \ll 1. \tag{5.27e}$$

In the large-λ limit, the resistance is overshadowed by the inertial radiation loading. The high-frequency limit of the impedance is the characteristic impedance of plane waves. Analytical expressions for the accession to inertia have been obtained for a variety of nonuniform circular radiators.[14] Complex impedances have been computed for various nonrigid axisymmetric pistons.[15,16]

We now turn to rectangular sound sources. We shall study them first in terms of Rayleigh's formula, and then of transforms.

5.5 The Far-Field of Rectangular Radiators Evaluated by Rayleigh's Formula; the Rigid Piston

An acceleration distribution $\ddot{w}(x_0, y_0)$ is prescribed over a baffled planar radiator located in the $(z_0 = 0)$ plane in the region $-L_x < x_0 < L_x$, $-L_y < y_0 < L_y$ (figure 5.7). For future reference, it is noted that this distribution, being aperiodic in both x_0 and y_0, must be expressed as a double

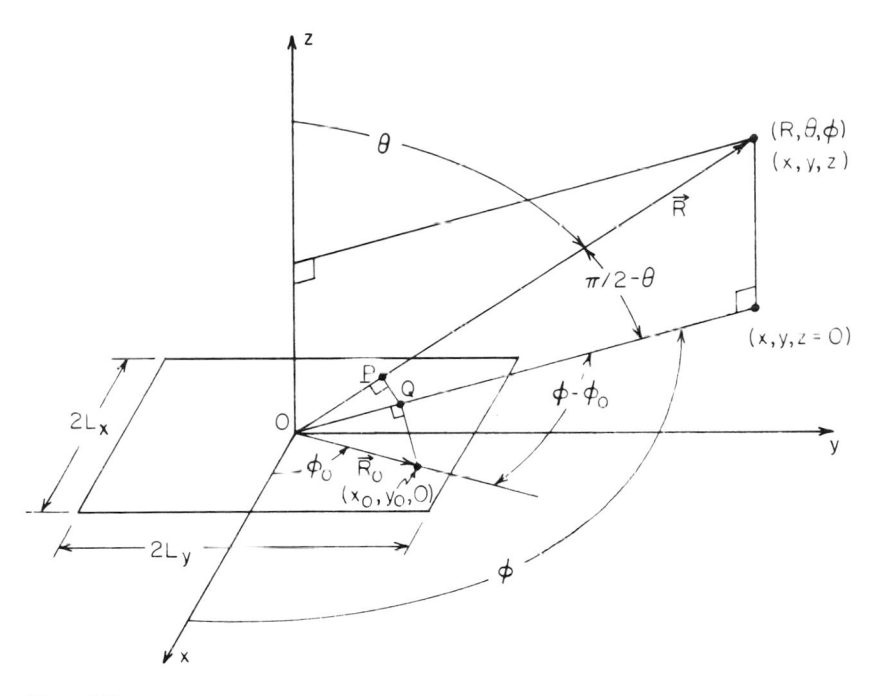

Figure 5.7
Construction of the far-field of a rectangular piston radiator.

Fourier transform[17] defined by the formulas

$$\tilde{f}(\gamma_x, \gamma_y) = \int\!\!\!\int_{-\infty}^{\infty} f(x, y) \exp[-i(\gamma_x x + \gamma_y y)] \, dx \, dy, \tag{5.28a}$$

$$f(x, y) = \frac{1}{(2\pi)^2} \int\!\!\!\int_{-\infty}^{\infty} \tilde{f}(\gamma_x, \gamma_y) \exp[i(\gamma_x x + \gamma_y y)] \, d\gamma_x \, d\gamma_y. \tag{5.28b}$$

The range of integration applicable to the acceleration transform is limited to the region occupied by the radiator, since $\ddot{w} = 0$ elsewhere:

$$\tilde{\ddot{w}}(\gamma_x, \gamma_y) = \int_{-L_x}^{L_x} \int_{-L_y}^{L_y} \ddot{w}(x_0, y_0) \exp[-i(\gamma_x x_0 + \gamma_y y_0)] \, dx_0 \, dy_0. \tag{5.29}$$

The far-field of this radiator is given by Rayleigh's formula, equation (4.29), with $dS(\mathbf{R}_0) = dx_0\, dy_0$. The path length, equation (5.7), must be similarly expressed in rectangular coordinates (figure 5.7). Expanding the cosine term in equation (5.7), and setting $r_0 \cos\phi_0 = x_0$ and $r_0 \sin\phi_0 = y_0$, one obtains the path length

$$|\mathbf{R} - \mathbf{R}_0| = R - x_0 \sin\theta \cos\phi - y_0 \sin\theta \sin\phi + O\left[\frac{(x_0^2 + y_0^2)}{R}\right].$$

Rayleigh's formula now takes the form

$$p(R, \theta, \phi)$$

$$= \frac{\rho e^{ikR}}{2\pi R} \int_{-L_x}^{L_x} \int_{-L_y}^{L_y} \ddot{w}(x_0, y_0) \exp[-ik \sin\theta\, (x_0 \cos\phi + y_0 \sin\phi)]\, dx_0\, dy_0.$$

Referring to equation (5.29), one notes that the above integral is the acceleration transform $\tilde{\ddot{w}}(\gamma_x, \gamma_y)$ specialized to the values $\gamma_x = \bar{\gamma}_x$, $\gamma_y = \bar{\gamma}_y$ of the transform parameters:

$$\bar{\gamma}_x = k \sin\theta \cos\phi,$$

$$\bar{\gamma}_y = k \sin\theta \sin\phi. \tag{5.30}$$

Rayleigh's formula can thus be formally integrated without specifying the acceleration distribution:

$$p(R, \theta, \phi) = \frac{\rho e^{ikR}}{2\pi R} \tilde{\ddot{w}}(k \sin\theta \cos\phi, k \sin\theta \sin\phi). \tag{5.31}$$

To illustrate the use of this formula, consider a rectangular piston over which a constant acceleration amplitude $\ddot{W}$ is prescribed. This configuration, being even in x_0 and y_0, the complex Fourier transforms, equation (5.29), reduce to cosine transforms

$$\tilde{\ddot{w}}(\gamma_x, \gamma_y) = 4\ddot{W} \int_0^{L_x} \int_0^{L_y} \cos\gamma_x x_0 \cos\gamma_y y_0\, dx_0\, dy_0$$

$$= \frac{4\ddot{W} \sin\gamma_x L_x \sin\gamma_y L_y}{\gamma_x \gamma_y}.$$

This transform is evaluated at the values $\bar{\gamma}_x$ and $\bar{\gamma}_y$ defined in equation (5.30). The result will be expressed concisely in terms of the spherical Bessel function, equation (2.45). When the resultant expression is substituted for $\tilde{\tilde{w}}$ in equation (5.31), the far-field of the rectangular piston is obtained:

$$p(R, \theta, \phi) = \frac{2 L_x L_y \ddot{W} \rho e^{ikR}}{\pi R} j_0(kL_x \sin \theta \cos \phi) j_0(kL_y \sin \theta \sin \phi). \qquad (5.32a)$$

Since $4 L_x L_y \ddot{W} = \ddot{Q}$, this is concisely expressed in terms of the pressure field $p_s(R)$ of the point source whose volume acceleration is that of the piston, equation (3.2):

$$
\begin{aligned}
p(R, \theta, \phi) &= 2 p_s(R) j_0(kL_x \sin \theta \cos \phi) j_0(kL_y \sin \theta \sin \phi) \\
&\approx 2 p_s(R), \qquad kL_x, kL_y \ll 1.
\end{aligned}
\qquad (5.32b)
$$

Thus, we have verified once again that an extended radiator can be replaced by a point source of equivalent volume acceleration provided the radiator is small in terms of acoustic wavelengths, in other words, in the small kL_x-, kL_y-limit. Whatever the dimensions of the piston, the pressure field takes on a maximum value of $2p_s$ on the piston axis, at $\theta = 0$. If the piston has a large aspect ratio, kL_y being small and kL_x large, the L_y-dependent factor in equation (5.32) can be set equal to unity, but the L_x factor cannot be thus simplified. The pressure field thus approaches that of the linear broadside array, equation (3.17), multiplied by 2 to account for the presence of the baffle. In fact, it can be verified that the field of the rectangular piston can be constructed by multiplying the directivity functions of two broadside line arrays oriented, respectively, along the x and y axes. A generalized formulation of this observation constitutes the *second product theorem* used in array theory.[18]

5.6 The Transform Solution of Rectangular Sound Radiators

The procedure parallels the transform technique for axisymmetric radiators. The formulation is somewhat more complex because the two-dimensional acceleration distribution $\ddot{w}(x_0, y_0)$ requires two Fourier transform parameters, γ_x and γ_y, equations (5.28). The pressure transform $\tilde{p}(\gamma_x, \gamma_y; z)$ must satisfy the doubly transformed boundary condition

$$\rho \ddot{\tilde{w}}(\gamma_x, \gamma_y) = \frac{-\partial \tilde{p}(\gamma_x, \gamma_y; z)}{\partial z}, \qquad z = 0, \qquad (5.33a)$$

and the similarly transformed Helmholtz equation

$$\int\int_{-\infty}^{\infty}\left(\frac{\partial^2}{\partial x^2} + \frac{\partial^2}{\partial y^2} + \frac{\partial^2}{\partial z^2} + k^2\right)p(x,y,z)$$

$$\cdot \exp[-i(\gamma_x x + \gamma_y y)]\,dx\,dy = 0.$$

(5.33b)

The z-derivative and the k^2 term in the above integrand, being independent of x and y, can be removed from under the integral sign, thus giving rise to an operator multiplying an integral in the form of equation (5.28a), which is recognized as the pressure transform:

$$\left[\left(\frac{\partial^2}{\partial z^2}\right) + k^2\right]\tilde{p}(\gamma_x,\gamma_y;z).$$

The x and y derivatives in equation (5.33b) integrate to

$$(-\gamma_x^2 - \gamma_y^2)\tilde{p}(\gamma_x,\gamma_y;z).$$

When these terms are added, the transformed Helmholtz equation becomes

$$\left(k^2 - \gamma_x^2 - \gamma_y^2 + \frac{\partial^2}{\partial z^2}\right)\tilde{p}(\gamma_x,\gamma_y;z) = 0.$$

(5.34)

The solution of this equation is

$$\tilde{p}(\gamma_x,\gamma_y;z) = A\exp[i(k^2 - \gamma_x^2 - \gamma_y^2)^{1/2}z].$$

(5.35)

To satisfy the Sommerfeld radiation condition, equation (4.13), the real part of the exponent must be negative. The undertermined coefficient is once again evaluated from the transformed boundary condition, equation (5.33a):

$$A = \frac{i\rho\tilde{\tilde{w}}(\gamma_x,\gamma_y)}{(k^2 - \gamma_x^2 - \gamma_y^2)^{1/2}}.$$

When this is combined with equation (5.35), one obtains an explicit expression for $\tilde{p}$ in terms of $\tilde{\tilde{w}}$. Substituting $\tilde{p}$ in place of $\tilde{f}$ in equation (5.28b),

we construct a double integral for the pressure field:

$$p(x, y, z)$$

$$= \frac{i\rho}{(2\pi)^2} \int\!\!\!\int_{-\infty}^{\infty} \frac{\tilde{\tilde{w}}(\gamma_x, \gamma_y) \exp[i\gamma_x x + i\gamma_y y + i(k^2 - \gamma_x^2 - \gamma_y^2)^{1/2} z]}{(k^2 - \gamma_x^2 - \gamma_y^2)^{1/2}} d\gamma_x \, d\gamma_y.$$

$$(5.36)$$

Like the inverse Hankel transform, this integral representation is generally not tractable in the near-field. It can be evaluated analytically in the near-field only for certain combinations of acceleration distribution and field points. For the rigid piston, the pressure can be evaluated at a piston corner by a rather laborious series expansion. By dividing the actual piston into four rectangles whose common point coincides with the desired field point, the surface pressure can be computed anywhere on the piston and, in fact, on the baffle.[19] Plots of the complex impedance of square pistons are available in the literature.[20]

The integral can always be evaluated in the far-field. Proceeding as in the far-field evaluation of the Hankel transform, we transform to spherical coordinates:

$$x = R \sin\theta \cos\phi, \qquad y = R \sin\theta \sin\phi, \qquad z = R \cos\theta.$$

The pressure integral now becomes

$$p(R, \theta, \phi)$$

$$= \frac{i\rho}{(2\pi)^2}$$

$$\cdot \int\!\!\!\int_{-\infty}^{\infty} \frac{\exp\{iR[(k^2 - \gamma_x^2 - \gamma_y^2)^{1/2} \cos\theta + \gamma_x \sin\theta \cos\phi + \gamma_y \sin\theta \sin\phi]\}}{(k^2 - \gamma_x^2 - \gamma_y^2)^{1/2}}$$

$$\cdot \tilde{\tilde{w}}(\gamma_x, \gamma_y) \, d\gamma_x \, d\gamma_y.$$

$$(5.37)$$

The integrand is in the exponential form of equation (5.18) and is thus suitable for stationary-phase evaluation, but with two integration parameters, γ_x and γ_y, in lieu of a single parameter γ. In the notation of equation (5.18), the modulus and phase of the integrand are, respectively,

$$\Phi(\gamma_x, \gamma_y) = \tilde{\tilde{w}}(\gamma_x, \gamma_y)\,(k^2 - \gamma_x^2 - \gamma_y^2)^{-1/2}, \tag{5.38a}$$

$$\Psi(\gamma_x, \gamma_y) = R[\,(k^2 - \gamma_x^2 - \gamma_y^2)^{1/2}\cos\theta + \gamma_x\sin\theta\cos\phi + \gamma_y\sin\theta\sin\phi\,]. \tag{5.38b}$$

The method of stationary phase is here based on the premise that the main contribution to the integral is associated with the region where the phase does not vary with either of the two integration parameters:

$$\frac{\partial\Psi}{\partial\gamma_x} = \frac{\partial\Psi}{\partial\gamma_y} = 0, \qquad \gamma_x = \bar{\gamma}_x, \qquad \gamma_y = \bar{\gamma}_y. \tag{5.39}$$

The stationary-phase approximation to the double integral is[21]

$$I = \frac{\pm i2\pi}{|D(\bar{\gamma}_x, \bar{\gamma}_y)|^{1/2}}\,\Phi(\bar{\gamma}_x, \bar{\gamma}_y)\,\exp[i\Psi(\bar{\gamma}_x, \bar{\gamma}_y)]. \tag{5.40}$$

The alternative signs are associated with the corresponding sign of the determinant D:

$$D(\gamma_x, \gamma_y) = \left(\frac{\partial^2\Psi}{\partial\gamma_x\,\partial\gamma_y}\right)^2 - \frac{\partial^2\Psi}{\partial\gamma_x^2}\frac{\partial^2\Psi}{\partial\gamma_y^2}. \tag{5.41}$$

The first partial derivative of the phase angle, equation (5.38b), is

$$\frac{\partial\Psi}{\partial\gamma_x} = R\cos\phi\sin\theta - \frac{\gamma_x R\cos\theta}{(k^2 - \gamma_x^2 - \gamma_y^2)^{1/2}},$$

$$\frac{\partial\Psi}{\partial\gamma_y} = R\sin\phi\sin\theta - \frac{\gamma_y R\cos\theta}{(k^2 - \gamma_x^2 - \gamma_y^2)^{1/2}}. \tag{5.42}$$

The simultaneous roots of equation (5.42), which define the point of stationary phase in the (γ_x, γ_y) plane, are precisely the values stated in equation (5.30), as verified by noting that

$$(k^2 - \bar{\gamma}_x^2 - \bar{\gamma}_y^2)^{1/2} = k\cos\theta.$$

The modulus and phase of the integrand, evaluated at the point of stationary phase, reduce to

$$\Phi(\bar{\gamma}_x, \bar{\gamma}_y) = \frac{\tilde{\tilde{w}}(\bar{\gamma}_x, \bar{\gamma}_y)}{k\cos\theta},$$

$$\Psi(\bar{\gamma}_x, \bar{\gamma}_y) = kR.$$

Second derivatives required in equation (5.41) are

$$\frac{\partial^2\Psi}{\partial\gamma_x\,\partial\gamma_y} = -\frac{\gamma_x\gamma_y R\cos\theta}{(k^2 - \gamma_x^2 - \gamma_y^2)^{3/2}}$$

$$= -\left(\frac{R}{k}\right)\tan^2\theta\cos\phi\sin\phi, \qquad \gamma_x = \bar{\gamma}_x,\quad \gamma_y = \bar{\gamma}_y,$$

$$\frac{\partial^2\Psi}{\partial\gamma_x^2} = \frac{R\cos\theta}{(k^2 - \gamma_x^2 - \gamma_y^2)^{1/2}} - \frac{\gamma_x^2 R\cos\theta}{(k^2 - \gamma_x^2 - \gamma_y^2)^{3/2}},$$

and similarly for $\partial^2\Psi/\partial\gamma_y^2$. The stationary-phase values of the latter two derivatives are

$$\left.\begin{array}{l} \dfrac{\partial^2\Psi}{\partial\gamma_x^2} = -\left(\dfrac{R}{k}\right)(1 + \tan^2\theta\cos^2\phi) \\[3mm] \dfrac{\partial^2\Psi}{\partial\gamma_y^2} = -\left(\dfrac{R}{k}\right)(1 + \tan^2\theta\sin^2\phi) \end{array}\right\} \quad \gamma_x = \bar{\gamma}_x,\quad \gamma_y = \bar{\gamma}_y.$$

Substituting these derivatives in equation (5.41), one reduces the determinant D at the point of stationary phase to

$$D(\bar{\gamma}_x, \bar{\gamma}_y) = -\left(\frac{R}{k\cos\theta}\right)^2.$$

When these results are substituted in equation (5.40), one obtains the stationary-phase approximation to the integral in equation (5.37):

$$\left. I = -2i\pi\tilde{\tilde{w}}(\bar{\gamma}_x, \bar{\gamma}_y)\frac{e^{ikR}}{R}\right\} \quad \begin{array}{l} \bar{\gamma}_x = k\sin\theta\cos\phi, \\[1mm] \bar{\gamma}_y = k\sin\theta\sin\phi. \end{array}$$

When this is substituted in place of the double integral in equation (5.37), the resulting expression for the pressure is identical to that evolved from Rayleigh's formula, equation (5.31). A physical interpretation of the equiv-

alence of the approximation inherent in the stationary-phase and Rayleigh's formulations will now be attempted.

5.7 Equivalence of Rayleigh's and the Stationary-Phase Formulations of the Far-Field

The reason for evolving two alternative solutions is that it will be found convenient, in chapter 8, to use the transform solution, approximated in the far-field by the stationary-phase formulation, while in other situations Rayleigh's formula is more practical. The approximations inherent in these two formulations are (1) in the transform technique, dropping contributions to the inverse transform integral associated with ranges of integration other than the region of stationary phase; (2) in Rayleigh's formula, dropping terms of order $k(x_0^2 + y_0^2)/R$ in the surface integral that are, by definition, negligible in the Fraunhofer far-field. Even though these approximations are quite different, they both result in associating the pressure at a given far-field point exclusively with the wave-number values in equation (5.30). This conclusion can be given a physical interpretation if one defines $\bar{\gamma}_x$ and $\bar{\gamma}_y$ as the x and y components of a resultant wave number $\bar{\gamma}$,

$$\bar{\gamma}_x \equiv \bar{\gamma} \cos\phi, \quad \bar{\gamma}_y \equiv \bar{\gamma} \sin\phi,$$

where

$$\bar{\gamma} = (\bar{\gamma}_x^2 + \bar{\gamma}_y^2)^{1/2} = k \sin\theta$$

$$= \frac{2\pi \sin\theta}{\lambda}. \tag{5.43}$$

This resultant stationary-phase wave number is not tied to rectangular radiators but was also found to apply to axisymmetric sound radiators, equation (5.21c). In both cases, the distance $\lambda/\sin\theta$ is the trace, on the radiating plane, of the wavelength λ of plane waves propagating in the direction of the field point (figure 5.8).

This can be expressed qualitatively, by stating that, *among the continuous spectrum of wave numbers inherent in the pressure transform, the stationary-phase wave number is best coupled to the acoustic medium, to the effective exclusion of contributions of the remainder of the spectrum.* The acoustic medium and its boundary thus exert a *spatial filtering action*, the filter characteristics being a function of field point location. The situation may be clarified by comparing it to an

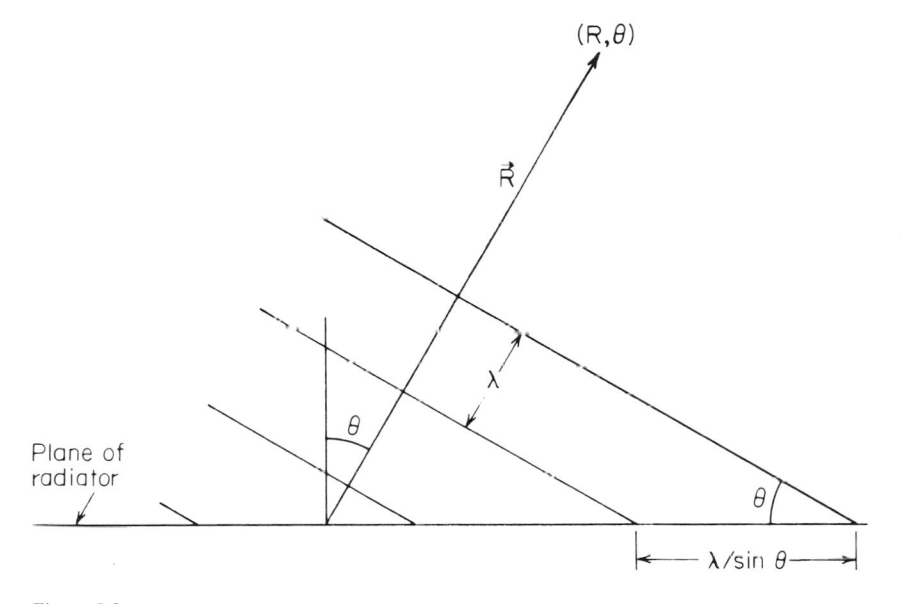

Figure 5.8
Interpretation of the stationary-phase value of the wave number.

analogous more familiar frequency filtering phenomenon in the field of mechanical vibrations.[22] Here, the response of a mass-spring oscillator to an impulsive force is expressed as an inverse Fourier transform over a continuous frequency spectrum. Only the frequency component that matches the natural frequency of the oscillator appears in the oscillator response. This frequency is thus analogous to the stationary-phase wave number that, in the stationary-phase approximation, determines the far-field of a rigid piston, to the exclusion of all other wave numbers in the wave-number spectrum. It is instructive to extend this parallel between the acoustical and vibrational situations to spectra embodying marked maxima. Thus, when, instead of an impulse, the transient excitation of an oscillator is a pulse of sine waves, the spectrum of the excitation displays a peak at the carrier frequency. If the natural frequency of the oscillator coincides with this peak of the frequency spectrum, the amplitude of the transient response at the oscillator natural frequency is enhanced. It grows indefinitely with the pulse duration measured in periods of oscillation. Returning to the acoustical problem, we note that the spatial analog of a sinusoidal pulse is a planar radiator whose dynamic configuration is characterized by sinusoidal stand-

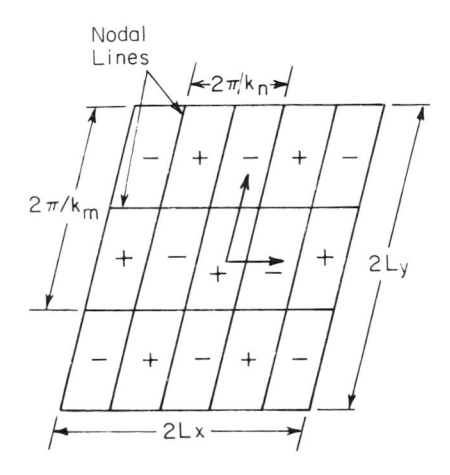

Figure 5.9
Even-even standing-wave configuration of a rectangular radiator.

ing waves of wavelength $(2\pi/k_s)$. The corresponding wavenumber spectrum of the acceleration distribution peaks at $\gamma = k_s$. We shall see in the next section that the coincidence of the structural wavenumber k_s with the stationary-phase wave number $\bar{\gamma}$ gives rise to a marked pressure peak whose amplitude tends to infinity with increasing radiator size measured in structural wavelengths $2\pi/k_s$.

5.8 The Far-Field of Rectangular Radiators Displaying a Sinusoidal Acceleration Distribution

Having applied the general far-field solution, equation (5.31), to rigid pistons, we shall illustrate it further by considering a spatially varying acceleration distribution specialized to a situation of practical interest, studied in chapter 7—the in vacuo modes of the simply supported rectangular plate. These modes are coupled by radiation loading when the plate is submerged, but the far-field pressures constructed here are a useful approximation to the field of the submerged plate, and an excellent approximation to the field of the plate vibrating in the atmosphere. We shall begin by considering a configuration even in both x and y (figure 5.9):

$$\ddot{w}(x, y) = \ddot{W} \cos k_n x \cos k_m y, \qquad |x| < L_x, \quad |y| < L_y$$
$$= 0, \qquad |x| \geq L_x, \quad |y| \geq L_y. \tag{5.44}$$

The boundary conditions are

$$\cos k_n L_x = \cos k_m L_y = 0,$$

$$\sin k_n L_x = (-1)^n, \qquad \sin k_m L_y = (-1)^m. \tag{5.45a}$$

These boundary conditions restrict the wave numbers k_n and k_m to the values

$$k_n L_x = (n + \tfrac{1}{2})\pi, \qquad n = 0, 1, 2\ldots,$$

$$k_m L_y = (m + \tfrac{1}{2})\pi, \qquad m = 0, 1, 2\ldots. \tag{5.45b}$$

The number of nodal lines displayed by this configuration is $2(n + m)$. The numbers (n, m) define the mode order. When this acceleration distribution is substituted in equation (5.29), one obtains the Fourier transform of $\ddot{w}$. Only the cosine terms of the exponentials contribute to the integral:

$$\tilde{w}(\gamma_x, \gamma_y) = 4\ddot{W} \int_0^{L_x} \int_0^{L_x} \cos \gamma_x x \cos k_n x \cos \gamma_y y \cos k_m y \, dx \, dy$$

$$= \frac{4\ddot{W} k_n k_m (-1)^{n+m} \cos \gamma_x L_x \cos \gamma_y L_y}{(k_n^2 - \gamma_x^2)(k_m^2 - \gamma_y^2)}. \tag{5.46}$$

The conciseness of this result is a result of the simply supported boundary conditions, equation (5.45), which simplify the upper limit of the integral. Evaluating this transform at the stationary-phase values of γ_x and γ_y, equation (5.30), and substituting the result in equation (5.31), one obtains an explicit expression for the pressure field:

$$p(R, \theta, \phi)$$

$$= \frac{2\rho \ddot{W} \exp(ikR) k_n k_m (-1)^{n+m} \cos(kL_x \sin\theta \cos\phi) \cos(kL_y \sin\theta \sin\phi)}{\pi R (k_n^2 - k^2 \sin^2\theta \cos^2\phi)(k_m^2 - k^2 \sin^2\theta \sin^2\phi)}. \tag{5.47}$$

In the long-wavelength limit, the k^2 terms in the denominator can be neglected. These terms vanish, whatever k, for $\theta = 0$. For this field-point location, the cosines in the numerator equal unity. This result can be further reduced by multiplying the numerator and denominator by $L_x L_y$, and introducing equation (5.45b). This pressure peak becomes

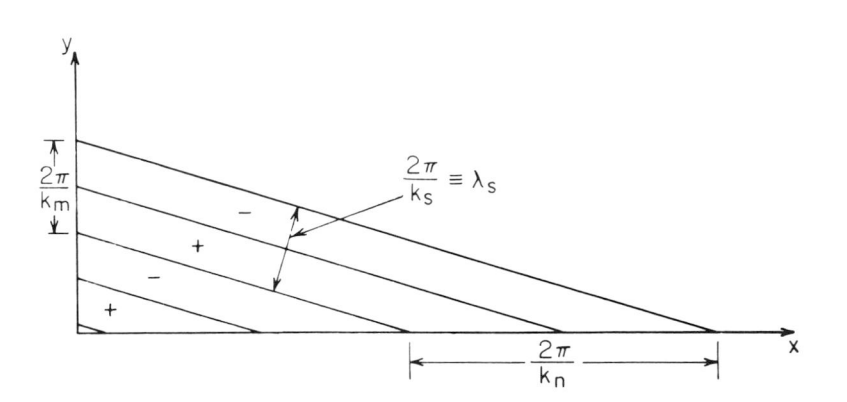

Figure 5.10
Effective structural wavelength of a radiator configuration embodying orthogonal nodal lines.

$$p(R, \theta = 0) = \frac{2\rho \ddot{W} L_x L_y (-1)^{n+m} \exp(ikR)}{\pi^3 R(n + \frac{1}{2})(m + \frac{1}{2})}, \qquad m, n = 0, 1 \ldots . \tag{5.48}$$

The higher the mode order, the smaller this maximum.

Other spatial maxima occur near values of θ and ϕ that cause the difference factors in the denominator of equation (5.47) to become zero, the exact locations of those maxima being given by the zeros of the derivative of this equation. Because both the location and level of the maxima are adequately approximated by the zeros of the denominator, we shall determine only the latter. The denominator vanishes for certain combinations of θ and ϕ if k_n, $k_m < k$, or, in physical terms, if both dimensions of the in-phase regions of the radiators exceed one-half acoustic wavelength (figure 5.9). If these conditions are met, pressure maxima occur at field points (θ_0, ϕ_0) where

$$k_n = k \sin \theta_0 \cos \phi_0,$$
$$k_m = k \sin \theta_0 \sin \phi_0. \tag{5.49a}$$

These two wave numbers can be combined, in the manner of equation (5.43). We thus define a resultant wave number whose x and y components are, respectively, k_n and k_m (figure 5.10):

$$k_s = (k_n^2 + k_m^2)^{1/2}.$$

In anticipation of the interaction analysis in chapter 8, the k_s can be associated with an effective structural wavelength $\lambda_s = 2\pi/k_s$. The relation in equation (5.49a) can now be expressed as the angular coordinates of four symmetrically located field points:

$$\theta_0 = \sin^{-1}\frac{k_s}{k} = \sin^{-1}\frac{\lambda}{\lambda_s},$$

$$(5.49b)$$

$$\phi_0 = \tan^{-1}\frac{k_m}{k_n}, \qquad \phi_1 = -\phi_0, \qquad \phi_{2,3} = \pi \pm \phi_0.$$

These are the *critical* or *coincidence angles*, for which equation (5.47) tends to the indeterminate form (zero/zero). Applying l'Hospital's rule, we have

$$\left.\begin{aligned}
\lim_{\substack{\theta\to\theta_0\\ \phi\to\phi_q}} \frac{\cos(kL_x\sin\theta\cos\phi)}{k_n^2 - (k\sin\theta\cos\phi)^2} &= \frac{(-1)^n L_x}{2k_n} \\[2ex]
\lim_{\substack{\theta\to\theta_0\\ \phi\to\phi_q}} \frac{\cos(kL_y\sin\theta\sin\phi)}{k_m^2 - (k\sin\theta\sin\phi)^2} &= \frac{(-1)^m L_y}{2k_m}
\end{aligned}\right\} \qquad q = 0, 1, 2, \text{ and } 3. \qquad (5.49c)$$

When this is substituted in equation (5.47), one obtains an expression for the maximum pressure:

$$p(R,\theta_0,\phi_q) = \frac{\rho\ddot{W}\exp(ikR)\,L_x L_y}{2\pi R}, \qquad k_n, k_m < k. \qquad (5.50)$$

This and other expressions for pressure maxima corresponding to different configurations are summarized in table 5.1. Unlike the peak at $\theta = 0$, this result is independent of the index numbers m and n.

It is of practical interest to determine whether the amplitude of this maximum exceeds the one on the radiator axis, at $\theta = 0$. It is seen that the coincidence peaks, equation (5.50), exceed the on-axis pressure, equation (5.48), except when both n and m are zero—in other words, except when the entire radiator vibrates in phase. These maxima are illustrated for n, $m \neq 0$ in the plots of the distribution in-angle in figures 5.11 and 5.12. As anticipated, the peak pressure exceeds slightly the value in equation (5.50), and occurs at angles somewhat different from the coincidence angles.

If $k_m > k$, but $k_n < k$, the x dimension of the in-phase regions of the

Table 5.1
Pressure maxima radiated by planar sources simulating in vacuo modes of simply supported rectangular plates[a] $[2L_x, 2L_y$ = plate dimensions; $k_n L_x = (n + \frac{1}{2})\pi$, $k_m L_y = (m + \frac{1}{2})\pi$, $m, n = 0, 1 \ldots]$

Effectively radiating area elements	Field point	
	On perpendicular through plate	At critical angle
Entire surface $(k_n, k_m < k)$	$\dfrac{2\rho \ddot{W} L_x L_y}{\pi^3 R (n + \frac{1}{2})(m + \frac{1}{2})}$	$\dfrac{\rho \ddot{W} L_x L_y}{2\pi R}$
Two edges parallel to x-axis $(k_n < k, k_m > k)$	Same	$\dfrac{\rho \ddot{W} L_x L_y}{\pi^2 R (m + \frac{1}{2})}$
Four corners $(k_n, k_m > k)$	Same	Not applicable

a. For modes even in x and y. For even-odd and odd-odd configurations, peak pressures are commensurate if distances between nodal lines are comparable, except in the small-kL_x, -kL_y limit, where the odd configurations generate pressures that are negligible compared with those tabulated here [see equations (5.53), (5.54), (5.56), and (5.57)].

radiator exceeds $\lambda/2$, but their y dimension is smaller than the acoustic half-wavelength. For these conditions, only one of the two factors in parentheses in the denominator of equation (5.47) can vanish. For a given ϕ, the pressure plotted as a function of θ has a maximum when the denominator vanishes, at a ϕ-dependent value of θ given by

$$\theta_n = \sin^{-1}\left(\frac{k_n}{k \cos \phi}\right), \qquad \phi \le \cos^{-1}\left(\frac{k_n}{k}\right).$$

The corresponding pressure is obtained by means of the former of equations (5.49c). The pressure maximum in this region can be envisioned as a ridge that peaks at $\phi = 0$, in the (x, z) plane, where $\cos(kL_y \sin \theta \sin \phi)$ reaches its maximum value of unity. A similar ridge is centered on $\phi = \pi$. The peak pressure at $\phi = 0$ (or π) is

$$p(R, \theta_n, \phi = 0) = \left. \frac{\rho \ddot{W} e^{ikR} (-1)^m L_x L_y}{\pi^2 R (m + \frac{1}{2})} \right\} \quad \begin{array}{l} k_m > k, \quad k_n < k, \\ m = 0, 1 \ldots. \end{array} \qquad (5.50a)$$

Except for $n = 0$ and 1, the pressure peak at $\theta = 0$, equation (5.48), is smaller than equation (5.50a) for the same value of m.

In the long-wavelength limit, the k^2 terms in the denominator of

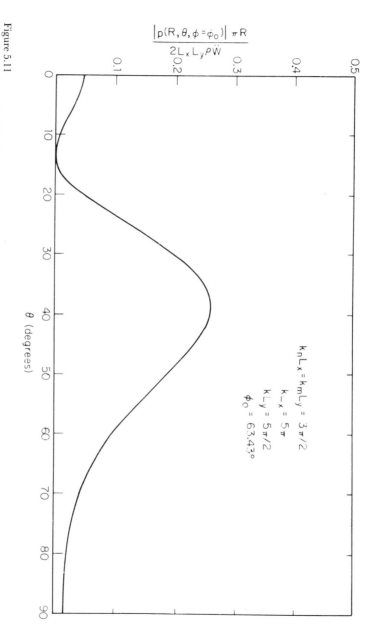

$$\frac{|p(R, \theta, \phi = \phi_0)| \ \pi R}{2 L_x L_y \rho \ddot{W}}$$

θ (degrees)

$k_n L_x = k_m L_y = 3\pi/2$

$k_{-x} = 5\pi$

$k L_y = 5\pi/2$

$\phi_0 = 63.43°$

Figure 5.11
Distribution in angle of the far-field pressure radiated by an even-even standing-wave configuration of a rectangular radiator in the ϕ plane containing the coincidence peak [equation (5.49b)] for ϕ_0] as a function of the polar angle θ [equation (5.47)].

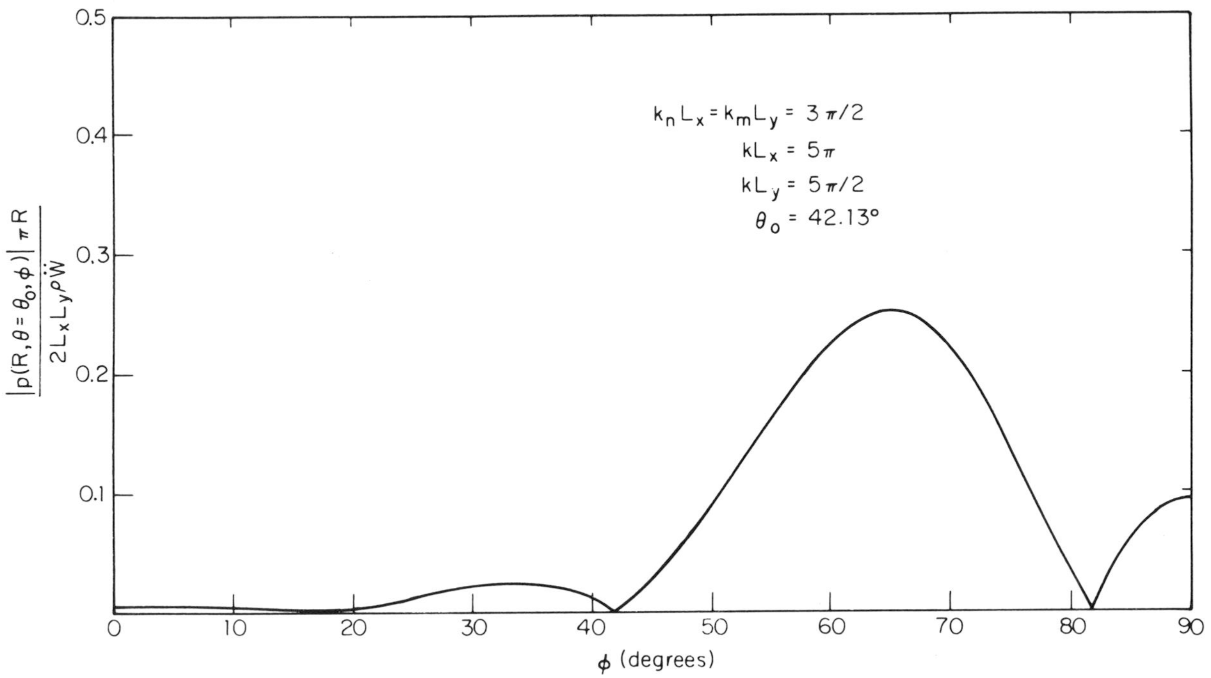

Figure 5.12
Same as figure 5.11, as a function of the circumferential angle ϕ, for field points on the conical surface of vertex angle θ containing the coincidence peaks [equation (5.49b) for θ_0].

$k_n L_x = k_m L_y = 3\pi/2$

$kL_x = 5\pi$

$kL_y = 5\pi/2$

$\theta_0 = 42.13°$

equation (5.47) can be neglected. Multiplying the numerator and denominator by $L_x L_y$, and introducing equation (5.45b), the pressure field becomes

$p(R, \theta, \phi)$

$$= \frac{2\rho \ddot{W} L_x L_y \exp(ikR) (-1)^{n+m}}{\pi^3 R (n + \frac{1}{2}) (m + \frac{1}{2})}$$

(5.51a)

$$\left. \cdot \cos(kL_x \sin\theta \cos\phi) \cos(kL_y \sin\theta \sin\phi) \right\} \quad \begin{array}{l} k^2 \ll k_n^2, \, k_m^2, \\ m, n = 0, 1 \ldots . \end{array}$$

long wavelength limit.

The long-wavelength approximation may be applicable to only one of the two wave-number components that characterize the acceleration distribution. Thus, in the limit of $k_m^2 \gg k^2$, the pressure field can be approximated by dropping the $\sin\phi$ term from the denominator in equation (5.47), multiplying the numerator and denominator by L_y, and introducing the latter of equations (5.45b) to eliminate $k_m L_y$:

$p(R, \theta, \phi)$

$$\approx \frac{2\rho \ddot{W} \exp(ikR) (-1)^{n+m} k_n L_y \cos(kL_x \sin\theta \cos\phi) \cos(kL_y \sin\theta \sin\phi)}{\pi^2 R (k_n^2 - k^2 \sin^2\theta \cos^2\phi) (m + \frac{1}{2})},$$

(5.51b)

where $\quad k^2 \ll k_m^2,$

$\quad k > k_n,$

$\quad m = 0, 1 \ldots .$

The physical meaning of these results will now be discussed.

5.9 Physical Interpretation of This Solution; Coincidence

In the long-wavelength limit, equation (5.51a), the width of each in-phase region of the radiator measures much less than $\lambda/2$. Its volume acceleration can therefore be combined with that of its neighbors in computing their contribution to the far-field pressure. The volume accelerations of successive pairs of out-of-phase area elements cancel as one scans the radiator in, for example, the x direction, until one reaches the edge of the radiator. Thus, at the ends of a strip parallel to the x axis, one is left with regions of uncanceled volume acceleration of width $\pi/2k_n = L_x/(2n + 1)$. Similarly regions of width $L_y/(2m + 1)$ are encountered at the end of a y-oriented strip

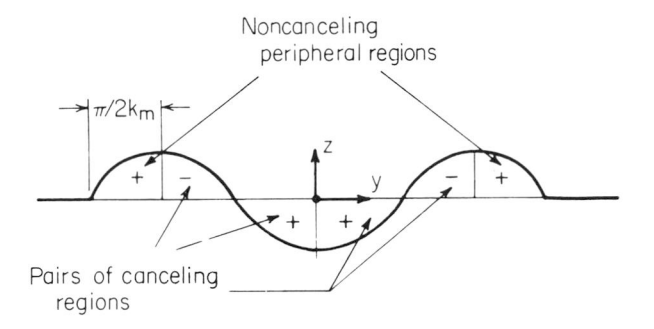

Figure 5.13
Cancellation of volume acceleration in the long-wavelength limit.

(figure 5.13). One thus obtains a checkerboard pattern of canceling rectangular regions, leaving only the regions at the four corners of the radiator as sources of volume acceleration. Other effectively radiating areas are obtained for other boundary conditions. In particular, more effective sound radiation associated with larger uncanceled regions results from "clamped" boundary conditions, whereby not only $\ddot{w}$, but its first spatial derivatives, vanish at the boundaries.[23]

The maximum of the pressure field described by equation (5.51a) occurs when the two cosines equal unity, for $\theta = 0$. This type of mode is described by some authors as a *corner mode*.[24] In the $\phi = 0$ (or x, z) plane, the second cosine becomes unity, and the first cosine takes on the form of the directivity factor of two point sources, separated by the distance $2L_x$, equation (3.7). In the $\phi = \pi$ (or y, z) plane, the directivity factor tends to that of two point sources separated by a distance $2L_y$. The directivity factor in equation (5.51a) is thus of a form associated with four point sources located at $x = \pm L_x, y = \pm L_y$, whose volume acceleration is

$$\ddot{Q} = \frac{\ddot{W} L_x L_y}{\left(n + \frac{1}{2}\right)\left(m + \frac{1}{2}\right)\pi^2}.$$

Effective cancellation between neighboring regions only occurs if the dimensions of the in-phase elements are small in terms of acoustic wavelengths. Thus, when $k^2 \ll k_m^2$, but $k > k_n$, for the pressure field in equation (5.51b), cancellation only takes place between elements located along a strip parallel to the y direction, but not between neighboring strips. This leaves a region of width $L_y/(2m + 1)$ and length $2L_x$, along each of the two

x-oriented radiator boundaries. Referring to a concept developed in section 3.6, one can see that the distribution in angle can be associated with two phased line sources, coinciding with the radiator boundaries at $y = \pm L_y$. This radiator configuration has been described as an *edge mode*.[24] These phased line sources differ from those whose pressure field is given in equation (3.18) in that they correspond to standing rather than traveling sinusoidal distributions of acceleration over the radiator.

Finally, there are configurations for which cancellation does not occur in either direction, when $k > k_n, k_m$. In this case, each in-phase region of the radiator measures more than $\lambda/2$ in both x and y directions and consequently cannot be combined with its neighbors. Each area element of the radiator thus makes a contribution to the far-field. Such a configuration is referred to as a *surface mode*.[24] The maxima occur at the field points defined by equation (5.49b). Referring to our earlier interpretation of the stationary-phase wave numbers, equation (5.43), we verify that pressure maxima occur at field points for which $2\pi/\lambda_s$ equals $(\bar{\gamma}_x^2 + \bar{\gamma}_y^2)^{1/2}$, or alternatively, $k_n = \bar{\gamma}_x, k_m = \bar{\gamma}_y$.

In summary, excluding the $(n = 0, m = 0)$ mode when the entire radiator vibrates in phase, the highest pressures are associated with coincidence effects of surface modes for which $k_n, k_m < k$, and the lowest pressures with corner modes for which $k_n, k_m > k$. Intermediate pressure peaks result for edge modes when $k_n < k$ and $k_m > k$, or vice versa. All even modes take on the same value at $\theta = 0$, on the perpendicular through the center of the radiator. We now turn to radiator configurations that are odd in one or both coordinates.

5.10 Other Standing-Wave Configurations of Rectangular Radiators

A radiator configuration compatible with simply supported boundary conditions that is even in one direction and odd in the other (figure 5.14) has an acceleration distribution

$$\ddot{w}(x, y) = \ddot{W} \cos k_n x \sin k_m y, \qquad |x| < L_x, |y| < L_y$$
$$= 0, \qquad\qquad\qquad |x| \geq L_x, |y| \geq L_y \tag{5.52}$$

with

$$k_n L_x = (n + \tfrac{1}{2})\pi, \qquad n = 0, 1 \ldots,$$
$$k_m L_y = m\pi, \qquad m = 1, 2 \ldots.$$

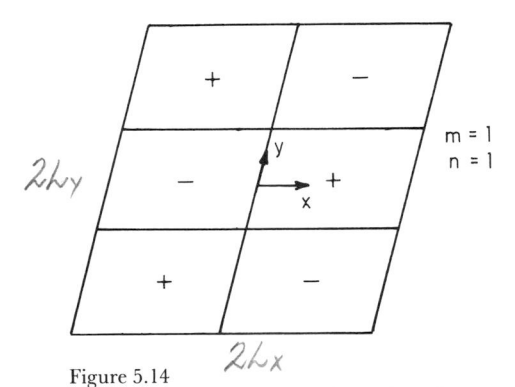

Figure 5.14
Even-odd standing-wave configuration of a rectangular radiator.

The transform of this acceleration distribution can be constructed from the transform of the even-even distribution, equation (5.46), by replacing $\cos \gamma_y L_y$ with $i \sin \gamma_y L_y$. When this transform is evaluated at the point of stationary phase and substituted in equation (5.31), the far-field pressure becomes

$$p(R, \theta, \phi)$$

$$= \frac{i2\rho \ddot{W} \exp(ikR) \, k_n k_m (-1)^{n+m} \cos(kL_x \sin\theta \cos\phi) \sin(kL_y \sin\theta \sin\phi)}{\pi R (k_n^2 - k^2 \sin^2\theta \cos^2\phi)(k_m^2 - k^2 \sin^2\theta \sin^2\phi)}.$$

$$(5.53)$$

The pressure pattern can again be interpreted in terms of radiating area elements located at the four corners of the radiating window. The two corners at $y = -L_y$ radiate out of phase with the two corners at $y = L_y$, resulting in a zero pressure in the ($\phi = 0$) plane. In the long-wavelength limit, for $k_n^2, k_m^2 \gg k^2$, the maximum occurs near the field points for which the argument of the sine in the numerator is $\pi/2$, and that of the cosine zero:

$$p\left(R, \theta_m = \sin^{-1}\frac{\pi}{2kL_y}, \phi = \pm\frac{\pi}{2}\right)$$

$$\approx \frac{2i\rho \ddot{W} L_x L_y (-1)^{n+m} \exp(ikR)}{\pi^3 R [m - (1/4m)](n + \frac{1}{2})} \left.\begin{array}{l} k^2 \ll k_n^2, k_m^2, \\ n = 0, 1 \ldots, \quad m = 1, 2 \ldots. \end{array}\right.$$

$$(5.54)$$

When the y dimension of the radiator is so small as to make $2kL_y < \pi$, the

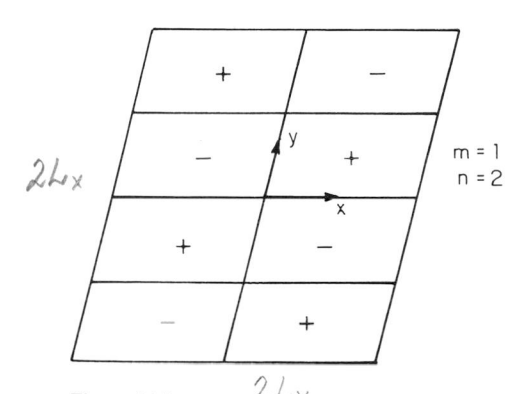

Figure 5.15 $2 L_y$
Odd-odd standing-wave configuration of a rectangular radiator.

pressure peaks at $\theta = \pi/2$. For the lowest $(m = 1, n = 0)$ mode, the peak amplitude is 2/3 of the on-axis pressure of the lowest even-even mode. When $k < k_n, k_m$, maxima obtain near the coincidence angles, equation (5.49). The corresponding pressure is again given by equation (5.50).

Finally, we consider a configuration antisymmetric in both x and y (figure 5.15):

$$\ddot{w}(x, y) = \ddot{W} \sin k_n x \sin k_m y, \qquad |x| < L_x |y| < L_y$$
$$- 0, \qquad\qquad |x| \geq L_x |y| \geq L_y, \tag{5.55}$$

with

$$k_n L_x = n\pi, \qquad n = 1, 2 \ldots,$$
$$k_m L_y = m\pi, \qquad m = 1, 2 \ldots.$$

The transform is constructed from equation (5.46), where the cosines are replaced by $(-\sin \gamma_x L_x \sin \gamma_y L_y)$. The corresponding pressure field is then obtained from equation (5.31):

$$p(R, \theta, \phi)$$
$$= -\frac{2\rho \ddot{W} \exp(ikR) k_n k_m (-1)^{n+m} \sin(kL_x \sin\theta\cos\phi) \sin(kL_y \sin\theta\sin\phi)}{\pi R (k_n^2 - k^2 \sin^2\theta\cos^2\phi)(k_m^2 - k^2 \sin^2\theta\sin^2\phi)}.$$
$$\tag{5.56}$$

When the k^2 terms in the denominator are negligible, the directivity factor is in the form of the product of two dipole factors, equation (3.9), associated with out-of-phase radiation from the two pairs of area elements located at the corners of the radiator. In the long-wavelength limit, the maxima occur at the four points for which the two sines in the numerator are of unit amplitude, their argument being $\pm \pi/2$:

$$
\begin{aligned}
p\left[R, \theta = \sin^{-1} \frac{\pi (L_x^2 + L_y^2)^{1/2}}{2kL_xL_y}, \phi = \pm \tan^{-1} \frac{L_x}{L_y} \right] & \\
= \left. \frac{2\rho L_x L_y \ddot{W}(-1)^{n+m} \exp(ikR)}{\pi^3 R[m - (1/4m)][n - (1/4n)]} \right\} & \quad \begin{aligned} & k^2 \ll k_n^2, k_m^2, \\ & m, n = 1, 2 \ldots . \end{aligned}
\end{aligned}
\tag{5.57}
$$

If the radiator is so small in terms of λ as to make the argument of the inverse sine in equation (5.57) larger than unity, a smaller peak pressure is obtained at $\theta = \pi/2$. For $k > k_n, k_m$, maxima again occur at the coincidence angles given by equation (5.49). These maxima are again given by equation (5.50).

5.11 The Infinite Planar Radiator with a Sinusoidal Acceleration Distribution

We shall now construct the pressure field of the effectively infinite plane radiator and compare it with that of the finite radiator. Consider a periodic acceleration distribution extending to infinity over a boundary coinciding with the $z = 0$ plane. The y axis is rotated parallel to the wave front, thus making the configuration independent of y and $k_n = k_s$:

$$
\ddot{w}(x) = \ddot{W} \cos k_s x.
\tag{5.58}
$$

The analysis below applies equally well to traveling waves, if we substitute $\exp(\pm ik_s x)$ for the cosine. In lieu of a continuous wave-number spectrum γ, the acceleration distribution in equation (5.58) is characterized by a single wave number. Since the prescribed boundary condition is spatially periodic, the pressure field can be constructed in the manner of equations (4.24). In the notation of that equation, this pressure field is

$$
p(x, z) = PR(z) \cos k_s x.
\tag{5.59}
$$

This pressure satisfies the two-dimensional Helmholtz equation, which for the assumed acceleration distribution takes the form

$$\left(\frac{\partial^2}{\partial x^2} + \frac{\partial^2}{\partial z^2} + k^2\right)p(x, z) = 0.$$

On the radiator, the pressure satisfies the boundary condition

$$\frac{\partial p}{\partial z} = P \cos k_s x \frac{dR}{dz}$$

$$= -\rho \ddot{W} \cos k_s x, \qquad z = 0.$$

In the notation of chapter 4, $k_s = k_\sigma$, $z = \zeta$. The z-dependent factor of the solution must therefore satisfy an ordinary differential equation in the form of the equation governing the radial function R_{mn} introduced in equation (4.25). For waves propagating in the positive z direction, the pressure field satisfying the two-dimensional Helmholtz equation and the boundary condition in the $(z = 0)$ plane can be verified to be of the form

$$p(x, z) = i\rho \ddot{W} \cos k_s x \frac{\exp[i(k^2 - k_s^2)^{1/2} z]}{(k^2 - k_s^2)^{1/2}}, \qquad k > k_s. \tag{5.60}$$

Acoustic waves propagating in the z direction, away from the boundary, arise only when $(k^2 - k_s^2)^{1/2}$ is real, that is, when $k > k_s$. The wave fronts are plane, even close to the boundary, and propagate exclusively in the coincidence directions whose angles with the (y, z) plane are, respectively, $\pm \sin^{-1} k_s/k$. The surface pressure can be expressed in terms of a specific acoustic resistance r_s as

$$p(x, z = 0) = \ddot{W} r_s \cos k_s x,$$

where

$$r_s = \frac{\rho c k}{(k^2 - k_s^2)^{1/2}}, \qquad k > k_s$$

$$\simeq \rho c, \qquad k^2 \gg k_s^2. \tag{5.61}$$

The resistance thus tends to the characteristic plane wave impedance when $k \gg k_s$. It becomes infinite as $k \to k_s$ (figure 5.16).

When the acoustic wavelength exceeds the structural wavelength,

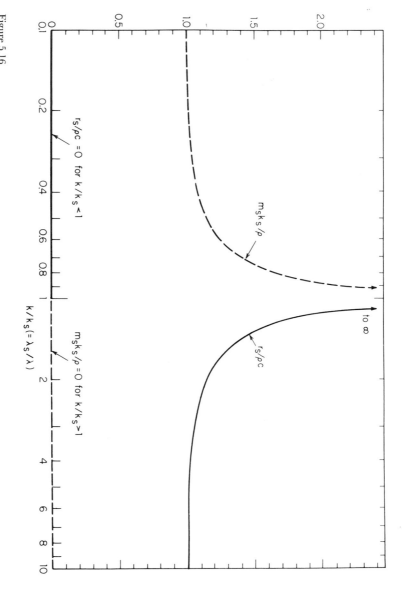

Figure 5.16
Specific acoustic resistance r_s [equation (5.61)] and accession to inertia m_s [equation (5.63)] of an infinite plane radiator whose acceleration distribution is in the form of straight-crested waves of wavelength $\lambda_s = 2\pi/k_s$.

$k_s > k$, and $(k^2 - k_s^2)^{1/2}$ is imaginary. The pressure field now takes the form

$$p(x,z) = \rho \ddot{W} \cos k_s x \frac{\exp[-(k_s^2 - k^2)^{1/2} z]}{(k_s^2 - k^2)^{1/2}}, \qquad k < k_s$$

$$\simeq \rho \ddot{W} e^{-k_s z} \frac{\cos k_s x}{k_s}, \qquad k^2 \ll k_s^2. \tag{5.62}$$

This pressure decays exponentially with distance from the plate. No energy is radiated. The surface pressure can be expressed in terms of an accession to inertia per unit area, m_s, as

$$p(x, z = 0) = \ddot{W} m_s \cos k_s x,$$

where

$$m_s = \frac{\rho}{(k_s^2 - k^2)^{1/2}}, \qquad k < k_s$$

$$\approx \frac{\rho}{k_s} = \frac{\rho \lambda_s}{2\pi}, \qquad k^2 \ll k_s^2. \tag{5.63}$$

The latter corresponds to the incompressible, hydrodynamic solution. Like the resistance, it tends to infinity as $k \to k_s$. In the long-wavelength limit, it takes on a constant value that corresponds to the mass of a fluid layer of thickness $\lambda_s/2\pi$ (figure 5.16). It differs from the accession in inertia of finite radiators in that it is precisely equal to zero when $k > k_s$ (or $\lambda < \lambda_s$), instead of gradually tending to zero as $k_s/k \to 0$ ($\lambda/\lambda_s \to 0$). Similarly, with varying frequency, the resistance of the infinite radiator vanishes abruptly when $\lambda > \lambda_s$, in contrast to the resistance of finite radiators, which approaches zero gradually as $\lambda_s/\lambda \to 0$. The fact that the specific acoustic resistance of the infinite plane radiator is rigorously zero in this range of λ is not surprising if one remembers that in the long-wavelength limit, the far-field pressure radiated by a finite radiator was associated exclusively with radiation from edges or corners, a concept that is meaningless for the infinite radiator.

The infinite radiator will be used to approximate the resistance of surface-radiating modes ($k_n, k_m < k$) of rectangular sources. Before we can proceed with these resistance calculations, we require the pressure field of one more radiator configuration, which will be used to approximate the

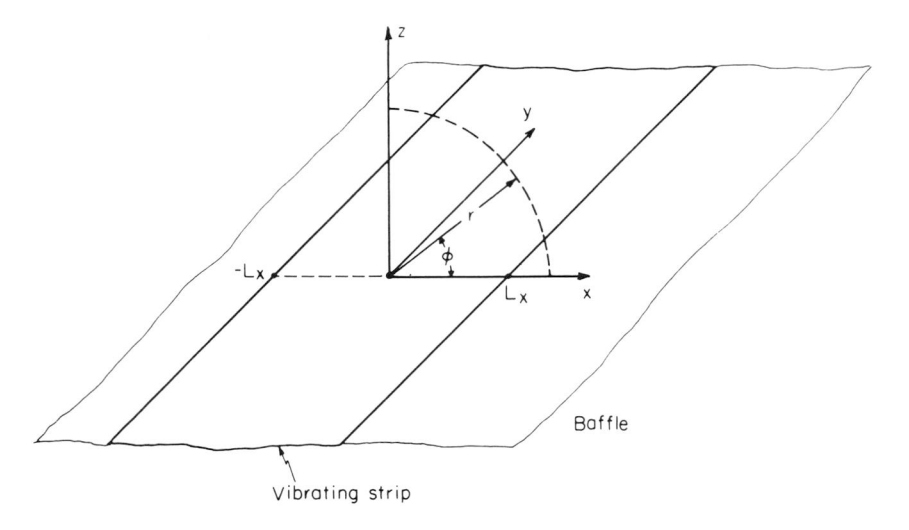

Figure 5.17
The radiating strip of infinite length.

resistance of edge-radiating modes $(k_m < k, k_n > k)$ of rectangular radiators. This application is, in fact, the reason for including this configuration in this chapter, as it is of little practical interest compared to the finite radiators analyzed earlier.

5.12 The Radiating Strip of Infinite Length and Finite Width

The source configuration under consideration is a strip located in the $(z = 0)$ plane extending to infinity in the y direction, and boundary by semiinfinite planar baffles at $x = \pm L_x$ (figure 5.17). The acceleration distribution is periodic or constant in the x direction:

$$\ddot{w}(x, y) = \ddot{W}f(x) \cos k_m y, \qquad |x| < L_x$$
$$= 0, \qquad\qquad\quad |x| \geq L_x,$$

(5.64a)

where $f(x)$ is a dimensionless function of unit amplitude whose transform is

$$\tilde{f}(\gamma_x) = \int_{-L_x}^{L_x} f(x) e^{-i\gamma_x x}\, dx.$$

(5.64b)

A y-independent configuration is simulated by setting $k_m = 0$. Like the axisymmetric configurations studied in section 5.3, this geometry requires a single transform parameter. The pressure transform satisfies the x-transformed boundary conditions and wave equation

$$\rho \ddot{W} \tilde{f}(\gamma_x) \cos k_m y = -\frac{\partial \tilde{p}(\gamma_x; y, z)}{\partial z}, \qquad z = 0,$$

$$\left[\frac{\partial^2}{\partial y^2} + \frac{\partial^2}{\partial z^2} + (k^2 - \gamma_x^2) \right] \tilde{p}(\gamma_x, y, z) = 0.$$

$$(5.65)$$

The solution of equation (5.65) is

$$\tilde{p}(\gamma_x; y, z) - \frac{i \ddot{W} \tilde{f}(\gamma_x) \rho \cos k_m y \exp[i(k^2 - \gamma_x^2)^{1/2} z]}{(k^2 - \gamma_x^2)^{1/2}}. \tag{5.66a}$$

The corresponding pressure field is given by the inverse transform

$$p(x, y, z) = \frac{1}{2\pi} \int_{-\infty}^{\infty} \tilde{p}(\gamma_x, y, z) e^{i\gamma_x x} \, d\gamma_x. \tag{5.66b}$$

To construct an integrand suitable for stationary-phase integration, we use cylindrical coordinates:

$$z = r \sin \phi, \qquad x = r \cos \phi.$$

If we apply this transformation to equations (5.66a) and (5.66b), the pressure field becomes

$$p(r, \theta, y)$$

$$= \frac{i \rho \ddot{W} \cos k_m y}{2\pi} \int_{-\infty}^{\infty} \frac{\tilde{f}(\gamma_x) \exp[ir(k^2 - \gamma_x^2)^{1/2} \sin \phi + i r \gamma_x \cos \phi]}{(k^2 - \gamma_x^2)^{1/2}} \, d\gamma_x.$$

$$(5.66c)$$

This integral is in the form of equation (5.18) and thus tractable by stationary-phase integration. The modulus $\Phi(\gamma)$ and the phase $\Psi(\gamma)$ of the integrand are

$$\Phi(\gamma_x) = \tilde{f}(\gamma_x)\,(k^2 - \gamma_x^2)^{-1/2}, \tag{5.67a}$$

$$\Psi(\gamma_x) = r[(k^2 - \gamma_x^2)^{1/2}\sin\phi + \gamma_x\cos\phi]. \tag{5.67b}$$

The derivative of the phase is

$$\frac{\partial\Psi}{\partial\gamma_x} = [-\gamma_x(k^2 - \gamma_x^2)^{-1/2}\sin\phi + \cos\phi]r.$$

The desired root is

$$\bar{\gamma}_x = k\cos\phi.$$

The second derivative of the phase angle is

$$\frac{\partial^2\Psi}{\partial\gamma_x^2} = -k^2 r\sin\phi\,(k^2 - \gamma_x^2)^{-3/2}. \tag{5.67c}$$

The values of the functions in equations (5.67a), (5.67b), and (5.67c) evaluated at $\gamma_x = \bar{\gamma}_x$ are

$$\Psi(\bar{\gamma}_x) = kr,$$

$$\frac{\partial^2\Psi}{\partial\gamma_x^2} = -\frac{r}{k\sin^2\phi}, \qquad \gamma_x = \bar{\gamma}_x,$$

$$\Phi(\bar{\gamma}_x) = \frac{\tilde{f}(k\cos\phi)}{k\sin\phi}.$$

When these values are substituted in the stationary-phase approximation, equation (5.21b) and the integral is multiplied by the coefficient outside the integral sign in equation (5.66c), the far-field pressure is obtained:

$$p(r,\phi,y) = \frac{i\rho\ddot{W}\tilde{f}(k\cos\phi)\cos k_m y}{(2\pi kr)^{1/2}}\exp\left(ikr - i\frac{\pi}{4}\right), \qquad kr \gg 1. \tag{5.68}$$

The $r^{-1/2}$ or "cylindrical" spreading loss is characteristic of two-dimensional pressure fields—for example, the field of infinite cylindrical sources with no axial dependence. More generally, it is characteristic of pressures radiated by sources with one infinite dimension associated with a periodic distribu-

tion, such as the cylindrical source (see section 6.8). It can be verified that this functional dependence ensures that the acoustic power spreading radially through concentric cylindrical boundaries of the same length is independent of r.

The field associated with various acceleration distributions can now be constructed. For a rigid strip, characterized by $k_m = 0, f(x) = 1$, the transform equation (5.64a) is

$$\tilde{f}(\gamma_x) = 2L_x j_0(\gamma_x L_x).$$

Hence,

$$|p(r, \phi)| = \left(\frac{2}{\pi kr}\right)^{1/2} \rho \ddot{W} L_x |j_0(kL_x \cos \phi)|, \qquad kr \gg 1. \tag{5.69}$$

For an even acceleration distribution $f(x) = \cos k_n x$ simulating simply supported boundary conditions, similar to those of the rectangular radiator, equations (5.45), the transform of the acceleration distribution, is

$$\tilde{f}(\gamma_x) = 2k_n(-1)^n \frac{\cos \gamma_x L_x}{(k_n^2 - \gamma_x^2)}. \tag{5.70a}$$

The corresponding pressure field is

$$|p(r, \phi, y)| = \frac{2^{1/2} \rho \ddot{W} k_n |\cos(kL_x \cos \phi) \cos k_m y|}{|k_n^2 - (k \cos \phi)^2| (\pi kr)^{1/2}}. \tag{5.70b}$$

This simulates the familiar directivity pattern of two line sources located along the edges of the radiating strip when $k_n^2 \gg k^2$, that is, for edge-radiating modes:

$$|p(r, \phi, y) \approx \frac{2^{1/2} \rho \ddot{W} |\cos(kL_x \cos \phi) \cos k_m y|}{k_n (\pi kr)^{1/2}}, \qquad k_n^2 \gg k^2. \tag{5.70c}$$

When $k_n < k$, the critical angle $\cos^{-1} k_n/k$ corresponds to a maximum evaluated in the manner of equation (5.49c). For an odd function $f(x) = \sin k_n x$, with simply supported boundary conditions similar to those in equation (5.55), the pressure field is

$$|p(r,\phi,y)| \approx \frac{2^{1/2}\rho\dot{W}|\sin(kL_x\cos\phi)\cos k_m y|}{k_n(\pi kr)^{1/2}}, \qquad k_n^2 \gg k^2. \qquad (5.71)$$

We now have the tools to compute the approximate specific acoustic resistance of various planar radiators.

5.13 Acoustic Resistance of Circular Pistons and of Surface-Radiating Rectangular Source Configurations with Sinusoidal Acceleration Distributions

A direct calculation of the total resistance or of the radiated power in terms of the surface pressure, such as performed for the circular piston, equation (5.27), is possible for some source configurations. In general, however, the power must be obtained by integrating the far-field intensity, as shown in equation (3.27). This integration can be performed numerically, if the directivity function is available. For some configurations, it can be performed analytically. The result can be conveniently expressed in the form of a specific acoustic resistance $\langle r \rangle$ averaged over the surface A of the radiator, thus eliminating the spatial fluctuations of the resistive component of the surface pressure. It is this space-averaged resistance that determines the radiation damping of a structural mode of vibration, keeping in mind that the concept of a mode is an approximation since the in vacuo modes are coupled by radiation damping. By definition, the specific acoustic resistance space averaged over the radiating surface equals the radiation resistance divided by the radiating area. For a velocity distribution that is uniform over the radiating surface, the radiation resistance can be expressed in terms of the time-averaged radiated power Π and the velocity amplitude $\dot{W}$ of the radiator as $2\Pi/\dot{W}^2$. One thus arrives at a relation between the space-averaged specific acoustic resistance and the time-averaged radiated power Π computed from the far-field pressure. If the velocity is constant over the radiator,

$$\langle r \rangle = \frac{2\Pi}{\dot{W}^2 A}. \qquad (5.72a)$$

In this notation, the resistance R_p of the circular piston, equation (5.27), equals $\pi a^2 \langle r_p \rangle$. To illustrate the procedure, we substitute equation (5.10a), with $\dot{W} = ck\dot{W}$, in Equation (3.27) to obtain the radiated power. Dividing by $\pi a^2 \dot{W}^2/2$, we note that the resistance becomes

$$\langle r_p \rangle = 2\rho c \int_0^{\pi/2} \frac{J_1^2(ka \sin\theta)}{\sin\theta} d\theta.$$

This integral is seen to be identical to the inverse transform formulation of the resistance, equation (5.27a), divided by the piston area. Analytical expressions for the radiation resistance of other circular radiators are available.[16,25]

If the velocity distribution varies over the radiator surface, it must also be space averaged, $\dot{W}^2$ being replaced by $\langle \dot{w}^2 \rangle$. Thus, for a rectangular radiator of area $4L_x L_y$ whose configuration simulates simply supported boundary conditions [equations (5.14) and (5.15)],

$$\langle r_{mn} \rangle = \frac{2\Pi}{L_x L_y \dot{W}^2}. \tag{5.72b}$$

As the emphasis in this chapter has been on pressure fields and their maxima, the derivation of the radiation resistance of rectangular radiators, which is the primary concern of much of the literature published in this area, will be rapidly reviewed. The general expression for the radiated pressure of a rectangular radiator whose configuration simulates a simply supported plate mode, equation (5.47), is too complicated to permit analytical integration of the power integral, equation (3.27). Integration can be circumvented for modal configurations for which cancellation effects between adjoining out-of-phase regions are unimportant. This includes the fundamental mode that does not embody nodal lines. For these configurations, the resistance can be approximated by considering a rigid rectangular piston of equivalent volume acceleration, or, in the limit of kL_x, $kL_y \ll 1$, by an equivalent point source whose volume acceleration equals that of the plate. For surface modes $(k_n, k_m < k)$, for which cancellation effects are unimportant except at points within approximately a half-acoustic wavelength from the boundary, each area element of the radiator radiates approximately as though it were part of an infinite radiator. Hence, the average impedance can be approximated by that of an infinite radiator of the same structural wavelength, equation (5.61), provided the source dimensions are sufficiently large in terms of acoustic wavelengths to make an error of $\lambda/2$ in the source dimensions unimportant:

$$\langle r_{mn} \rangle \simeq \frac{\rho c k}{[k^2 - (k_m^2 + k_n^2)]^{1/2}}, \qquad k_n, k_m < k, \quad k_n l_x, k_m l_y \gg 1.$$

This and subsequent results are summarized in table 5.2. A more rigorous expression is derived by Davies,[26] who published a systematic asymptotic analysis of various combinations of acoustic and structural wave numbers, based on analytical integration of equation (3.27). Other asymptotic integrations are available for clamped[23] and unrestrained[27] boundary conditions.

5.14 Acoustic Resistance of Edge-Radiating Rectangular Source Configurations with Sinusoidal Acceleration Distributions

The space-averaged specific acoustic resistance of an edge-radiating source configuration $(k_n > k, k_m < k)$ simulating the in vacuo mode of a simply supported rectangular plate can be approximated in the $(k_n^2 \gg k^2)$ limit, when $kL_y \gg 1$. The derivation is based on the same assumption as equation (5.72) whereby the boundary only affects the surface pressure and hence the far-field contribution of source points located within a distance of approximately $\lambda/2$ of the boundary or less. When this assumption is applied to an edge mode, one can compute the far-field contributions of source points located within the span $-L_y + (\lambda/2) < y < L_y - (\lambda/2)$ as though they were part of the infinite strip analyzed in the preceding section. Consequently, the power radiated by an area element $dA = 2L_x \, dy$ is approximately

$$d\Pi \approx \frac{1}{2\rho c} \int_0^\pi |p(1, \phi, y)|^2 \, d\phi \, dy, \qquad kL_y \gg 1,$$

where $|p|$ is given in equation (5.70) or (5.71). If the dimension $2L_y$ of the plate is sufficiently large to make the contribution of the poorly radiating extremities of width $\lambda/2$ negligible in comparison with the resultant power, the space-averaged acoustic resistance can be obtained from the last expression for $d\Pi$, from equation (5.72b). Because the y dependence is effectively the same for the pressure as for the radiator velocity configuration, the latter need not be averaged over y but only over x:

$$\langle r_{mn} \rangle \approx \frac{1}{\dot{W}^2 \cos^2(k_m y) L_x} \frac{d\Pi}{dy}, \qquad kL_y \gg 1.$$

Substituting equation 5.70c, we see that the resistance for a mode which is even in x becomes

Table 5.2
Asymptotic expressions for the space-averaged specific acoustic resistance $\langle r_{mn} \rangle$ of planar sources simulating in vacuo modes of simply supported rectangular plates [$2L_x, 2L_y$ = plate dimensions; $k_n L_x = (n + \frac{1}{2})\pi$, $k_m L_y = (m + \frac{1}{2})\pi$, $m, n = 0, 1, \ldots$]

Source configuration	Effectively radiating area		
	Entire surface	Two edges parallel to y-axis	Four corners
	$(k_n, k_m < k, kL_x, kL_y \gg 1)$	$(k_m < k, k_n^2 > k^2)$	$(k_n^2, k_m^2 > k^2)$
$\cos k_n x \cos k_m y$	$\langle r_{mn} \rangle = \dfrac{\rho c k}{[k^2 - (k_m^2 + k_n^2)]^{1/2}}$	$\langle r_{mn} \rangle = \dfrac{\rho c k L_x}{2\pi^2 (n + \frac{1}{2})^2} F'(kL_x),$ $F'(kL_x)$ $\equiv 1 - J_0(2kL_x)$ $\approx 2, \quad k^2 L_x^2 \ll 1$ $\approx 1, \quad (kL_x)^{1/2} \gg 1$	$\langle r_{mn} \rangle = \dfrac{2\rho c k^2 L_x L_y}{\pi^5 (n + \frac{1}{2})^2 (n + \frac{1}{2})^2} F'(kL_x, kL_y)$ $F'(kL_x, kL_y)$ $\equiv 1 + j_0(2kL_x)$ $\quad - j_0((2kL_x)) + j_0[2k\langle L_x^2 + L_y^2)^{1/2}]$ $\approx 4, \quad k^2 L_x^2, k^2 L_y^2 \ll 1$ $\approx 1, \quad kL_x, kL_y \gg 1$
$\sin k_n x \cos k_m y$	Same	Same $F(kL_x)$ $\equiv 1 - j_0(2kL_x)$ $\approx k^2 L_x^2, \quad k^2 L_x^2 \ll 1$ $\approx 1, \quad (kL_x)^{1/2} \gg 1$	Same $F(kL_x, kL_y)$ $\equiv 1 - j_0(2kL_x)$ $+ j_0(2kL_y) - j_0[2k(L_x^2 + L_y^2)^{1/2}]$ $\approx 2k^2 L_x^2/3, \quad k^2 L_x, k^2 L_y^2 \ll 1$ $\approx 1, \quad kL_x, kL_y \gg 1$
$\sin k_n x \sin k_m y$	Same	Same	Same $F(kL_x, kL_y)$ $\equiv 1 - j_0(2kL_x)$ $- j_0(2kL_y) + j_0[2k(L_x^2 + L_y^2)^{1/2}]$ $\approx 4k^4 L_x^2 L_y^2/15, \quad k^2 L_x^2, kL_y^2 \ll 1$ $\approx 1, \quad kL_x, kL_y \gg 1$

$$\langle r_{mn} \rangle \approx \frac{2\rho ck I}{\pi k_n^2 L_x} \left.\begin{array}{c} \\ \\ \end{array}\right\} \quad kL_y \gg 1, \quad k_m < k, \tag{5.73a}$$
$$\left.\approx \frac{2\rho ck L_x I}{\pi^3 \left(n + \frac{1}{2}\right)^2}\right\}$$

where

$$I = \int_0^\pi \cos^2\left(kL_x \cos\phi\right) d\phi, \qquad k_n^2 \gg k^2$$
$$= \int_0^\pi \left[\tfrac{1}{2} + \tfrac{1}{2}\cos\left(2kL_x \cos\phi\right)\right] d\phi. \tag{5.73b}$$

The first term yields $\pi/2$. The second term is recognized as the integral representation of $\mathcal{J}_0$, equation (5.9). Hence

$$\langle r_{mn} \rangle \approx \frac{\rho ck L_x}{\pi^2 \left(n + \frac{1}{2}\right)^2} \left[1 + \mathcal{J}_0\left(2kL_x\right)\right], \qquad k_m < k, \quad k_n^2 \gg k^2, \quad kL_y \gg 1. \tag{5.74}$$

Making use of the asymptotic relations

$$\mathcal{J}_0(x) \approx 1 - \left(\frac{x^2}{4}\right), \qquad x^4 \ll 1,$$
$$\mathcal{J}_0(x) \approx O\left(x^{-1/2}\right), \qquad x \gg 1,$$

we obtain the low- and high-frequency limits of the resistance:

$$\langle r_{mn} \rangle \approx \frac{2\rho ck L_x}{\pi^2 \left(n + \frac{1}{2}\right)^2}, \quad k^2 L_x^2 \ll 1 \left.\begin{array}{c} \\ \\ \end{array}\right. \tag{5.75a}$$
$$\approx \frac{\rho ck L_x}{\pi^2 \left(n + \frac{1}{2}\right)^2}, \quad (kL_x)^{1/2} \gg 1 \left.\begin{array}{c} \\ \\ \end{array}\right\} \quad k_n^2 \gg k^2, \; k_m < k, \; kL_y \gg 1. \tag{5.75b}$$

The resistance is limited by the number of nodal lines $2n$ parallel to the two uncanceled, effectively radiating edges. For small radiators, $(kL_x, kL_y < \pi)$, edge modes can be shown to radiate like corner modes $(k_n, k_m > k)$.

For a modal configuration that is odd in x, the power integral corre-

sponding to equation (5.73b) is constructed by means of equation (5.71):

$$
\begin{aligned}
I &= \int_0^\pi \sin^2 (kL_x \cos \phi) \, d\phi \\
&= \frac{1}{2} \int_0^\pi [1 - \cos (2kL_x \cos \phi)] \, d\phi \\
&= \frac{\pi}{2} [1 - \mathcal{J}_0 (2kL_x)].
\end{aligned}
\tag{5.76}
$$

The corresponding small-kL_y resistance is

$$
\left.
\begin{aligned}
\langle r_{mn} \rangle &\simeq \frac{\rho c k L_x}{\pi^2 n^2} [1 - \mathcal{J}_0 (2kL_x)] \\
&\approx \frac{\rho c k^3 L_x^3}{\pi^2 n^2}, \qquad k^2 L_x^2 \ll 1 \\
&\approx \frac{\rho c k L_x}{\pi^2 n^2}, \qquad (kL_x)^{1/2} \gg 1
\end{aligned}
\right\}
\qquad k_n^2 \gg k^2, \quad k_m < k, \quad kL_y \gg 1.
\tag{5.77}
$$

In contrast to even modes [see equation (5.74)], which radiate with mono-polelike efficiency at all frequencies, odd-edge modes display the markedly frequency-dependent small resistance of dipoles when the spacing of the radiating edges is small in terms of wavelengths. A recent detailed analysis of the limiting $k_m \to 0$ situation provides insight into the factors that control acoustic power.[28]

We conclude this chapter by analyzing corner-radiating modes characterized by $k_n^2, k_m^2 \gg k^2$.

5.15 The Resistance of Corner-Radiating Rectangular Source Configurations with Sinusoidal Source Configurations

The pressure field of these modes is given in equation (5.51a). Combining this equation with equations (5.72) and (3.27), the space-averaged specific acoustic resistance can be written in the long-wavelength limit as

$$
\langle r_{mn} \rangle = \frac{4 \rho c k^2 L_x L_y I}{\pi^6 (n + \frac{1}{2})^2 (m + \frac{1}{2})^2}, \qquad k^2 \ll k_n^2, k_m^2.
\tag{5.78}
$$

For a mode even in x and y, the power integral is constructed by squaring equation (5.51a):

$$I = \int_0^{2\pi} \left[\int_0^{\pi/2} \cos^2 (kL_x \sin \theta \cos \phi) \cos^2 (kL_y \sin \theta \sin \phi) \sin \theta \, d\theta \right] d\phi. \quad (5.79)$$

When the integrand is transformed in the manner of equations (5.73b), this integral becomes

$$I = \tfrac{1}{4} \int_0^{2\pi} \left[\int_0^{\pi/2} (1 + \cos X + \cos Y + \cos X \cos Y) \sin \theta \, d\theta \right] d\phi, \quad (5.80)$$

where

$$X \equiv 2kL_x \sin \theta \cos \phi,$$
$$Y \equiv 2KL_y \sin \theta \sin \phi.$$

Performing the ϕ integration first, the unit term yields 2π. The second term is integrated by making use of the integral representation of J_0, equation (5.9), where x is now set equal to $2kL_x \sin \theta$:

$$\int_0^{2\pi} \cos X \, d\phi = 2\pi J_0 (2kL_x \sin \theta).$$

To integrate the third term of the integrand in equation (5.80), we note that the integral representation of the Bessel function in equation (5.9) applies when $\sin \phi_0$ is substituted for $\cos \phi_0$. Setting $x = kL_y \sin \theta$ in that equation, the third term of the integrand of equation (5.80) integrates to

$$\int_0^{2\pi} \cos Y \, d\phi = 2\pi J_0 (2kL_y \sin \theta).$$

The fourth term is evaluated by means of the expansion[29]

$$\cos (x \sin \phi) = \sum_{l=0,1}^{\infty} \varepsilon_l J_{2l}(x) \cos (2l\phi), \quad (5.81a)$$

where $\varepsilon_l = 1$ for $l = 0$ and 2 for $l > 0$. A variant of this expansion is

$$\cos(x \cos \phi) = \sum_{l=0,1}^{\infty} \varepsilon_l (-1)^l \mathcal{J}_{2l}(x) \cos(2l\phi).$$ (5.81b)

Applying the latter expansion to the Y factor, and the former to the X factor, we can now write the cross product in equation (5.80) as

$$\cos X \cos Y = \left[\sum_{l=0,1}^{\infty} \varepsilon_l (-1)^l \mathcal{J}_{2l}(2kL_x \sin \theta) \cos(2l\phi) \right]$$
$$\cdot \left[\sum_{q=0,1}^{\infty} \varepsilon_q \mathcal{J}_{2q}(2kL_y \sin \theta) \cos(2q\phi) \right].$$

When we integrate this series with respect to ϕ, products with $l \neq q$ drop out. The other terms yield $2\pi/\varepsilon_l$:

$$\int_0^{2\pi} \cos X \cos Y \, d\phi = 2\pi \sum_{l=0,1}^{\infty} \varepsilon_l (-1)^l \mathcal{J}_{2l}(2kL_x \sin \theta) \mathcal{J}_{2l}(2kL_y \sin \theta).$$

If one sums the four ϕ integrals thus obtained, the double integral, equation (5.80), is reduced to the ϕ integral

$$I = \frac{\pi}{2} \int_0^{\pi/2} \left[1 + \mathcal{J}_0(2kL_x \sin \theta) + \mathcal{J}_0(2kL_y \sin \theta) \right.$$
$$\left. + \sum_{l=0,1}^{\infty} \varepsilon_l (-1)^l \mathcal{J}_{2l}(2kL_x \sin \theta) \mathcal{J}_{2l}(2kL_y \sin \theta) \right] \sin \theta \, d\theta.$$ (5.82)

The unit term in the above integrand yields unity. For large kL_x and kL_y, this term makes the most important contribution to the integral. The second and third terms can be integrated by using the integral[30]

$$\int_0^{\pi/2} \mathcal{J}_0(2x \sin \theta) \sin \theta \, d\theta = \frac{\pi}{2} \mathcal{J}_{-1/2}(x) \mathcal{J}_{1/2}(x)$$ (5.83)
$$= j_0(2x).$$

The latter form of the result is constructed by multiplying the functions

$$\mathcal{J}_{1/2}(x) \equiv \left(\frac{2\pi}{x}\right)^{1/2}(\sin x),$$

$$\mathcal{J}_{-1/2}(x) \equiv \left(\frac{2\pi}{x}\right)^{1/2}(\cos x),$$

and replacing the resulting product, $(\sin 2x)/2x$, with $j_0(2x)$. Substituting $x = 2kL_x$ and $2kL_y$, we integrate the second and third terms in the integrand of equation (5.82) in the forms $j_0(2kL_x)$ and $j_0(2kL_y)$. Finally, the l series can be integrated by means of Neumann's addition theorem[31,32]

$$\mathcal{J}_0(z) = \sum_{l=0,1}^{\infty} \varepsilon_l \mathcal{J}_l(x)\mathcal{J}_l(y)\cos l\alpha, \tag{5.84}$$

where $z \equiv (x^2 + y^2 - 2xy\cos\alpha)^{1/2}$. Setting $\alpha = \pi/2$, $x = 2kL_x \sin\theta$, and $y = 2kL_y \sin\theta$, equation (5.83) takes the form of the l series in equation (5.82) with

$$z = 2k(L_x^2 + L_y^2)^{1/2}\sin\theta.$$

The series in the integrand of equation (5.82) thus becomes $\mathcal{J}_0(z)$. This function can be integrated in the manner of equation (5.83):

$$\int_0^{\pi/2} \mathcal{J}_0[2k(L_x^2 + L_y^2)^{1/2}\sin\theta]\sin\theta\,d\theta = j_0[2k(L_x^2 + L_y^2)^{1/2}].$$

Adding these various results, we obtain the θ integral in equation (5.82), and hence the double integral in equation (5.80):

$$I = \frac{\pi}{2}\{1 + j_0(2kL_x) + j_0(2kL_y) + j_0[2k(L_x^2 + L_y^2)^{1/2}]\}$$

$$= \frac{\pi}{2}, \qquad kL_x, kL_y \gg 1 \tag{5.85}$$

$$= 2\pi, \quad k^2L_x^2, k^2L_y^2 \ll 1.$$

The resistance is finally constructed by substituting this integral in equation (5.78) (see table 5.2). Since the integral tends to a constant as L_x, $L_y \to \infty$ while k_m and k_n are maintained constant, the resistance tends to zero. The

smooth transition to the vanishing resistance of the infinite radiator in the range $k_s > k$ is thus verified.

For a source configuration that is even in y and odd in x, equation (5.53) yields a power integral that differs from equation (5.79) only in that $\sin^2(kL_x \sin\theta \cos\phi)$ is used instead of cosine squared. This power integral is integrated to

$$I = \frac{\pi}{2}\{1 - j_0(2kL_x) + j_0(2kL_y) - j_0[2k(L_x^2 + L_y^2)^{1/2}]\}.$$

The large-kL_x, -kL_y limit is the same as for equation (5.85), but the long-wavelength limit is of higher order than for the even-even mode. To evaluate it, we use the small-argument expansion of j_0, equation (2.45). The lower-order terms cancel, leaving

$$I = \frac{\pi k^2 L_x^2}{3}, \qquad (kL_x)^2, (kL_y)^2 \ll 1.$$

Finally, for a mode that is odd in both x and y [equation (5.56)], the power integral is constructed by substituting two sine-squared factors in equation (5.79). The resulting integrand integrates to

$$I = \frac{\pi}{2}\{1 - j_0(2kL_x) - j_0(2kL_y) + j_0[2k(L_x^2 + L_y^2)^{1/2}]\}.$$

In the short-wavelength limit, this becomes

$$I = \frac{2\pi k^4 L_x^2 L_y^2}{15}, \qquad (kL_x)^2, (kL_y)^2 \ll 1.$$

This is even smaller than the corresponding resistance of the even-odd configuration, the cancellation of volume flow being more nearly perfect, In contrast, the even-even configuration does not display any cancellation between the contributions from the uncanceled area elements at its four corners. The fact that all three configurations display the same large kL_x, kL_y resistance is consistent with the observation stated in chapter 3 to the effect that small sources separated by a distance of several wavelengths do not interact, thus making their relative phase unimportant in computing

the power they radiate. In contrast, the *pressure* in the far-field depends on the phase of the individual contributions at a given field point.

The techniques developed in this chapter will be applied to sound radiation by elastic plates in Chapter 8. Before that, however, we shall explore nonplanar sound sources.

References

1. C. J. Tranter, *Integral Transforms in Mathematical Physics*, 3rd ed. (New York: John Wiley, 1966), p. 16.

2. Ibid., p. 48, equation (4.15).

3. M. Abramowitz and I. A. Stegun, eds., *Handbook of Mathematical Functions* (Washington, D.C.: NBS, Supt. of Doc., 1964), p. 484, formula (11.3.20) with $v = 1$.

4. Ibid., p. 360, formula (9.1.18).

5. Tranter, *Integral Transforms*, pp. 47–48.

6. Abramowitz and Stegun, *Handbook*, p. 358, formulas (9.1.3), (9.1.4), and (9.1.6).

7. Ibid., p. 364, formula (9.2.3).

8. E. T. Copson, *Asymptotic Expansions* (Cambridge: University Press, 1967), pp. 27–35. For a brief introduction to this technique see H. Lamb, *Hydrodynamics*, 6th ed. (New York: Dover, 1932), pp. 395–396.

9. Y. L. Luke, *Integrals of Bessel Functions* (New York: McGraw-Hill, 1962, p. 328, formula (7), with $v = 1$, $t = \gamma a$, and $z = ka$.

10. Abramowitz and Stegun, *Handbook*, pp. 496–497.

11. N. W. McLachlan, *Bessel Functions for Engineers*, 2nd ed. (Oxford: Clarendon Press, 1955), p. 74.

12. Ibid., p. 64.

13. Ibid., p. 191.

14. N. W. McLachlan, *Loudspeakers*, (New York: Dover, 1960), p. 90, table 5A.

15. D. T. Porter, "Self- and Mutual-Radiation Impedances and Beam Patterns for Flexural Disks in a Rigid Plane," *J. Acoust. Soc. Am.* 36:1154–1161 (1964).

16. M. Greenspan, "Piston Radiator: Some Extensions of the Theory," *J. Acoust. Soc. Am.* 65:608–621 (1979).

17. Tranter, *Integral Transforms*, pp. 6–11, 31–33.

18. V. M. Albers, *Underwater Acoustics Handbook—II* (University Park, PA: Pennsylvania State University Press, 1965), pp. 161–163.

19. H. Stenzel and O. Brosze, *Leitfaden zur Berechnung von Schallvorgängen* (Berlin: Springer Verlag, 1958), pp. 99–105; also, Stenzel, *Acustica* 2:263–281 (1952).

20. G. W. Swenson, Jr., and W. E. Johnson, "Radiation Impedance of a Rigid Square Piston in an Infinite Baffle," *J. Acoust. Soc. Am.* 24:84 (1952).

21. D. S. Jones and M. Klein, *J. Math. Phys.* 37:1–28 (1958).

22. P. M. Morse and K. U. Ingard, *Theoretical Acoustics* (New York: McGraw-Hill, 1968), pp. 53, 54.

23. P. W. Smith, Jr., "Coupling of Sound and Panel Vibration below the Critical Frequency," *J. Acoust. Soc. Am.* 36:1516–1520 (1964).

24. For the basis of the classification of modes in this physically evocative manner see G. Maidanik, "Response of Ribbed Panels to Reverberant Acoustic Fields," *J. Acoust. Soc. Am.* 34:809–826 (1961), as well as Smith, reference 23 above and references 29–30 in chapter 1.

25. McLachlan, *Loudspeakers*, pp. 128–130, tables 9 and 10.

26. H. G. Davies, "Low Frequency Random Excitation of Water-Loaded Plates," *J. Sound Vib.* 15:107–136 (1971).

27. E. Skudrzyk, *Simple and Complex Vibratory Systems* (University Park, PA: Pennsylvania State University Press, 1968), pp. 390–398.

28. H. Levine, "A Note on Sound Radiation from Distributed Sources," *J. Sound Vib.* 68:203–207 (1980).

29. Abramowitz and Stegun, *Handbook*, p. 361, formulas (9.1.42) and (9.1.44).

30. Luke, *Integrals*, p. 294, formula (2), with $\beta = 1/2$, $\nu = 0$.

31. Abramowitz and Stegun, *Handbook*, p. 363, formula (9.1.97).

32. G. N. Watson, *Theory of Bessel Functions*, 2nd ed. (Cambridge, U.K.: Cambridge University Press, 1952), p. 358.

6

6.1 Characteristics of Convex Boundaries

The major portion of this chapter, through section 6.13, deals with spherical
and cylindrical radiators. The difference between geometries compatible
with separation of the wave equation and arbitrary source configurations
is explored in terms of the vector intensity field (section 6.14). Having
thoroughly familiarized the reader with cylindrical wave harmonics, these
functions will be used to construct asymptotic solutions for bodies of revolution whose geometry does not permit separation of the wave equation
(section 6.15).

Under steady-state conditions, a convex radiating surface completely
enclosing a volume perforce displays an acceleration that is *spatially periodic*.
Consequently, its dependence on the surface coordinate is expressible in
series form and does not require the integral transform representation developed in chapter 5. Such a series formulation is associated with a discrete
wave-number spectrum. If the boundary geometry is compatible with separation of the wave equation, the pressure field is expressible in terms of this
same series [see equation (4.24)].

For *infinite cylindrical* radiators this spatial periodicity may be confined
to the circumferential coordinate ϕ. Both the boundary configuration and
the sound field can therefore be expressed as a Fourier series in ϕ. Thus the
boundary configuration takes the form

$$\ddot{w}(z, \phi) = \sum_n \ddot{W}_n f_m(z) \cos n\phi \tag{6.1}$$

If the ϕ axis cannot be oriented to make the configuration even in ϕ, a sine
series will also be required. Because the corresponding development parallels
the procedure presented here for even configurations, it is not included. To
avoid unnecessarily burdening this presentation, a single z dependence $f_m(z)$
is assumed whenever possible. Unlike the two-dimensional acceleration
distribution over the plane baffle, which for nonperiodic configurations
requires two continuous wave-number spectra, the acceleration distribution
over the cylinder embodies at least one discrete wave number spectrum, n/a.
The z dependence of the configuration may embody a continuous spectrum,
if $f_m(z)$ is spatially nonperiodic and represented by a Fourier transform in
z (figure 6.1):

$$\ddot{w}(z, \phi) - \frac{1}{2\pi} \sum_n \ddot{W}_n \cos n\phi \int_{-\infty}^{\infty} \tilde{f}_m(\gamma) e^{i\gamma z} \, d\gamma. \tag{6.2}$$

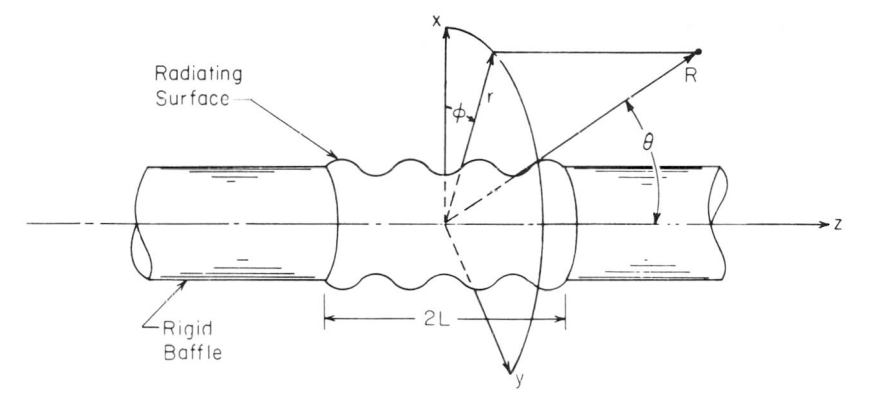

Figure 6.1
Infinite cylindrical boundary with a standing-wave acceleration distribution restricted to a finite length of cylinder.

If f_m is periodic in z, it can be represented by a Fourier series. The acceleration distribution thus becomes a surface-harmonic series in the form of equation (4.26):

$$\ddot{w}(z,\phi) = \sum_{n,m} \ddot{W}_{mn} \cos k_m z \cos n\phi. \tag{6.3}$$

This configuration is associated with two discrete spectra of wave numbers. The evaluation of the coefficients $\ddot{W}_{mn}$ is described in section 6.8.

    In contrast to the infinite cylinder, spherical or spheroidal boundaries allow the representation of all acceleration distributions as a double series. For the sphere, the acceleration is represented as a double series in *Legendre functions*[1] of the polar angle (or latitude) θ, and in cosines of the circumferential angle (or longitude) ϕ (figure 6.2):

$$\ddot{w}(\theta,\phi) = \sum_{n=0}^{\infty} \sum_{m=0}^{n} \ddot{W}_{mn} P_n^m (\cos\theta) \cos m\phi. \tag{6.4}$$

Here, the functions of index $m = 0$,

$$P_n^0(\eta) = P_n(\eta), \qquad \eta \equiv \cos\theta,$$

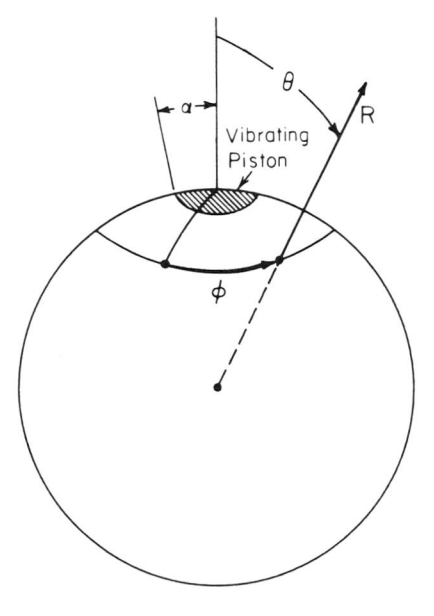

Figure 6.2
Sphere with radiating spherical cap.

are the *Legendre polynomials*. For $m \geq 1$, P_n^m are the *associated Legendre functions*, defined as[2]

$$P_n^m(\eta) \equiv (1 - \eta^2)^{m/2} \frac{d^m P_n(\eta)}{d\eta^m}. \qquad (6.5)$$

To visualize the radiator configuration it is noted that the subscript n indicates the number of nodal circles coaxial with the $\theta = 0$ axis. The superscript m is the number of nodal circles that are great circles intersecting the $\theta = 0$ axis. It is seen that the expansion of axisymmetric, ϕ-independent acceleration distributions, which are associated with the index $m = 0$, only requires Legendre polynomials. For configurations with spherical symmetry, equation (6.4) reduces to the $n = m = 0$ term, $P_0(\theta) = 1$. The normalization constants used in computing the coefficients $\ddot{W}_{mn}$ will be stated in the next section.

The two convex boundary configurations to be studied in detail are the sphere and the cylinder. The theory will be illustrated for some acceleration

distributions of practical importance. These results will be applied in chapter 9 to the vibrations of submerged shells and in chapters 10 and 11 to the sound scattering action of respectively, rigid and elastic scatterers.

6.2 The Green's Function for the Spherical Radiator

This source configuration will be used to illustrate the construction of the pressure field by the Green's function technique outlined in chapter 4. It is recalled that the primary requirement for obtaining an analytical solution to the Helmholtz integral equation is the construction of a Green's function, equation (4.23), that satisfies Neumann boundary conditions on the radiating surface — in this case, the sphere $R_0 = a$. The first step in constructing such a Green's function is to express the free-space Green's function, equation (4.8), in spherical coordinates. For this purpose the exponential in that expression is divided into its real and imaginary components:

$$\exp(ik|\mathbf{R} - \mathbf{R}_0|) = \cos(k|\mathbf{R} - \mathbf{R}_0|) + i\sin(k|\mathbf{R} - \mathbf{R}_0|). \tag{6.6}$$

From plane trigonometry, the distance $|\mathbf{R} - \mathbf{R}_0|$ from source to field point can be written as

$$|\mathbf{R} - \mathbf{R}_0| = (R^2 + R_0^2 - 2RR_0\cos\psi)^{1/2},$$

where ψ is the angle between $\mathbf{R}$ and $\mathbf{R}_0$. We can now apply an addition theorem[3] to the real and imaginary terms in equation (6.6), divided by $k|\mathbf{R} - \mathbf{R}_0|$ in anticipation of the construction of the Green's function:

$$\left.\begin{aligned}
\frac{\cos(k|\mathbf{R} - \mathbf{R}_0|)}{k|\mathbf{R} - \mathbf{R}_0|} &= -\sum_{n=0}^{\infty} y_n(kR)\,(2n + 1)\,j_n(kR_0)\,P_n(\cos\psi) \\
\frac{\sin(k|\mathbf{R} - \mathbf{R}_0|)}{k|\mathbf{R} - \mathbf{R}_0|} &= \sum_{n=0}^{\infty} j_n(kR)\,(2n + 1)\,j_n(kR_0)\,P_n(\cos\psi)
\end{aligned}\right\} \quad R_0 \le R, \tag{6.7}$$

where j_n and y_n are, respectively, the *spherical Bessel functions of the first and second kinds*. The above expressions are not yet sufficiently general for our purposes, because they restrict the polar axis of the coordinate system to

coincide with one or the other of the two position vectors $\mathbf{R}$ and $\mathbf{R}_0$. To generalize equation (6.7) to arbitrary coordinate axes, the angle ψ subtended by two points on a sphere is expressed in terms of the spherical coordinates (θ, ϕ) and (θ_0, ϕ_0) of these two points:

$$\cos \psi = \cos \theta \cos \theta_0 + \sin \theta \sin \theta_0 \cos(\phi - \phi_0). \tag{6.8}$$

The Legendre polynomials in equations (6.7) can now be expressed in terms of θ, θ_0, ϕ and ϕ_0 by applying the addition theorem[4]

$$P_n(\cos \psi) = \sum_{m=0}^{n} \varepsilon_m \frac{(n-m)!}{(n+m)!} P_n^m(\cos \theta) P_n^m(\cos \theta_0) \cos m(\phi - \phi_0), \tag{6.9}$$

where ε_m is the Neumann factor, which equals 1 for $m = 0$ and 2 for $m \geq 1$. Combining equations (6.6), (6.7, and (6.9) with the expression for the free-space Green's function, equation (4.8), one obtains this function as a series in spherical wave harmonics:

$$g(|\mathbf{R} - \mathbf{R}_0|) = -\frac{ik}{4\pi} \sum_{n=0}^{\infty} \sum_{m=0}^{n} \varepsilon_m \frac{(n-m)!}{(n+m)!} (2n+1) \cos m(\phi - \phi_0)$$

$$\cdot P_n^m(\cos \theta) P_n^m(\cos \theta_0) j_n(kR_0) h_n(kR), \qquad R_0 \leq R. \tag{6.10}$$

Here, h_n is the *spherical Hankel function of the first kind*, also referred to as *spherical Bessel function of the third kind*:[5]

$$h_n(kR) \equiv j_n(kR) + i y_n(kR) \tag{6.11a}$$

$$\approx -1 \cdot 3 \cdot 5 \cdots (2n-1)(kR)^{-(n+1)}, \qquad n \geq 1, \quad (kR)^2 \ll 2|2n-1| \tag{6.11b}$$

$$\approx \frac{1}{kR} \exp\left[i\left(kR - \frac{n+1}{2}\pi\right)\right], \qquad kR \gg n^2 + 1. \tag{6.11c}$$

The Green's function whose normal derivative vanishes on the surface $R_0 = a$ can now be constructed. The function $\Gamma(\mathbf{R}|\mathbf{R}_0)$ in equation (4.23) whose derivative $\partial\Gamma/\partial R_0$ cancels $\partial g/\partial R_0$ on the sphere $R_0 = a$ must have the same (ϕ, ϕ_0) and (θ, θ_0) dependence as equation (6.10). It can be verified that the R_0 derivative of the product

$$-\frac{h_n(kR)\,h_n(kR_0)\,j_n'(ka)}{h_n'(ka)} \tag{6.12}$$

cancels the R_0 derivative of the (R, R_0)-dependent terms in equation (6.10), when $R_0 = a$. The ratio in equation (6.12) thus embodies the (R, R_0) dependence required for the function Γ. Reducing the (R, R_0)-dependent factors of g and Γ to the same denominator $h_n'(ka)$, one obtains the (R, R_0)-dependent term of the Green's function:

$$\frac{h_n(kR)}{h_n'(ka)}\left[\,j_n(kR_0)\,h_n'(ka) - h_n(kR_0)\,j_n'(ka)\,\right]. \tag{6.13}$$

For $R_0 = a$, all the terms in square brackets have the same argument ka and can be simplified if one expands h_n and h_n' into their real and imaginary components:

$$j_n(ka)\,h_n'(ka) - h_n(ka)\,j_n'(ka) = i[\,j_n(ka)\,y_n'(ka) - y_n(ka)\,j_n'(ka)\,].$$

The terms in brackets can be reduced further by means of the Wronskian relation[6]

$$j_n(ka)\,y_n'(ka) - y_n(ka)\,j_n'(ka) = (ka)^{-2}. \tag{6.14}$$

When one substitutes $(ka)^{-2}$ for the terms in brackets in equation (6.13), the (R, R_0)-dependent term of the desired Green's function $G(\mathbf{R}|\mathbf{R}_0)$, specialized to $R_0 = a$, takes the form

$$\frac{ih_n(kR)}{(ka)^2 h_n'(ka)}.$$

The Green's function $G(\mathbf{R}|\mathbf{R}_0)$, specialized to $R_0 = a$, can now be constructed by substituting this ratio in lieu of $h_n(kR)$ in the expression for the free-space Green's function, equation (6.10):

$$G(R, \theta, \phi|a, \theta_0, \phi_0) = \frac{1}{4\pi k a^2} \sum_{n=0}^{\infty} \sum_{m=0}^{n} \varepsilon_m \frac{(n-m)!}{(n+m)!} (2n+1)\cos m(\phi - \phi_0)$$

$$\cdot P_n^m(\cos\theta)\, P_n^m(\cos\theta_0)\, \frac{h_n(kR)}{h_n'(ka)}. \tag{6.15}$$

6.3 The Pressure Field of a Spherical Radiator

The integral representation of the pressure, equation (4.22), specialized to spherical coordinates, now becomes

$$p(R, \theta, \phi) = -\rho a^2 \int_0^{2\pi} \int_0^{\pi} G(R, \theta, \phi | a, \theta_0, \phi_0) \ddot{w}(\theta_0, \phi_0) \sin \theta_0 \, d\theta_0 \, d\phi_0. \quad (6.16)$$

An arbitrary surface displacement distribution, expanded in series form, equation (6.4), can now be substituted in this integral. When the product of the two double series in equations (6.4) and (6.15) is integrated over the spherical surface, cross products drop out because of the orthogonality of the functions:

$$\int_{-1}^{1} P_n^m(\eta) P_l^m(\eta) \, d\eta = \frac{2\delta_{nl}}{2n+1} \frac{(n+m)!}{(n-m)!}, \qquad \eta \equiv \cos\theta, \quad -d\eta \equiv \sin\theta \, d\theta,$$
$$(6.17)$$

$$\int_0^{2\pi} \cos m\phi \cos l\phi \, d\phi = \frac{2\pi\delta_{ml}}{\varepsilon_m},$$

where δ_{nl} is the Kronecker delta, which equals one when $n = l$ and zero when $n \neq l$. These orthogonality relations are also used in determining the coefficients $\ddot{W}_{mn}$ in the expansion of an arbitrary acceleration distribution, equation (6.4):

$$\ddot{W}_{mn} = \frac{1}{\mathcal{N}_{mn}} \int_0^{2\pi} \left[\int_{-1}^{1} P_n^m(\eta) \ddot{w}(\eta, \phi) \, d\eta \right] \cos m\phi \, d\phi, \qquad \eta \equiv \cos\theta, \quad (6.18a)$$

where $\mathcal{N}_{mn}$ is the normalization factor obtained from equation (6.17):

$$\mathcal{N}_{mn} = \int_0^{2\pi} \cos^2 m\phi \, d\phi \int_{-1}^{1} [P_n^m(\eta)]^2 \, d\eta$$
$$= \frac{2\pi}{\varepsilon_m} \frac{2}{2n+1} \frac{(n+m)!}{(n-m)!}. \quad (6.18b)$$

The pressure equation (6.16) can now be written explicitly as

$$p(R, \theta, \phi) = -\frac{\rho}{k} \sum_{n=0}^{\infty} \sum_{m=0}^{n} \ddot{W}_{mn} P_n^m (\cos \theta) \cos m\phi \frac{h_n(kR)}{h_n'(ka)}. \qquad (6.19)$$

The coefficients multiplying the products of wave harmonics can be expressed in terms of the velocity amplitude as $i\rho c \dot{W}_{mn}$. Noting that

$$h_0(x) = -\frac{i}{x} e^{ix},$$

$$h_0'(x) = \frac{e^{ix}}{x^2}(x + i) = \left(i - \frac{1}{x}\right) h_0, \qquad (6.20)$$

one retrieves the pressure field of the uniformly pulsating sphere, equation (2.38). The far-field is obtained by using the large-argument asymptotic expansion [equation (6.11c)] for $h_n(kR)$:

$$p(R, \theta, \phi) = \frac{\rho e^{ikR}}{k^2 R} \sum_{n=0}^{\infty} \sum_{m=0}^{n} \frac{(-i)^{n-1}}{h_n'(ka)} \ddot{W}_{mn} P_n^m (\cos \theta) \cos m\phi, \qquad kR \gg n^2 + 1.$$

$$(6.21)$$

The low-frequency limit is evaluated by replacing $h_n'(ka)$ with the derivative of the small-argument asymptotic limit of h_n, equation (6.11b):

$$h_n'(x) \approx i \cdot 1 \cdot 3 \cdots (2n - 1)(n + 1) x^{-(n+2)}, \qquad x^2 \ll 2|2n - 1|. \qquad (6.22)$$

Because of the rapid convergence of the reciprocal of these functions, only the $(n = 0)$ and $(n = 1)$ terms need be retained in equation (6.21) in the small-ka limit. For these two terms, the Legendre functions are

$$P_0 = 1, \qquad P_n^m \equiv 0,$$

$$P_1 = \cos \theta, \qquad P_1^1 = \sin \theta, \qquad P_1^m \equiv 0 \quad \text{for } m > 1.$$

The low-frequency expression for the far-field pressure thus becomes

$$p(R, \theta, \phi)$$

$$\approx \frac{e^{ikR} \rho a^2}{R} \left[\ddot{W}_{00} - i\frac{ka}{2} (\ddot{W}_{01} \cos \theta + \ddot{W}_{11} \sin \theta \cos \phi) \right], \qquad k^2 a^2 \ll 1.$$

$$(6.23)$$

The monopole term equals equation (3.1) and can therefore be expressed in terms of the volume acceleration, equation (3.2). The dipole terms are proportional to the volume $\cdot V$ of the sphere and, in fact, to the inertia associated with the rigid-body translation of the fluid:

$$p(R, \theta, \phi) \simeq \frac{\rho e^{ikR}}{4\pi R} \left[\ddot{Q} - \frac{i3kV}{2} (\ddot{W}_{01} \cos \theta + \ddot{W}_{11} \sin \theta \cos \phi) \right]. \tag{6.24}$$

It will be found in section 6.6 that $3\rho V/2$ is the sum of the displaced mass and of the mass of fluid entrained by the sphere's reciprocating motion. If the monopole and dipole accelerations are comparable, the two θ-dependent terms can be dropped in the low-frequency limit.

　　The high-frequency asymptotic limit cannot be conveniently computed from equation (6.21) because of the slow convergence of the waveharmonic series. A high-frequency formulation in terms of series whose convergence accelerates with increasing ka is presented in chapter 12.

6.4 Circular Piston and Point Sources on a Spherical Baffle

A source of particular interest is a piston subtending a small polar angle $0 \le \theta \le \alpha$ on a rigid sphere and vibrating with uniform surface acceleration $\ddot{W}$ (figure 6.2):

$$\ddot{w}(\theta) = \ddot{W}, \qquad 0 \le \theta \le \alpha$$
$$= 0, \qquad \alpha < \theta < \pi.$$

Since $\ddot{w}(\theta) = 0$ for $\theta > \alpha$, the η integral in equation (6.18a) is reduced to the range $\cos \alpha \le \eta \le 1$. When the above ϕ-independent expression for $\ddot{w}(\theta)$ is substituted in equation (6.18a), only the $m = 0$ coefficients are nonzero:

$$\ddot{W}_{0n} = (n + \tfrac{1}{2}) \ddot{W} \int_{\cos \alpha}^{2} P_n(\eta) \, d\eta$$

$$= \frac{\ddot{W}}{2} [P_{n-1}(\cos \alpha) - P_{n+1}(\cos \alpha)]. \tag{6.25a}$$

If the piston is reduced to a point source ($\alpha^2 \ll 1$), the η integration can be simplified by taking the functions $P_n(\eta)$ out from under the integral sign, and assigning to them the value $P_n(1) = 1$. The η integral thus reduces to

$$
\left.
\begin{aligned}
\ddot{W}_{0n} &\approx (n + \tfrac{1}{2}) \ddot{W} \int_{\cos \alpha}^{1} d\eta \\
&\approx (n + \tfrac{1}{2})(1 - \cos \alpha) \ddot{W} \\
&\approx \frac{(2n + 1)\alpha^2 \ddot{W}}{4}
\end{aligned}
\right\} \qquad \alpha^2 \ll 1.
$$

This is conveniently expressed in terms of the volume acceleration $\pi a^2 \alpha^2 \ddot{W}$:

$$
\ddot{W}_{0n} = \frac{(2n + 1)\ddot{Q}}{4\pi a^2}. \tag{6.25b}
$$

The far-field of the piston is obtained by substituting equation (6.25a) in equation (6.21):

$$
p(R, \theta) = \frac{e^{ikR} \rho \ddot{W}}{2k^2 R} \sum_{n=0}^{\infty} \frac{(-i)^{n-1}}{h_n'(ka)} [P_{n-1}(\cos \alpha) - P_{n+1}(\cos \alpha)] P_n(\cos \theta). \tag{6.26}
$$

The pressure field of the point source is obtained by substituting the coefficients equation (6.25b) in equation (6.21):

$$
p(R, \theta) = \frac{\rho \ddot{Q} e^{ikR}}{4\pi R} \sum_{n=0,1}^{\infty} \frac{(2n + 1)(-i)^{n-1}}{k^2 a^2 h_n'(ka)} P_n(\cos \theta). \tag{6.27a}
$$

The coefficient multiplying the series is recognized as the pressure $p_s(R)$ radiated by a point source, equation (3.2). In the low-frequency limit, from equation (6.23),

$$
p(R, \theta) \approx p_s(R)\left(1 - \frac{i3ka}{2}\cos \theta\right), \qquad (ka)^2 \ll 1
$$
$$
\approx p_s(R), \qquad ka \ll 1. \tag{6.27b}
$$

This last result indicates that, when *the spherical baffle is small in terms of λ, is does not alter the far-field radiated by a point source.*

This pressure field is controlled by the size of the spherical baffle, as well as by that of the radiating surface element, which was assumed small. The effect of the dimensions of the sphere on the directivity factor is illustrated by noting that $p(R, \theta = 0)$ differs from $p(R, \theta = 180°)$ only by ap-

proximately 1% for $ka = 1/10$, while for $ka = 10$, the drive-point pressure is almost twice the pressure at the antipode: $|p(R, \theta = 0)| \approx 1.97|p_s(R)|$ whereas $|p(R, \theta = 180°)| \approx 1.04|p_s(R)|$. Even though the pressure at $\theta = 180°$ is small compared with the pressure at $0°$, it is enhanced by a focusing effect discussed further in section 12.2. Consequently, the pressure field in the shadow zones, $90° < \theta$, drops to considerably lower levels at points not in the vicinity of the antipode, provided the circumference measures one wavelength or more ($ka > 3$). In summary, a spherical baffle casts a shadow when its diameter is commensurate with or larger than a wavelength. More generally, it will be verified repeatedly that an object does not significantly distort the far-field if it is small in terms of wavelengths.

6.5 The Radiation Loading of Spherical Radiators

As in chapter 3, the surface pressure obtained by setting $R = a$ in equation (6.19) can be expressed in terms of modal specific acoustic impedances z_n,

$$p(a, \theta, \phi) = \sum_{n=0}^{\infty} \sum_{m=0}^{n} z_n \dot{W}_{mn} P_n^m (\cos \theta) \cos m\phi, \tag{6.28}$$

where

$$z_n \equiv i\rho c \frac{h_n(ka)}{h_n'(ka)}. \tag{6.29}$$

Separating the complex ratio h_n/h_n' into its real and imaginary components, we can write the impedance in terms of a resistance and an accession to inertia:

$$z_n \equiv r_n - i\omega m_n. \tag{6.30}$$

Combining equations (6.20) and (6.29), one obtains the impedance components found in equation (2.41) for the breathing mode of the sphere.

More generally, the resistance can be written as

$$r_n = \rho c \operatorname{Re} \left[\frac{ih_n(ka)}{h_n'(ka)} \right].$$

Using equations (6.11a) and (6.14), we obtain

$$r_n = \rho c [ka|h'_n(ka)|]^{-2}$$

$$\approx \frac{\rho c (ka)^{2n+2}}{(n+1)^2 [1 \cdot 3 \cdots (2n-1)]^2}, \qquad (ka)^2 \ll |2(2n-1)| \qquad (6.31)$$

$$\approx \rho c, \qquad ka \gg n^2 + 1.$$

The modal accession to inertia per unit area is

$$m_n = -\frac{\rho c}{\omega} \mathrm{Im} \left[\frac{ih_n(ka)}{h'_n(ka)} \right]$$

$$= -\frac{\rho a}{ka} \frac{j'_n(ka) j_n(ka) + y'_n(ka) y_n(ka)}{|h'_n(ka)|^2} \qquad (6.32)$$

$$\approx \rho a (n+1)^{-1}, \qquad (ka)^2 \ll 2n+3$$

$$\approx \rho a (ka)^{-2}, \qquad ka \gg n^2 + 1.$$

The modal resistances and accessions to inertia are plotted in figures 6.3[8] and 6.4, respectively. These results display the by now familiar features: When the dimension of the in-phase regions of the radiator—in other words, the distance between nodal circles, $\pi a/n$—equals or exceeds the acoustic wavelength—that is, when $ka > n$—the resistance is commensurate with the characteristic, plane-wave impedance. At low frequencies, the resistance is a second-order effect compared with the reactance. In this range, the accession to inertia displays a frequency-independent value. It peaks when the distance between nodal circles equals $\lambda/2$, and tends to zero at high frequencies.

These results can be used to compute the impedance of a circular piston located on a spherical baffle. The surface pressure is obtained by substituting the coefficients, equation (6.25a), expressed in terms of $\dot{W}$ instead of $\ddot{W}$, in equation (6.28):

$$p(a, \theta) = \frac{\dot{W}}{2} \sum_{n=0}^{\infty} z_n [P_{n-1}(\cos \alpha) - P_{n+1}(\cos \alpha)] P_n(\cos \theta).$$

The resultant force on the piston is computed by integrating the pressure over the spherical cap extending from $\theta = 0$ to α, using the integral in equation (6.25a). The resultant radiation impedance $\mathcal{Z}_p$ is finally obtained by dividing this force by $\dot{W}$:

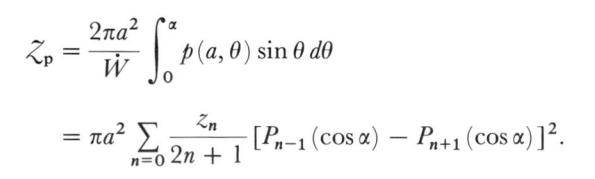

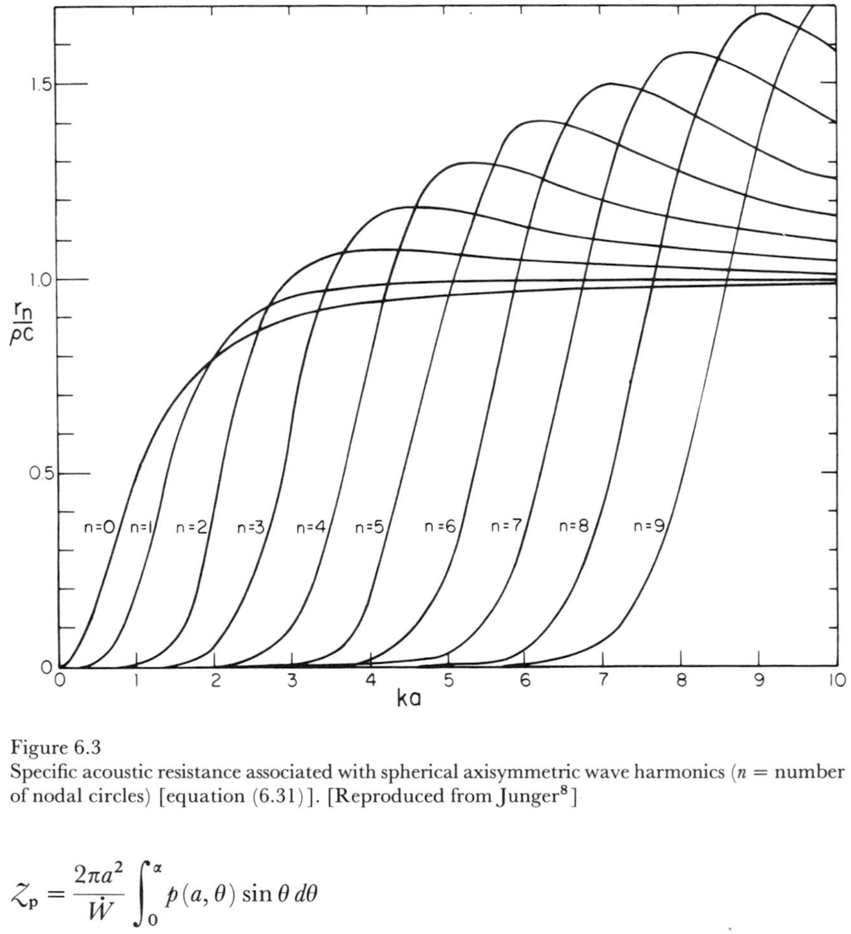

Figure 6.3
Specific acoustic resistance associated with spherical axisymmetric wave harmonics (n = number of nodal circles) [equation (6.31)]. [Reproduced from Junger[8]]

$$\mathscr{Z}_p = \frac{2\pi a^2}{\dot{W}} \int_0^\alpha p(a,\theta)\sin\theta\, d\theta$$

$$= \pi a^2 \sum_{n=0}^{} \frac{z_n}{2n+1}[P_{n-1}(\cos\alpha) - P_{n+1}(\cos\alpha)]^2.$$

The resistance can be approximated as the sum of those terms whose order number n does not exceed $2ka$. Thus, with increasing ka, a larger number of modes contributes to the resistance. For small ka, the resistance falls below that of the piston in a plane baffle by a factor of two. This is anticipated from the fact that, in the small-ka limit, the latter was shown in chapter 3 to radiate a far-field pressure $2p_s(R)$ into a half-space, in contrast to the small

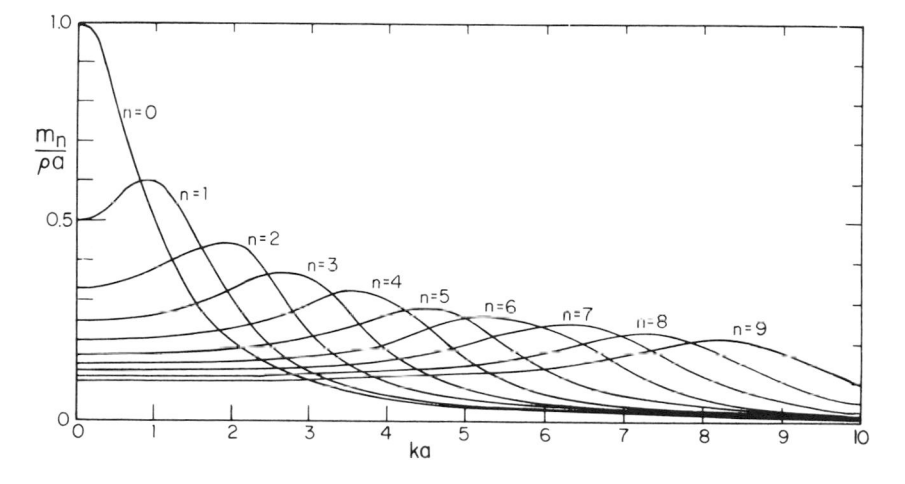

Figure 6.4
Accession to inertia per unit area corresponding to the resistance in figure 6.3[8] [equation (6.32)]. [Reproduced from Junger[8]]

piston in a spherical baffle, which in the small-ka limit, equation (6.27b), radiates a pressure $p_s(R)$ into the infinite medium.

6.6 Concentrated Force Applied to the Acoustic Medium

The results of the preceding section can be used to verify the interpretation given in chapter 4 for the $\partial g/\partial \xi_0$ term in the Helmholtz integral equation as the pressure field radiated by a force of unit amplitude applied to the fluid medium. Consider a liquid sphere of infinitesimal radius translating with an acceleration amplitude $\ddot{W}_{01}$ under the effect of a reciprocating force of amplitude F. The corresponding acceleration distribution is described by the single term $\ddot{W}_{01} P_1 (\cos\theta) \equiv \ddot{W}_{01} \cos\theta$. The force balances the inertia force associated with the effective mass, specifically, the sum of the mass of displaced fluid ρV, and the resultant mass of entrained fluid, M_1:

$$M_1 = 2\pi a^2 m_1 \int_0^\pi \cos^2\theta \sin\theta \, d\theta.$$

The integral yields $2/3$. The entrained mass per unit area, m_1, from equation (6.32) specialized to long wavelengths, is $\rho a/2$. One thus computes

$$M_1 = \frac{\rho V}{2}. \tag{6.33}$$

This being half the displaced mass of fluid, the total effective mass is $3M_1/2$. The acceleration of the fluid sphere equals the force F divided by this effective mass:

$$\ddot{W}_{01} = \frac{2F}{3\rho V}.$$

Substituting this result in equation (6.24) with $\ddot{W}_{00} = \ddot{W}_{11} = 0$, one obtains the pressure generated by a force applied directly to the fluid:

$$p(R, \theta) = -\frac{ikFe^{ikR}}{4\pi R} \cos \theta. \tag{6.34a}$$

The peak pressure amplitude, located at $\theta = 0$ (or π), is conveniently expressed in terms of the wavelength:

$$|p(R, 0)| = \frac{F}{2\lambda R}. \tag{6.34b}$$

For practical purposes, it is desirable to express this result as a sound pressure level (SPL), defined in section 2.2. When the acoustic medium is water,

$$\text{SPL} = 20 \log fF - 20 \log R + \dot{A},$$

where F is the rms force; f is frequency in Hz; R is range in meters; and $A = -36$ dB re 1 μbar for F in pounds, -149 dB re 1 μbar for F in dynes, and 51 dB re 1 μPa for F in newtons.

Equation (6.34) can be generalized by setting $R = |\mathbf{R} - \mathbf{R}_0|$. The pressure can now be expressed in terms of the free-space Green's function, equation (4.8):

$$p(|\mathbf{R} - \mathbf{R}_0|, \theta) = -ikFg(|\mathbf{R} - \mathbf{R}_0|) \cos \theta. \tag{6.34c}$$

Comparing the result with equation (4.16), we verify the interpretation of

the derivative $\partial g/\partial \xi_0$ as the far-field sound pressure generated by a unit force acting in the direction $\hat{\xi}_0$.

We now turn our attention to cylindrical boundaries.

6.7 Cylindrical Radiators with Spatially Periodic Configurations

Having illustrated the Green's function technique, we shall construct the sound field of cylindrical sources by means of the boundary-matching technique used in chapter 5. The boundary condition relating the pressure field to the source configuration is

$$\frac{\partial p(r, z, \phi)}{\partial r} = -\rho \ddot{w}(z, \phi), \qquad r = a. \tag{6.35}$$

For the simple configuration, equation (6.3), whereby the acceleration distribution is periodic in the axial coordinate z, the coefficients of the acceleration series are obtained from the latter of the two orthogonality relations in equation (6.17):

$$\ddot{W}_{mn} = \frac{\varepsilon_n \varepsilon_m}{(2\pi)^2} \int_0^{2\pi} \left[\int_0^{2\pi/k_m} \ddot{w}(\phi, z) \cos k_m z \, dz \right] \cos n\phi \, d\phi. \tag{6.36}$$

Because the wave equation is separable in cylindrical coordinates, the sound field is in the form of equation (4.24), with $\xi = r$, $\eta = z$, and $\zeta = \phi$. To satisfy the boundary condition, equation (6.35), the function S_{mn} in equation (4.25) must embody the z and ϕ dependence of the source configuration, equation (6.3). The pressure field is therefore expressible as the series

$$p(r, z, \phi) = \sum_{m, n} A_{mn} R_{mn}(r) \cos k_m z \cos n\phi, \tag{6.37}$$

where A_{mn} are coefficients to be determined from the boundary condition, equation (6.35). Previously, however, the radial functions R_{mn} must be constructed. The differential equation governing the surface harmonics in equation (6.37) is constructed specializing the general procedure developed in connection with equations (4.25) to cylindrical coordinates. When the surface Laplace operator in cylindrical coordinates, equation (2.61), is substituted in the former of these equations, one obtains

$$\left(\frac{1}{r^2} \frac{\partial^2}{\partial \phi^2} + \frac{\partial^2}{\partial z^2} \right) \cos k_m z \cos n\phi = -\left(k_m^2 + \frac{n^2}{r^2} \right) \cos k_m z \cos n\phi.$$

The radial harmonic satisfies the ordinary differential equation

$$\left[\frac{\partial^2}{\partial r^2} + \frac{1}{r}\frac{\partial}{\partial r} - \frac{n^2}{r^2} + (k^2 - k_m^2)\right] R_{mn}(r) = 0. \tag{6.38}$$

This is *Bessel's differential equation*, whose solutions are linear combinations of Bessel functions of the first and second kinds:[9]

$$R_{mn}(r) = J_n[(k^2 - k_m^2)^{1/2}r] + BY_n[(k^2 - k_m^2)^{1/2}r]. \tag{6.39}$$

To determine B, we consider the large-argument asymptotic limit[10] of equation (6.39), with $k_m \to 0$:

$$\lim_{\substack{k/k_m \\ kr \to \infty}} R_{mn}(r) = \frac{2^{1/2}}{(\pi kr)^{1/2}}\left[\cos\left(kr - \frac{2n+1}{4}\pi\right) + B\sin\left(kr - \frac{2n+1}{4}\pi\right)\right].$$

The proper value of the coefficient B can be determined by noting that, when the wave front is nearly plane, i.e., when $k_m \ll k$ and $kr \gg 1$, $R_{mn}(r)$ must embody the propagation characteristics of a plane wave traveling outward, in the positive r direction, with the velocity of sound. In this asymptotic, large-kr limit, the function R_{mn} must therefore be proportional to a complex exponential $\exp(ikr - i\omega t)$, whereby increasing time requires a radial coordinate increasing at the speed of sound to maintain the phase invariant. The desired exponential dependence on kr is achieved when the coefficient B is selected equal to i. For outgoing waves, the solution in equation (6.39) must therefore be specialized to

$$\begin{aligned} R_{mn}(r) &= J_n[(k^2 - k_m^2)^{1/2}r] + iY_n[(k^2 - k_m^2)^{1/2}r] \\ &= H_n[(k^2 - k_m^2)^{1/2}r], \end{aligned} \tag{6.40}$$

where H_n is the *Hankel function of the first kind*.[9] This function is normally identified, as in equations (5.19), by the superscript (1), but since we shall not have occasion in what follows to use the function of the second kind, which is the complex conjugate of the function of the first kind, we shall omit superscripts. After equations (6.40) and (6.37) are combined, the final task of determining the coefficients A_{mn} can be performed by substituting the general solution in the boundary condition, equation (6.35), and solving for A_{mn}:

$$A_{mn} = \frac{\rho \ddot{W}_{mn}}{(k^2 - k_m^2)^{1/2} H_n'[(k^2 - k_m^2)^{1/2} a]}.$$

The pressure field is thus finally found to be

$$p(r, z, \phi) = -\rho \sum_{n, m} \frac{\ddot{W}_{mn} H_n[(k^2 - k_m^2)^{1/2} r]}{(k^2 - k_m^2)^{1/2} H_n'[(k^2 - k_m^2)^{1/2} a]} \cos k_m z \cos n\phi. \qquad (6.41)$$

Derivation of the asymptotic far-field and low-frequency limits will be delayed until the more realistic case of a cylindrical source of finite length is analyzed.

6.8 Radiation Loading of Infinite Cylinders with Standing-Wave Configurations

The surface pressure obtained from equation (6.41) can be written in terms of modal specific acoustic impedances z_{mn} as

$$p(a, z, \phi) = \sum_{n, m} \dot{W}_{mn} z_{mn} \cos k_m z \cos n\phi, \qquad (6.42a)$$

where

$$z_{mn} = \frac{i\rho c k H_n[(k^2 - k_m^2)^{1/2} a]}{(k^2 - k_m^2)^{1/2} H_n'[(k^2 - k_m^2)^{1/2} a]}. \qquad (6.42b)$$

Because the boundary configuration, equation (6.3), is spatially periodic in the entire domain $-\infty < z < \infty$, this impedance can be expected to display some of the features of the impedance found for infinite trains of waves on a plane boundary (section 5.11). Specifically, one expects that no energy will be radiated when $k < k_m$—in other words, when $(k^2 - k_m^2)^{1/2}$ is imaginary. In this regime, the Hankel function of imaginary argument is conveniently replaced by a *modified Hankel function* of real argument:[11]

$$K_n(x) \equiv \frac{\pi}{2} i^{n+1} H_n(ix). \qquad (6.43)$$

When this function is substituted in equation (6.42b), one obtains a purely masslike impedance

$$z_{mn} = -i\omega m_{mn}, \qquad k_m > k,$$

where the mass per unit area is

$$m_{mn} = -\frac{\rho K_n[(k_m^2 - k^2)^{1/2}a]}{(k_m^2 - k^2)^{1/2} K_n'[k_m^2 - k^2)^{1/2}a]}, \qquad k_m > k \qquad (6.44a)$$

$$\left.\begin{aligned} &\approx -\rho a \ln(|k_m^2 - k^2|^{1/2}a), && n = 0 \\ &\approx \frac{\rho a}{n}, && n \geq 1 \end{aligned}\right\} \qquad (k_m^2 - k^2)a^2 \ll 2n + 1 \qquad (6.44b)$$

$$\approx \frac{\rho}{(k_m^2 - k^2)^{1/2}}, \qquad (k_m^2 - k^2)^{1/2}a \gg n^2 + 1. \qquad (6.44c)$$

A peculiarity of the infinite cylinder, which it shares with infinite planar radiators (figure 5.16) but distinguishes it from convex boundaries enclosing a finite volume, is that the accession to inertia goes to infinity for the axially symmetric ($n = 0$) mode as $(k_m^2 - k^2)^{1/2}a \to 0$ (figures 6.5 and 6.6).[12] When the argument of the Hankel function in equation (6.42b) is real, the impedance has both a real and an imaginary component:

$$\left.\begin{aligned} m_{mn} &= \frac{-\rho(\mathcal{J}_n\mathcal{J}_n' + Y_n Y_n')}{(k^2 - k_m^2)^{1/2}|H_n'|^2} \\ r_{mn} &= \frac{\rho c k(\mathcal{J}_n Y_n' - Y_n\mathcal{J}_n')}{(k^2 - k_m^2)^{1/2}|H_n'|^2} \end{aligned}\right\} \quad k_m < k, \qquad \begin{matrix}(6.45a) \\ \\ (6.45b)\end{matrix}$$

where the suppressed argument of the cylinder functions is $[k^2 - k_m^2)^{1/2}a]$. Asymptotic values of the accession to inertia for small argument are given by (6.44b). For large argument

$$m_{mn} \approx \frac{\rho a}{2(k^2 - k_m^2)a^2}, \qquad (k^2 - k_m^2)^{1/2}a \gg n^2 + 1. \qquad (6.45c)$$

Before evaluating the limiting values of the resistive component, r_{mn} will be simplified by noting that the terms in the parentheses are the Wronskian,[13]

$$\mathcal{J}_n(x)Y_n'(x) - Y_n(x)\mathcal{J}_n'(x) = \frac{2}{\pi x}. \qquad (6.45d)$$

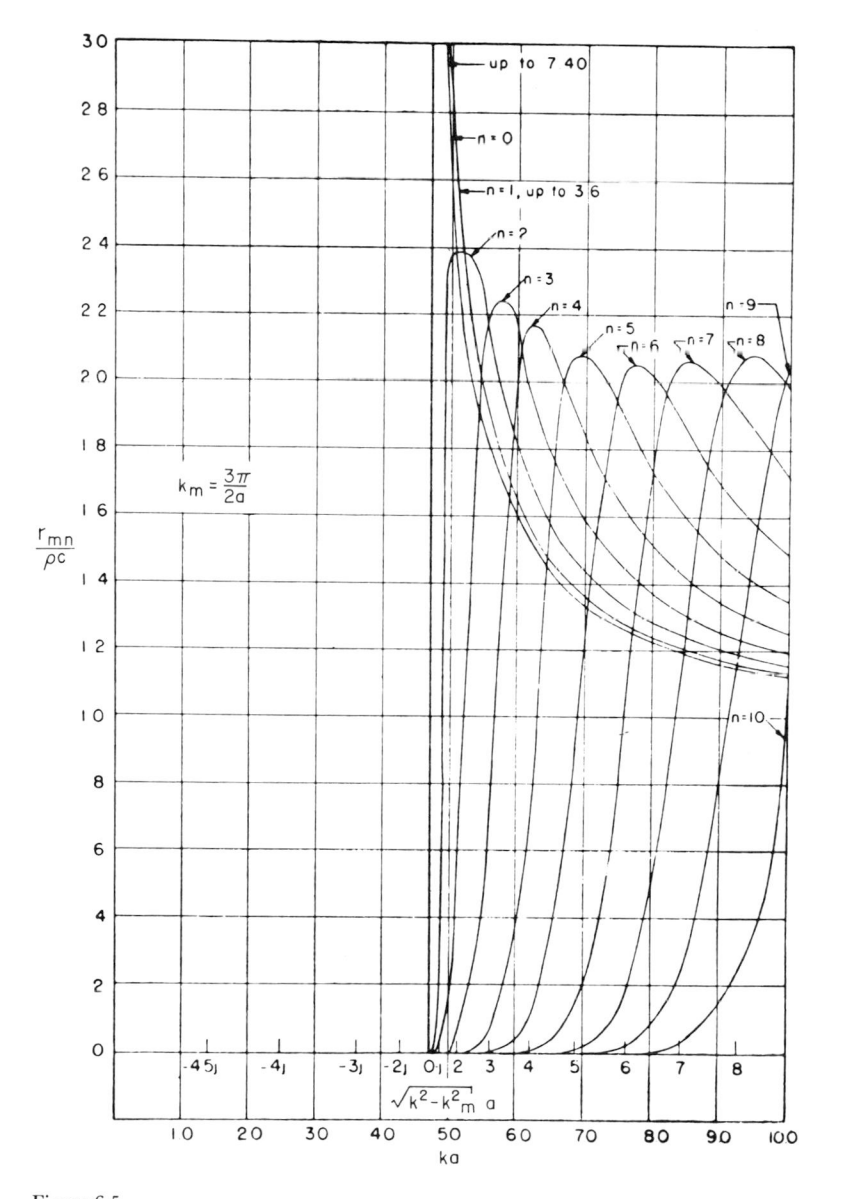

Figure 6.5
Specific acoustic resistance associated with cylindrical wave harmonics [equation (6.46)].
[Reproduced from Junger[12]]

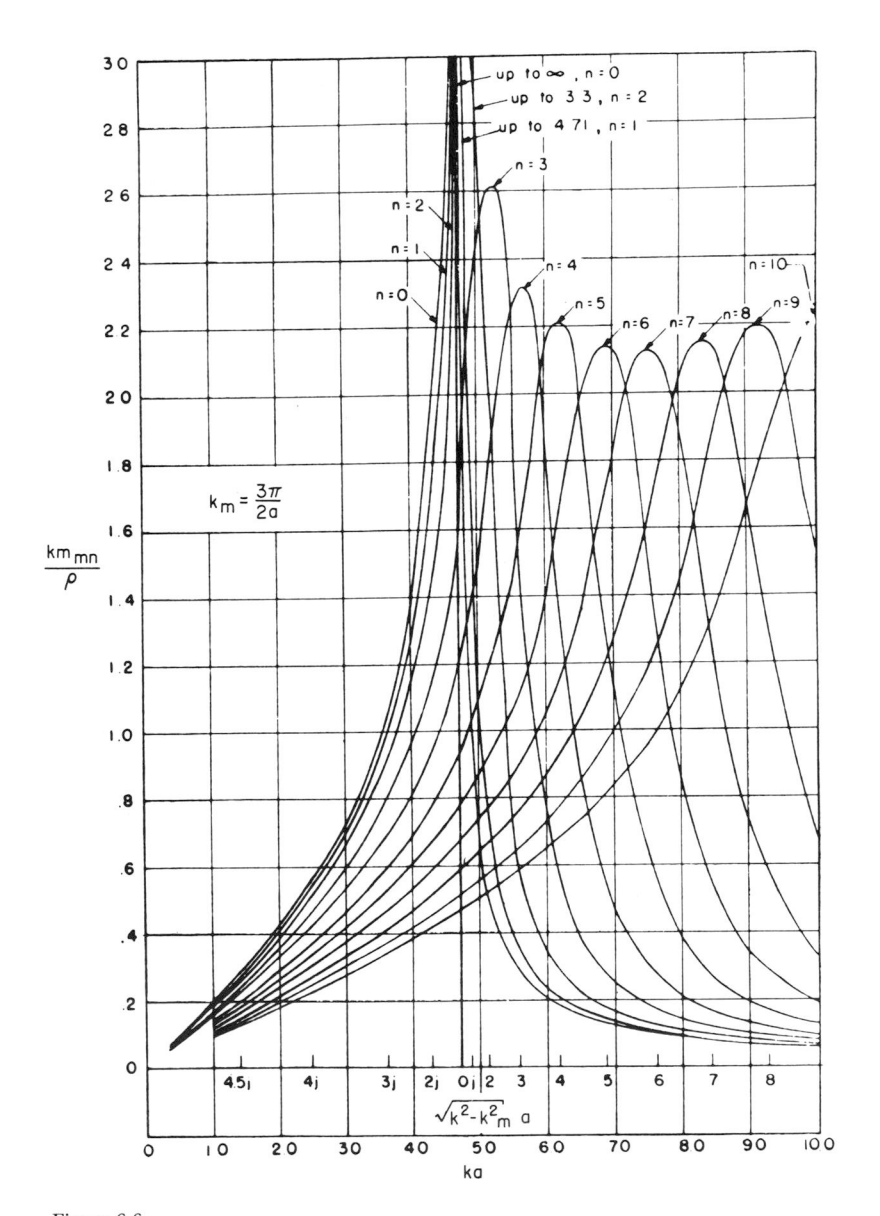

Figure 6.6
Specific acoustic reactance (defined as circular frequency times accession to inertia per unit area) corresponding to the resistance in figure 6.5 [equations (6.44) and (6.45a)]. [Reproduced from Junger[12]]

The resistance now becomes

$$
\left.
\begin{aligned}
r_{mn} &= \frac{2\rho c k}{\pi (k^2 - k_m^2) a |H_n'[(k^2 - k_m^2)^{1/2}a)]|^2} \\[2mm]
&\approx \frac{\rho c \pi k a}{(n!)^2} \frac{(k^2 - k_m^2)^n a^{2n}}{2^{|2n-1|}}, \qquad 0 < (k^2 - k_m^2)a^2 \ll 2n + 1 \\[2mm]
&\approx \rho c, \qquad (k^2 - k_m^2)^{1/2}a \gg n^2 + 1 \\[2mm]
&= 0, \qquad k_m > k.
\end{aligned}
\right\} \quad k_m < k.
$$

$$(6.46a)$$
$$(6.46b)$$
$$(6.46c)$$
$$(6.46d)$$

The physical interpretation of these analytical results, specifically of the reactive impedance in the range $k_m > k$, is effectively the same as for the plane boundary of periodic configuration (section 5.11). For a small argument, the resistance, equation (6.46b), converges rapidly with increasing n, so that only the lowest value of n, corresponding to the $n = 0$ breathing mode, need be considered, for comparable velocity coefficients in equation (6.42).

Setting $k_m = 0$ in the above impedance expressions, one obtains the impedance components of the z-independent cylindrical radiator. The resistances take the simple form

$$r_{0n} = 2\rho c / \pi k a |H_n'(ka)|^2.$$

A feature peculiar to the cylinder, as opposed to the planar radiator, is that, when $k > k_m$, the impedance is complex, rather than purely resistive. The reactance in this range can be associated with acceleration of the fluid in the circumferential direction, the pressure field being in the nature of an edge mode. It is only when $k \gg n/a$ as well as $k \gg k_m$ that the impedance tends asymptotically to the purely resistive impedance characteristic of surface modes. In this limit, where the fluid particle velocity is almost purely radial, the pressure field can be envisioned as consisting of cylindrical, radially spreading wave fronts (figure 6.7a). When k is only moderately larger than k_m, the wave fronts propagate along conical surfaces of vertex angle $\theta_m = \cos^{-1} k_m/k$ and $(\pi - \theta_m)$ (figure 6.7b). When $k < k_m$, the impedance is purely reactive, the axial dependence of the pressure field being that of standing waves (figure 6.7c). The pressure decays exponentially in the radial direction, no power being radiated. Circumferential standing

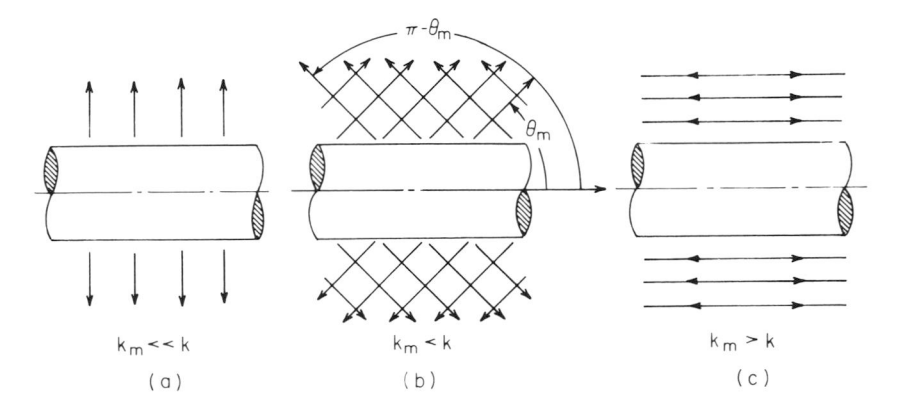

Figure 6.7
Configuration of the pressure field of cylindrical radiators with spatially periodic acceleration distributions, in three ranges of ratio of structural to acoustic wave numbers: (a) cylindrical waves; (b) conical waves; (c) and standing waves. [Reproduced from Junger[12]]

waves are of importance in the range $k < n/a$, where they are associated with rapid convergence of the wave-harmonic series for the far-field.

6.9 Transform Formulation of the Pressure Field of Cylindrical Radiators

As in the case of the plane boundary, some acoustic power is radiated by a *finite* train of waves, such as an acceleration distribution $\ddot{w}(z) \propto \cos k_m z$ limited to the region $|z| < L$, even when $k_m > k$. This configuration will now be studied using the same Fourier transform technique as in section 5.6, confined, however, to a single coordinate. When a Fourier transform in z is applied to the Helmholtz equation whose Laplace operator is formulated in cylindrical coordinates, equation (2.61), the partial differential equation governing the pressure transform $\tilde{p}(r, \phi; \gamma)$ is obtained:

$$\left(\frac{\partial^2}{\partial r^2} + \frac{1}{r} \frac{\partial}{\partial r} + k^2 - \gamma^2 + \frac{1}{r^2} \frac{\partial^2}{\partial \phi^2} \right) \tilde{p}(r, \phi; \gamma) = 0. \tag{6.47}$$

By analogy with the radial dependence of the spatially periodic pressure field, equation (6.40), solutions of this equation representing outward traveling waves must display the r dependence embodied in the Hankel function of the first kind:

$$\tilde{p}(r,\phi;\gamma) = \sum_n A_n H_n[(k^2 - \gamma^2)^{1/2}r] \cos n\phi. \tag{6.48}$$

The coefficients A_n are determined from the transform of the boundary condition, equation (6.35), the nonperiodic acceleration distribution in equation (6.1) being used for $\ddot{w}(z,\phi)$:

$$\frac{\partial \tilde{p}(r,\phi;\gamma)}{\partial r} = -\rho \sum_n \ddot{W}_n \tilde{f}_m(\gamma) \cos n\phi, \qquad r = a.$$

Substituting equation (6.48), one can solve for A_n:

$$A_n = \frac{-\rho \ddot{W}_n \tilde{f}_m(\gamma)}{(k^2 - \gamma^2)^{1/2} H_n'[(k^2 - \gamma^2)^{1/2}a]}.$$

When this is substituted in equation (6.48), one obtains $\tilde{p}$, whose inverse Fourier transform finally yields the pressure field:

$$p(r,\phi,z) = -\frac{\rho}{2\pi} \sum_n \ddot{W}_n \cos n\phi \int_{-\infty}^{\infty} \frac{e^{i\gamma z} H_n[(k^2 - \gamma^2)^{1/2}r] \tilde{f}_m(\gamma)\, d\gamma}{(k^2 - \gamma^2)^{1/2} H_n'[(k^2 - \gamma^2)^{1/2}a]}. \tag{6.49}$$

In the near-field, specifically on the surface $r = a$, this integral must be evaluated numerically,[14] except in the asymptotic large-ka limit. Here Watson's transformation presents a concise analytical formulation of the surface pressure. This is illustrated in chapter 12.

For an acceleration distribution in the form of a standing wave, say, $\ddot{w}(z) \propto \cos k_m z$, confined to the region $|z| < L$, the accession to inertia can, in the range $k_m > k$, be approximated by the one derived for a boundary displacement truly periodic in z, equations (6.44) and (6.45a). The basis for this approximation is that the accession to inertia in the long-wavelength range is effectively determined by the incompressible, hydrodynamic solution, whereby the local surface pressure is primarily a function of the local acceleration distribution. This hydrodynamic approximation does not apply when $k_m < k$. In this range, the accession to inertia is fortunately relatively unimportant compared with the radiation resistance.

Except in the short-wavelength limit, where the average radiation resistance tends to the characteristic plane-wave impedance, the resistance loading cannot be approximated without accounting for the finiteness of the length $2L$ of the active portion of the cylindrical boundary. The reason is

that radiation of acoustic power, in the range $k_m > k$, is associated with area elements located near the boundaries $z = \pm L$ of the active region of the cylinder. The corresponding acoustic resistance can therefore not be approximated by assuming boundary accelerations extending over the entire range $-\infty < z < \infty$. The resistance will be obtained in section 6.13, from the power computed from the far-field pressure. This pressure can be approximated in closed form by applying the method of stationary phase familiar from chapter 5.

6.10 Stationary-Phase Approximation to the Far-Field of Cylindrical Radiators

In the far-field the large-argument asymptotic value of the Hankel function, equation (5.19b), is applicable in equation (6.49). Furthermore, the cylindrical coordinates in equation (6.49) are transformed into spherical coordinates by setting $r = R \sin\theta$, $z = R \cos\theta$ (figure 6.1). The inverse transform, equation (6.49), now becomes

$$p(R, \theta, \phi) = \frac{-\rho}{(2\pi^3 R \sin\theta)^{1/2}} \sum_n \ddot{W}_n (-i)^n e^{-i\pi/4} \cos n\phi$$

$$\cdot \int_{-\infty}^{\infty} \frac{\tilde{f}_m(\gamma) \exp\{iR[(k^2 - \gamma^2)^{1/2} \sin\theta + \gamma \cos\theta]\}}{(k^2 - \gamma^2)^{3/4} H_n'[(k^2 - \gamma^2)^{1/2}a]} \cdot d\gamma. \tag{6.50}$$

The integrand is in the form $\Phi \exp(i\Psi)$ and, like equation (5.18), suitable for stationary-phase integration. The modulus and phase of the integrand are

$$\Phi(\gamma) = \frac{\tilde{f}_m(\gamma)}{(k^2 - \gamma^2)^{3/4} H_n'[(k^2 - \gamma^2)^{1/2}a]}, \tag{6.51a}$$

$$\Psi(\gamma) = R[(k^2 - \gamma^2)^{1/2} \sin\theta + \gamma \cos\theta]. \tag{6.51b}$$

The derivative of the phase,

$$\frac{\partial \Psi}{\partial \gamma} = R\left[\cos\theta - \frac{\gamma \sin\theta}{(k^2 - \gamma^2)^{1/2}}\right],$$

vanishes for

$$\bar{\gamma} = k \cos\theta, \qquad (k^2 - \bar{\gamma}^2)^{1/2} = k \sin\theta. \tag{6.52}$$

The second derivative of the phase is

$$\frac{\partial^2 \Psi}{\partial \gamma^2} = -R \sin \theta \left[\frac{1}{(k^2 - \gamma^2)^{1/2}} + \frac{\gamma^2}{(k^2 - \gamma^2)^{3/2}} \right]$$

$$= -R/k \sin^2 \theta, \qquad \gamma = \bar{\gamma}.$$

The stationary-phase value of the modulus of the integrand is

$$\Phi(\bar{\gamma}) = \frac{\tilde{f}_m(k \cos \theta)}{(k \sin \theta)^{3/2} H_n'(ka \sin \theta)}.$$

The stationary-phase approximation to the integral can now be constructed by substituting the preceding results in equation (5.21b):

$$I = \frac{(2\pi)^{1/2} \tilde{f}_m(k \cos \theta) e^{(ikR - i\pi/4)}}{k(R \sin \theta)^{1/2} H_n'(ka \sin \theta)}.$$

When this is substituted for the integral in the inverse transform, the far-field pressure is finally obtained:

$$p(R, \theta, \phi) = \frac{\rho e^{ikR}}{\pi k R \sin \theta} \sum_{n=0}^{\infty} \ddot{W}_n \frac{\tilde{f}_m(k \cos \theta)(-i)^{n+1}}{H_n'(ka \sin \theta)} \cos n\phi. \qquad (6.53)$$

To determine the small-θ (or long-wavelength) behavior of this solution, we require the small-argument asymptotic expression[10] for H_n':

$$H_n'(x) \approx \frac{i 2^{n+1} n!}{\pi \varepsilon_n x^{n+1}}, \qquad x^2 \ll 2n + 1, \qquad (6.54)$$

where $\varepsilon_0 = 1$ and $\varepsilon_n = 2$ for $n \geq 1$. When this is used in equation (6.53), one obtains the small-$(ka \sin \theta)$ expression for the far-field:

$$p(R, \theta, \phi)$$

$$= \frac{-\rho a e^{ikR}}{R} \qquad (6.55)$$

$$\cdot \sum_{n=0}^{\infty} \frac{\varepsilon_n \ddot{W}_n \tilde{f}_m(k \cos \theta)(-ika \sin \theta)^n \cos n\phi}{n! 2^{n+1}}, \qquad (ka \sin \theta)^2 \ll 2n + 1.$$

6.11 Piston in a Cylindrical Baffle

The stationary-phase solution was originally formulated by Laird and Cohen[15] to study the field of a rectangular piston in a cylindrical baffle. If the piston subtends a circumferential angle 2α, the Fourier coefficients in (6.2) are

$$\ddot{W}_n = \frac{\ddot{W}\varepsilon_n}{2\pi} \int_0^\alpha \cos n\phi \, d\phi$$

$$= \frac{\alpha \ddot{W}\varepsilon_n}{\pi} j_0 (n\alpha).$$

For a rigid piston of axial dimension $2L$ the Fourier transform of the axial dependence of the dynamic configurations is

$$\tilde{f}_m(\gamma) = \int_{-L}^{L} \cos \gamma z \, dz \tag{6.56}$$

$$= 2L j_0 (kL \cos \theta), \qquad \gamma = k \cos \theta.$$

These results can be substituted in (6.53) to yield the pressure radiated by a piston in a cylindrical baffle:

$$p(R, \theta, \phi) = \frac{2\rho L\alpha \ddot{W} j_0 (kL \cos \theta) e^{ikR}}{\pi^2 kR \sin \theta} \sum_{n=0}^{\infty} \frac{\varepsilon_n j_0 (n\alpha) (-i)^{n+1} \cos n\phi}{H_n' (ka \sin \theta)}.$$

To verify the algebra, we now retrieve the sound field of the line source: Setting $\alpha = \pi$, all terms $n > 0$ vanish. Specializing the result to the small-ka asymptotic limit, (6.54), and introducing the volume acceleration $\ddot{Q} = 4\pi a L \ddot{W}$,

$$p(R, \theta) = \frac{\rho e^{ikR} \ddot{Q} j_0 (kL \cos \theta)}{4\pi R}.$$

This is precisely the solution constructed by elementary methods for the broadside array [equation (3.17)].

We shall now study acceleration distributions that simulate the in vacuo modes of cylindrical shells.

6.12 Far-Field of Cylinders with Standing-Wave Configurations of Finite Axial Extent

A configuration of particular interest is in the form of standing waves $\cos k_m z$ confined to the finite region $|z| < L$, the product $k_m L$ being selected to simulate the boundary conditions of a cylindrical shell simply supported at $z = \pm L$, figure 6.1. Such an acceleration is in the form of equation (6.3), with

$$
\begin{aligned}
f_m(z) &= \cos k_m z, \qquad |z| < L \\
&= 0, \qquad\qquad |z| > L,
\end{aligned}
\tag{6.57a}
$$

where

$$
k_m = \frac{\left(\frac{1}{2} + m\right)\pi}{L}, \qquad m = 0, 1, \ldots,
$$

$$
\cos k_m L = 0, \qquad \sin k_m L = (-1)^m.
$$

The transform of this acceleration distribution can be taken from equation (5.70a) if one writes m and γ in lieu of n and γ_x, respectively:

$$
\begin{aligned}
\tilde{f}_m(\gamma) &= \frac{2k_m(-1)^m \cos \gamma L}{k_m^2 - \gamma^2}. \\
&= L, \qquad k_m = \gamma.
\end{aligned}
\tag{6.57b}
$$

Evaluating this transform at the stationary-phase point $\bar{\gamma} = k \cos\theta$ and substituting the result in equation (6.53), we obtain the far-field pressure:

$$
\begin{aligned}
p(R, \theta, \phi) &= \frac{2\rho e^{ikR} k_m(-1)^m \cos(kL \cos\theta)}{\pi k R \sin\theta (k_m^2 - k^2 \cos^2\theta)} \\
&\quad \cdot \sum_n \frac{\ddot{W}_n(-i)^{n+1} \cos n\phi}{H_n'(ka \sin\theta)}, \qquad kR \gg n^2 + 1.
\end{aligned}
\tag{6.58}
$$

When $k_m < k$, the denominator displays a root at

$$
\theta_m = \cos^{-1}\frac{k_m}{k},
$$

and at $(\pi - \theta_m)$. Here θ_m is the coincidence angle that was discussed at length in section 5.9. The corresponding pressure is obtained in the manner of equation (5.49c), by means of l'Hospital's rule. Using these results and replacing $k \sin \theta_m$ by $(k^2 - k_m^2)^{1/2}$, equation (6.58) yields the pressure on the coincidence cone:

$$p(R, \theta_m, \phi) = \frac{\rho L e^{ikR}}{\pi (k^2 - k_m^2)^{1/2} R} \sum_n \frac{\ddot{W}_n (-i)^{n-1} \cos n\phi}{H_n' [(k^2 - k_m^2)^{1/2} a]}. \tag{6.59}$$

The coincidence angle does not locate the pressure maximum precisely, because it does not account for the θ dependence of H_n' in the denominator of equation (6.58). The discrepancy is important when H_n' takes on large values as $\theta \to \theta_m$ for $n > 0$. From equation (6.54), this is seen to occur when $\theta_m \ll 1$, i.e., when $k \approx k_m$. The dependence of individual terms on the wave number $(k^2 - k_m^2)^{1/2}$ can be formulated in terms of the square root of the corresponding modal resistance, equation (6.46a).

To explore the value taken by the solution when $ka \sin \theta \to 0$, the transform in equation (6.57b) is substituted in equation (6.54):

$$p(R, \theta, \phi) = \frac{-\rho a e^{ikR} k_m (-1)^m \cos(kL \cos \theta)}{R(k_m^2 - k^2 \cos^2 \theta)}$$
$$\cdot \sum_{n=0} \frac{\ddot{W}_n \varepsilon_n (-ika \sin \theta)^n \cos n\phi}{n! 2^n}, \qquad (ka \sin \theta)^2 \ll 1. \tag{6.60}$$

For comparable coefficients $\ddot{W}_n$, the $(n = 0)$ term is seen to dominate the pressure field. In this long-wavelength limit, the k^2 term in the denominator can be ignored. Consequently, the θ dependence is entirely embodied in the $\cos(kL \cos \theta)$ term. It may be recalled that this θ dependence is characteristic of the pressure field of two point sources radiating in phase and separated by a distance $2L$, equation (3.7). Hence, in the long-wavelength limit, the far field can be associated with ring sources located at the edges of the radiating surfaces. The physical interpretation of this result parallels the discussion of a similar result obtained for plane radiators in section 5.9, whereby only peripherally located elements contribute to the far-field.

Whatever k_m/k, the θ dependence of the pressure displays a maximum at $\theta = \pi/2$. The corresponding pressure is

$$p\left(R, \theta = \frac{\pi}{2}, \phi\right) = \frac{2\rho e^{ikR} (-1)^m}{\pi k k_m R} \sum_n \frac{\ddot{W}_n (-i)^{n+1} \cos n\phi}{H_n' (ka)}. \tag{6.61}$$

We now consider a configuration that is odd in z, the boundary conditions being still those of a shell simply supported at $z = \pm L$:

$$f_m(z) = \sin k_m z, \qquad |z| \leq L$$
$$= 0, \qquad\qquad |z| > L, \tag{6.62}$$

with

$$k_m = \frac{m\pi}{L}, \qquad m = 1, 2, \ldots,$$

$$\sin k_m L = 0, \qquad \cos k_m L = (-1)^m.$$

The transform $\tilde{f}_m(\gamma)$ reduces to a sine transform

$$\tilde{f}_m(\gamma) = -2i \int_0^L \sin k_m z \sin \gamma z \, dz$$

$$= 2ik_m(-1)^m \frac{\sin \gamma L}{(k_m^2 - \gamma^2)}$$

$$= -iL, \qquad k_m = \gamma.$$

When this is substituted in equation (6.53), with $\bar{\gamma} = k \cos \theta$, the pressure field becomes

$$p(R, \theta, \phi) = \frac{2\rho e^{ikR} k_m(-1)^m \sin(kL \cos \theta)}{\pi k R \sin \theta (k_m^2 - k^2 \cos^2 \theta)} \sum_{n=0}^{\infty} \frac{\ddot{W}_n(-i)^n \cos n\phi}{H_n'(ka \sin \theta)}. \tag{6.63}$$

For $k_m < k$, coincidence peaks occur at the same angles as for equation (6.58).

In the small-ka or small-θ limits, an asymptotic expression can be constructed:

$$\left.
\begin{aligned}
p(R, \theta, \phi) &= \frac{\rho k_m a e^{ikR}(-1)^m \sin(kL \cos \theta)}{R(k_m^2 - k^2 \cos^2 \theta)} \\
&\quad \cdot \sum_{n=0}^{\infty} \frac{\varepsilon_n \ddot{W}_n(-i)^{n+1}(ka \sin \theta)^n \cos n\phi}{2^n n!}
\end{aligned}
\right\}
\quad (ka \sin \theta)^2 \ll 1.
\tag{6.64}$$

The $k^2 \cos^2 \theta$ term in the denominator of the coefficient multiplying this

series need be retained only if the asymptotic limit is approached by letting $\theta \to 0$, and can be dropped when $ka \to 0$. Comparison with equation (3.9) indicates that the θ-dependent factor of the dominant ($n = 0$) term is of a form associated with the pressure field of two out-of-phase ring sources located at $z = \pm L$. For the ($n = 0$) wave harmonic, the maximum occurs when the argument of the $\sin(kL\cos\theta)$ term is $(\pi/2)(2q + 1)$, or $\theta = \cos^{-1} \pm (\pi/2kL)(2q + 1)$, where $q = 0, 1, \ldots$, and assuming $kL > \pi/2$.

6.13 Comparison of Planar and Cylindrical Standing-Wave Radiators; Specific Acoustic Resistance

In the small-ka limit, where the pressure field is given by equations (6.60) and (6.64), for commensurate coefficients $\ddot{W}_n$, the pressure amplitude of a wave harmonic of order $n \geq 1$ compares with the ($n = 0$) term as $(ka \sin\theta)^n/2^n n!$. This ratio is much smaller than unity for $ka \sin\theta \ll 1$. The prevalence of the ($n = 0$) term results from the fact that it does not display any circumferential phase cancellation, however small ka. For $ka > 1/2$, this in-phase region exceeds one-half acoustic wavelength. If this condition occurs when $k_m > k$, the $n = 0$ term is the equivalent of an edge mode of the planar standing-wave radiator. In fact, edge-type radiation, embodying cancellation in the axial direction only, is displayed by higher modes of order $n < ka$, the ratio (n/a) being analogous to the wave number k_n of the planar standing-wave radiator. The series in equation (6.63) does not begin to converge until n exceeds ka, that is, until all the edge modes have been accounted for. When $n > (ka)^2/2$, the wave-harmonic terms take the asymptotic form shown in equation (6.64), the modal radiation characteristics being those of a corner mode of a planar rectangular radiator, with cancellation occurring in both the axial and circumferential directions. The series in equations (6.58) and (6.63) can, in practice, be truncated when $n > 2ka$.

For $k_m < k$, the $n = 0$ mode in the frequency range $\frac{1}{2} < ka$, and other modes in the range $n < ka$, radiate with the effectiveness of surface modes, there being no cancellation in either the circumferential or axial directions. Under these conditions, the resistance can be approximated as though the standing-wave pattern were extended to infinity and equations (6.46) can be used. A detailed theoretical description and results of an experimental study of the radiation characteristics of various dynamic configurations are available in the literature.[16]

As in the case of planar standing-wave radiators, the space-averaged specific acoustic impedance of edge modes ($k_m > k$) can be evaluated analytically in the long-wavelength limit by integrating the far-field intensity.

It was already pointed out, in connection with equations (6.60) and (6.64), that in this low-frequency range, and assuming commensurate coefficients $\ddot{W}_n$, the only significant term is the $(n = 0)$ wave harmonic. Starting with standing-wave configurations even in z, we compute the power by substituting equation (6.60) in equation (3.27). For the axisymmetric $(n = 0)$ term, the ϕ integration reduces to multiplication by 2π. The power takes the form

$$\Pi = \frac{1}{2}\rho c \left(\dot{W}_0 \frac{ka}{k_m} \right)^2 I, \qquad k_m^2 \gg k^2, \; k^2 a^2 \ll 1, \tag{6.65}$$

where

$$\begin{aligned} I &= 2\pi \int_0^\pi \cos^2(kL\cos\theta)\sin\theta \, d\theta \\ &= \pi \int_0^\pi [1 + \cos(2kL\cos\theta)]\sin\theta \, d\theta \\ &= 2\pi[1 + j_0(2kL)]; \end{aligned} \tag{6.66}$$

here use has been made of the integral representation of the spherical Bessel function.[17] Alternatively, for the $(n = 0)$ mode, this power can be expressed in terms of the specific acoustic resistance and a velocity squared, both averaged over the length $2L$.

$$\begin{aligned} \Pi &= \tfrac{1}{2}\langle r_{m0}\rangle \dot{W}_0^2 (2\pi a) \int_{-L}^{L} \cos^2 k_m z \, dz \\ &= \pi a L \langle r_{m0}\rangle \dot{W}_0^2. \end{aligned} \tag{6.67}$$

When this is equated to the former integral representation of the power, we can solve for the specific acoustic resistance

$$\left. \begin{aligned} \langle r_{m0}\rangle &\approx \rho c \left(\frac{k}{k_m}\right)^2 \left(\frac{a}{L}\right)[1 + j_0(2kL)] \\ &\approx 2\rho c \left(\frac{k}{k_m}\right)^2 \left(\frac{a}{L}\right), \qquad (kL)^2 \ll 1 \\ &\approx \rho c \left(\frac{k}{k_m}\right)^2 \left(\frac{a}{L}\right), \qquad (kL)^{1/2} \gg 1 \end{aligned} \right\} \begin{aligned} &(ka)^2 \ll 1, \\ &k^2 \ll k_m^2. \end{aligned} \tag{6.68}$$

If $\ddot{W}_0 = 0$, a procedure for evaluating the acoustic power radiated by one of the higher harmonics is required. For $n \geq 1$ the specific acoustic resistance $\langle r_{mn} \rangle$ is expressed in terms of integrals of the form

$$\int_0^\pi (\sin \theta)^{2n+1} [1 + \cos (kL \cos \theta)] \, d\theta.$$

These definite integrals are available in standard tables.[18]

For odd configurations whose z dependence is given by equation (6.62), the integral in equation (6.66) is constructed from the $n = 0$ term in equation (6.64), with $k^2 \ll k_m^2$:

$$I = 2\pi \int_0^\pi \sin^2 (kL \cos \theta) \sin \theta \, d\theta$$

$$= \pi \int_0^\pi [1 - \cos (2kL \cos \theta)] \sin \theta \, d\theta.$$

(6.69)

Except for the sign of the trigonometric function, this integral is identical to the one in equation (6.66). The value of the integral is therefore constructed by using a minus in lieu of the plus sign. The space-averaged resistance thus becomes

$$\langle r_{m0} \rangle \approx \rho c \left(\frac{k}{k_m} \right)^2 \left(\frac{a}{L} \right) [1 - j_0 (2kL)], \qquad \left(\frac{k}{k_m} \right)^2, \ (ka)^2 \ll 1. \qquad (6.70a)$$

The large-kL value is the same as for the even configuration. The small-kL value is obtained from the small-argument asymptotic expression, equation (2.45). When this is substituted in equation (6.70a), the resistance becomes

$$\langle r_{m0} \rangle = \frac{2\rho c k^4 a L}{3 k_m^2}, \qquad \left(\frac{k}{k_m} \right)^2, \ (ka)^2, \ (kL)^2 \ll 1. \qquad (6.70b)$$

A situation relevant to the $(n = 0)$ and the higher $(n = 1)$ resonances of submerged cylindrical shells involves cancellation in the axial direction $(k_m > k)$, but not circumferentially $(ka > 2)$. An analytical expression for the radiation resistance of these modes can be generated by using the large-ka asymptotic expression for $|H_n'| \approx |H_n|$ in equation (6.50). The resistance now becomes, for even modes,

$$\langle r_{mn} \rangle = \frac{\rho c k}{\pi^2 k_m^2 L} \, \mathrm{I}.$$

Here,

$$\mathrm{I} = 2\pi \int_0^\pi \cos^2 (kL \cos \theta) \, d\theta$$

$$= \pi^2 [1 + J_0(2kL)],$$

where the integral was evaluated in the manner of equation (5.74). Hence

$$\langle r_{mn} \rangle = \frac{\rho c k}{k_m^2 L} [1 + J_0(2kL)], \qquad k_m^2 \gg k^2, \quad ka \gg n^2 + 1. \tag{6.70c}$$

The large- and small-kL limits are obtained as in equation (6.68). The resistance corresponding to an odd mode is obtained, by replacing the plus sign by a minus sign [see equation (5.77)].

Having discussed the spherical and cylindrical radiators in detail, we shall formulate some features of the spatial configuration of a pressure field radiated by convex boundaries in general.

6.14 Nodal Planes and Acoustic Intensity in a Three-Dimensional Pressure Field

Consider the sound field derived earlier for a spherical source, equations (6.19) and (6.21). This solution is in the form of a double series of wave harmonics. Each wave harmonic is a *complex* function of the field-point position-vector **R**. A complex function of a real variable displays real roots if its real and imaginary components have coincident zeros. In this case a common real or imaginary factor displaying real roots can be factored out. Thus, in equation (6.19), the two functions of the surface coordinates θ and ϕ are real. For $m, n \neq 0$, they display real roots that define surfaces on which the sound pressure vanishes. Thus the function $\cos m\phi$ forms nodal planes defined by

$$\phi = \frac{\pi}{2m} (2q + 1), \qquad q = 0, \ldots, 2m - 1.$$

No such planes exist for $m = 0$. However, the corresponding surface har-

monic, $P_n(\eta)$, where $\eta \equiv \cos\theta$, defines nodal surfaces. The surface harmonic of order one, $P_1 = \eta$, defines a nodal surface that coincides with the equatorial plane $\theta = \pi/2$. The surface harmonic of order two,

$$P_2(\eta) = \tfrac{1}{4}(3\eta^2 - 1),$$

has roots $\eta = \pm(1/3)^{1/2}$, which define the conical surface $\theta \approx 55°$ (or $\theta \approx 180° - 55°$). Only the $(n = 0, m = 0)$ harmonic characterized by $P_0(\eta) \equiv 1$ does not display nodal surfaces. This is to be expected from the fact that this mode represents outgoing waves that are uniform over spherical surfaces concentric with the radiating sphere. The corresponding pressure can therefore not vanish for any combination of θ and ϕ unless it is zero everywhere. Let us now consider the flow of acoustic energy associated with the individual wave harmonics.

Whether they display nulls or not, the wave harmonics in equation (6.19) have a pressure gradient whose θ and ϕ components,

$$\frac{1}{R}\frac{\partial p}{\partial \theta} = \frac{\partial P_n^m(\cos\theta)/\partial\theta}{R P_n^m(\cos\theta)} p(R, \theta, \phi), \tag{6.71a}$$

$$\frac{1}{R\sin\theta}\frac{\partial p}{\partial \phi} = \frac{m\sin m\phi}{R\cos m\phi \sin\theta} p(R, \theta, \phi), \tag{6.71b}$$

are in phase or 180° out of phase with the pressure, since the ratios multiplying the pressure are real quantities. Referring to the relation between the pressure gradient and the velocity vector given by equation (2.28), it is concluded that the θ and ϕ components of the velocity are in phase quadrature with the pressure. The radial component of the pressure gradient,

$$\frac{\partial p}{\partial R} = \frac{k h_n'(kR)}{h_n(kR)} p(R, \theta, \phi), \tag{6.72}$$

displays a kR-dependent phase with the pressure because the ratio (h_n'/h_n) is complex. Consequently, by virture of equation (2.28), the radial velocity has a component in phase with pressure. The intensity vector (see section 3.8) therefore has a radial component but no tangential components.

The two tangential components of the particle velocity can be interpreted as standing waves entailing no power flow, while the radial com-

ponent is in the form of a traveling wave associated with energy radiation. The standing-wave characteristics of the spherical surface harmonics, their discrete wave-number spectrum, and their orthogonality, equation (6.17), suggest their interpretation as the *modes of propagation* of waveguide theory that permit energy flow only in the direction of the infinite dimension of the waveguide. More generally, whenever the boundary is compatible with a formulation of the pressure field as a series of products of a real surface harmonic multiplying a complex radial function, equation (4.24), the intensity vector is normal to the radiating surface.

There are two types of pressure fields representable by a wave-harmonic series for which the amplitude of this intensity vector vanishes. One of the two types is obtained when an infinite radiator—for example, a cylindrical or planar boundary—displays a spatially periodic acceleration distribution characterized by a structural wave number k_m that exceeds the acoustic wave number k. In this situation, the pressure field was found to consist entirely of tangential standing waves attenuated exponentially in the direction normal to the boundary. This is illustrated for the cylinder in figure 6.7, in the sketch labelled $k < k_m$. The frequency below which this inequality obtains corresponds to the cutoff frequency of waveguide theory.

A second type of pressure field divorced from power flow arises when the vibrating boundary encloses a finite fluid volume, such as a fluid-filled sphere. The $(n = 0)$ spherically symmetric pressure field of this system was formulated in equation (2.46). When the prescribed velocity distribution, instead of being uniform over the spherical surface, is in the form of the θ- and ϕ-dependent series in equation (6.4), the solution can still be constructed by the procedure that led to equation (2.46). The pressure field in the enclosed fluid sphere is of the form

$$p(r, \theta, \phi) = -\rho c k \sum_{n=0}^{\infty} \sum_{m=0}^{n} \dot{W}_{mn} P_n^m(\cos \theta) \cos m\phi \frac{j_n(kR)}{j_n'(ka)}. \tag{6.73}$$

The θ and ϕ components of the pressure gradient are still given by equations (6.71a) and (6.71b). The R component differs from equation (6.72) in that it is proportional to the ratio $j_n(kR)/j_n'(ka)$, which is a real function. All three components of the particle velocity vector are therefore in phase quadrature with the pressure. Consequently, the pressure field is not associated with an intensity vector. The field consists of three-dimensional standing waves, which include nodal surfaces in the form of planes, the roots of $\cos m\phi$; cones, the roots of $P_n^m(\cos \theta)$; and spheres, the roots of $j_n(kR)$.

When the radiator configuration does not permit a solution in the form of a wave-harmonic series—for example, for two sources vibrating in phase, equation (3.7)—the velocity, equation (3.8), has a tangential component associated with acoustic power cyclically circulating in alternate directions around the source. Only in the far-field does the intensity vector tend to its radial component, which alone is associated with radiated power.

For the infinite cylinder vibrating over a finite length, as for other boundary configurations compatible with separation of the wave equation, but displaying an acceleration distribution that is spatially nonperiodic, the pressure field cannot be formulated as a wave-harmonic series but requires an integral representation. These acceleration distributions entail acoustic power flowing tangentially over the radiating surfaces. Such a pressure field precludes the existence of normal modes of vibration in a finite cylindrical shell as it does in finite planar radiators. The existence of a tangential component of the intensity vector that gives rise to intermodal coupling can be anticipated from the fact that, for corner or edge modes of finite cylindrical or planar radiators in the frequency range $(k_m > k)$, acoustic radiation is associated with the peripheral regions of the standing-wave field of the acceleration distribution. The periphery of the vibrating surface thus acts as an energy sink absorbing energy from the entire vibrating surface and reradiating it as acoustic energy.

When the boundary itself, as opposed to the acceleration distribution, is of finite extent—for example, for a finite cylinder not bracketed between semiinfinite cylindrical baffles—the pressure field is necessarily spatially periodic over a surface enclosing or coinciding with the radiator. However, unless the surface geometry is compatible with a wave-harmonic series representation of the pressure field, the tangential velocity component will again be found to be associated with acoustic power flow parallel to the radiating surface. This configuration is less tractable than that of an infinite cylindrical boundary displaying a spatially nonperiodic acceleration distribution, because even the far-field cannot be represented explicitly. The finite cylindrical radiator, like other boundaries incompatible with separation of the wave equation, must be analyzed by numerical methods, or by the asymptotic approaches described in section 4.6. Fortunately, for practical purposes, when a finite cylinder has stationary ends, thus radiating with its cylindrical surface only, its far-field pressure can be approximated by the pressure field of an infinite cylindrical boundary whose active portion coincides with the finite cylinder.[19]

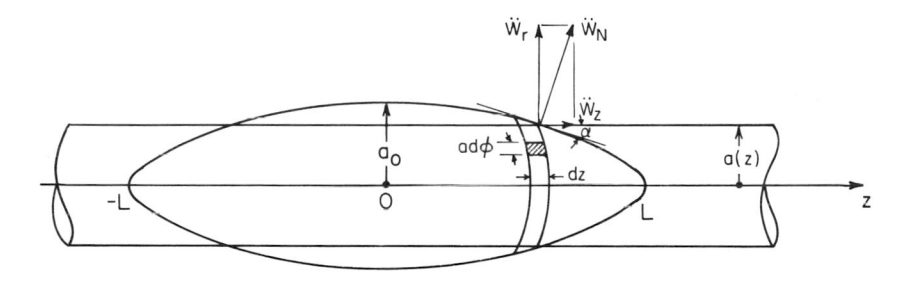

Figure 6.8
Mathematical model of a body of revolution used in the cylindrical-baffle asymptotic analysis. A differential area element at cross-section z is modeled as a source on an infinite cylindrical baffle of radius $a(z)$.

6.15 Far-Field of Slender Bodies of Revolution

The only finite, analytically tractable source geometries are ellipsoids and their degenerate forms, spheroids and spheres. Spheroidal wave harmonics, being rather inconvenient, will not be covered here. The interested reader is referred to the literature.[20,21,22] Instead, an asymptotic theory will be developed for slender bodies of revolution whose maximum radius a_0 is small in terms of wavelengths, no restriction being placed on kL. For the sake of conciseness, instead of evaluating the Helmholtz integral asymptotically, the development will be carried out as a heuristic extension of the theory of cylindrical radiators (sections 6.10–6.12). The underlying approximation is that each length element dz of the source radiates as though it were bracketed by two semiinfinite cylinders of the same radius $a(z)$ (see figure 6.8). It will be verified that as long as the radiator radius changes slowly, the far-field is adequately predicted by this formulation, even on the axis of symmetry. The inactive cylindrical baffles' relative unimportance is apparent from the similarity of the far-fields of baffled cylinders and of paraboloids near their respective axes.[23] It is further borne out by an observation made earlier, viz., that the infinite cylinder provides a basis for estimating the radiation loading of the equivalent finite cylinder.[19]

　　The source's dynamic configuration is described in cylindrical coordinates, equation (6.1). Since $f_m(z)\exp(-i\bar\gamma z)$ is effectively constant over the infinitesimal length dz, the far-field contribution of a differential area element is obtained by substituting

$$d\tilde{f}_m(\bar\gamma) \equiv f_n(z)\,e^{-i\bar\gamma z}\,dz, \qquad \bar\gamma = k\cos\theta \tag{6.74}$$

in equation (6.55) and integrating these differential contributions over the length of the body of revolution:

$$p(R, \theta, \phi)$$

$$\simeq \frac{\rho e^{ikR}}{2R} \sum_{n=0} \frac{\varepsilon_n \ddot{W}_n (-ik \sin \theta)^n \cos n\phi}{2^n n!}$$

$$\cdot \int_{-L}^{L} f_m(z) a^{n+1} \exp(-ikz \cos \theta) \, dz,$$

$$(ka_0 \sin \theta)^2 \ll 2n + 1.$$

(6.75)

On the axis, only the $n = 0$ term contributes. The results thus generated are consistent with a low-frequency slender-body theory for longitudinally vibrating bodies of revolution[24] and more general asymptotic theories,[25,26] effectively expansions of the solution of the Helmholtz integral in negative powers of sound velocity, c^{-n}, the zero-order term being the solution of the Laplace equation. The solutions are formulated in terms of the distributed inertia force of the entrained mass of fluid, which can be envisioned as the contribution of the surface pressure to the Helmholtz integral.

The z dependence of the effective radial cylindrical acceleration in equation (6.75) is selected so that the volume acceleration of an infinitesimal area element of the cylinder matches that of the body of revolution. If the latter is expressed in terms of its radial and axial components, respectively, $\ddot{W}_{rn} \cos n\phi$ and $\ddot{W}_{zn} \cos n\phi$, the volume acceleration is

$$d\ddot{Q}_n(z, \phi) = [\ddot{W}_{rn} - \ddot{W}_{zn}(da/dz)] a \cos n\phi \, d\phi \, dz.$$

(6.76)

The minus sign is required because positive longitudinal accelerations, pointing by definition in the positive z direction, contribute to the volume acceleration if the radiator cross section decreases with increasing z, i.e., if $da/dz < 0$. Comparison with (6.1) yields the equivalent cylindrical acceleration:

$$\ddot{W}_n f_m(z) = \ddot{W}_{rn} - \ddot{W}_{zn}(da/dz).$$

(6.77)

If the dynamic configuration is stated in terms of the acceleration $\ddot{W}_{Nn} \cos \phi$ normal to the radiating surface, the acceleration components are

$$\ddot{W}_{rn} = \ddot{W}_{Nn}\cos\alpha,$$
$$\ddot{W}_{zn} = \ddot{W}_{Nn}\sin\alpha,$$

where α is the angle between the normal and the z axis (figure 6.8). The trigonometric functions can be expressed in terms of $\tan\alpha$, which equals da/dz:

$$\cos\alpha = [1 + (da/dz)^2]^{-1/2},$$
$$\sin\alpha = (da/dz)[1 + (da/dz)^2]^{-1/2}.$$

When these equations are combined with (6.77) the equivalent cylindrical acceleration is obtained in the form

$$\ddot{W}_n f_m(z) = [1 + (da/dz)^2]^{1/2}\ddot{W}_{Nn}. \tag{6.78}$$

These results will now be illustrated for the prolate spheroid defined in terms of the dimensionless axial coordinate $\zeta \equiv z/L$:

$$a(\zeta) = a_0(1 - \zeta^2)^{1/2},$$
$$da/dz = -(a_0/L)\zeta(1 - \zeta^2)^{-1/2}. \tag{6.79}$$

For the purpose of obtaining a comparison with the prolate spheroidal wave-harmonic solution, a dynamic configuration studied by Chertock[20] is selected, viz., rigid-body vibration along the axis of symmetry. The acceleration is axial and z independent. Setting $\ddot{W}_{r0} = 0$ and $\ddot{W}_{z0} = \ddot{W}$, and substituting the latter in (6.77), the equivalent cylindrical acceleration in (6.78) becomes

$$\ddot{W}_0 f_0(\zeta) = \ddot{W}(a_0/L)\zeta(1 - \zeta^2)^{-1/2}.$$

With this result and the radius in the former equations (6.79), the $n = 0$ term in (6.75) becomes

$$p_0(R,\theta) = \frac{-i\rho e^{ikR}a_0^2\ddot{W}}{R}\int_0^1 \zeta\sin(\zeta kL\cos\theta)\,d\zeta.$$

The integral can be evaluated analytically by parts,

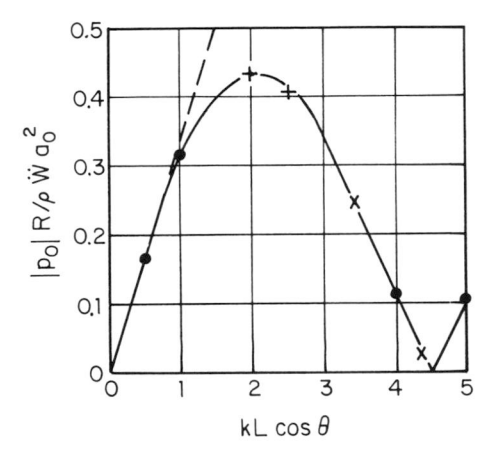

Figure 6.9
Sound pressure radiated by a prolate spheroid undergoing rigid-body axial vibration. The solid curve is the small-ka asymptotic solution in equation (6.80); Chertock's prolate spheroidal wave-harmonic results[20] are represented by discrete points, where 0, x, and + stand, respectively, for $\theta = 0°$, $30°$, and $60°$; the dotted line is the asymptotic small-kL and small-ka solution, equation (6.82). [Reproduced from Strasberg[24]]

$$\int_0^1 \zeta \sin (\zeta x) \, d\zeta = x^{-2} \sin x - x^{-1} \cos x$$

$$\equiv j_1 (x)$$

$$\simeq x/3, \qquad x^3/30 \ll 1,$$

where use has been made of (2.45). The sound pressure finally becomes

$$p_0 (R, \theta) = -i\rho e^{ikR} (a_0^2/R) \, \ddot{W} j_1 (kL \cos \theta). \tag{6.80}$$

This coincides with Strasberg's solution,[24] which is compared with the wave-harmonic results[20] in figure 6.9.

The small-argument asymptotic solution applicable when $\theta \to \pi/2$ or when the length is small in terms of wavelengths can be formulated concisely in terms of the spheroid's volume:

$$V = 4\pi a_0^2 L/3. \tag{6.81}$$

Using the small-argument expression for j_1, the resulting dipole pressure field is

$$p_0(R, \theta) = -i\rho e^{ikR} k V \ddot{W} \cos \theta / 4\pi R, \qquad k^2 L^2 \cos^2 \theta \ll 1. \tag{6.82}$$

Now consider a spheroid undergoing rigid-body vibrations in a direction normal to the axis of symmetry. The acceleration components now become $\ddot{W}_{z1} = 0$, $\ddot{W}_{r1} = \ddot{W}$, a constant; $\sin\alpha/f_1(\zeta) = 1$, equation (6.77), reduces to $\ddot{W}$. The $n = 1$ term in equation (6.75) yields

$$p_1(R, \theta, \phi) = \frac{-i\rho e^{ikR} k L a_0^2 \ddot{W}}{R} \sin\theta\cos\phi \int_0^1 (1 - \zeta^2) \cos(\zeta k L \cos\theta)\, d\zeta. \tag{6.83}$$

The integral can be evaluated analytically:[27]

$$\int_0^1 (1 - \zeta^2) \cos(\zeta x)\, d\zeta = 2j_1(x)/x \tag{6.84}$$

$$\simeq 2/3, \qquad x^2 \ll 1.$$

The pressure finally becomes

$$p_1 = \frac{-2i\rho e^{ikR} \ddot{W} a_0^2}{R} j_1(kL\cos\theta) \tan\theta\cos\phi \tag{6.85}$$

$$\simeq 2p_0(R, \theta), \qquad k^2 L^2 \cos^2\theta \ll 1,$$

where p_0 is given in (6.82) with $\cos\theta$ set equal to $\sin\theta\cos\phi$ to account for the 90° rotation of the θ axis.

To explore further the significance of the factor-two discrepancy between the maximum values of p_1 and p_0, these results will be compared with the pressure radiated by the translational mode of the sphere, where the distinction between major and minor axes does not exist. The $(0, 1)$ term in (6.24) yields $1.5p_0$, a result intermediate between equations (6.82) and (6.85). The numerical coefficients in these equations can be interpreted in terms of the entrained mass of fluid associated with the reciprocating motion of the source. If the entrained mass is expressed as $\gamma\rho V$, where ρV is the mass of displaced fluid, then $\gamma = 1/2$ for the sphere, equation (6.33); $\gamma \simeq 0$ for longitudinal vibrations of the slender prolate spheroid;[28] and $\gamma \simeq 1$ for its transverse vibrations.[29] One thus arrives at Strasberg's formulation of sound fields radiated by translational vibrations of bodies small in terms of wave-lengths:[25]

$$p(R,\theta) = -i\rho kV(1 + \gamma)\ddot{W}e^{ikR}\cos\theta/4\pi R, \qquad (kV^{1/3}\cos\theta)^2 \ll 1. \qquad (6.86)$$

This low-frequency acoustic solution merely requires the solution of the Laplace, rather than the Helmholtz, equation to obtain γ. A related result will be constructed in the discussion of small movable scatterers (section 11.9).

References

1. See A. Sommerfeld, *Partial Differential Equations in Physics* (New York: Academic Press, 1949), pp. 124–129, for a concise theory of these functions. Explicit polynominal expressions are given by M. Abramowitz and I. A. Stegun, *Handbook of Mathematical Functions* (Washington, D.C.: NBS Supt. of Doc., 1969), p. 303.

2. This definition is used by Sommerfeld, *Partial Differential Equations*, p. 128; P. M. Morse and H. Feshbach, *Methods of Theoretical Physics, Part I* (New York: McGraw-Hill, 1953), p. 549; and Jahnke and Emde, *Table of Functions*, p. 110. Other authors—for example, I. N. Sneddon, *Special Functions of Mathematical Physics and Chemistry* (New York: Interscience Publishers, 1956), p. 74, and Abramowitz and Stegun, *Handbook*—give a definition whereby the derivative in the foregoing equation is multiplied by $(\eta^2 - 1)^{m/2}$. This latter definition requires a factor of $(-1)^m$ in (6.9) and (6.10).

3. Abramowitz and Stegun, *Handbook*, p. 440, formulas (10.1.45) and (10.1.46); G. N. Watson, *A Treatise on the Theory of Bessel Functions*, 2nd ed. (Cambridge, U.K.: Cambridge University Press, 1952), p. 366.

4. Sommerfeld, *Partial Differential Equations*, p. 134, or Sneddon, *Special Functions*, p. 80. The $(-1)^m$ factor given by the latter author must be deleted if one uses our definition of P_n^m, equation (6.5).

5. Abramowitz and Stegun, *Handbook*, pp. 437 et seq. Tables of h_n' in terms of the amplitude and phase of these functions are given by P. M. Morse and K. U. Ingard, *Theoretical Acoustics* (New York: McGraw-Hill, 1968), p. 900.

6. Abramowitz and Stegun, *Handbook*, p. 439, combining formulas (10.1.21) and (10.1.31).

7. Ibid., p. 338, formulas (8.14.11) and (8.14.13).

8. M. C. Junger, "Radiation Loading of Cylindrical and Spherical Surfaces," *J. Acoust. Soc. Am.* 24:288–289 (1952).

9. See, for example, Abramowitz and Stegun, *Handbook*, p. 358. Tables of H_n', conveniently expressed in terms of amplitude and phase angle, are found in Morse and Ingard, *Theoretical Acoustics*, p. 897.

10. Abramowitz and Stegun, *Handbook*, p. 360, formulas (9.1.8) and (9.1.9).

11. Ibid., p. 375, formula (9.6.4).

12. M. C. Junger, "The Physical Interpretation of the Expression for an Outgoing Wave in Cylindrical Coordinates," *J. Acoust. Soc. Am.* 25:40–47 (1953).

13. Abramowitz and Stegun, *Handbook*, p. 360, formula (9.1.16).

14. J. E. Greenspon and C. H. Sherman, "Mutual Radiation Impedance and Near-Field Pressure for Pistons on a Cylinder," *J. Acoust. Soc. Am.* 36:153 (1964).

15. D. T. Laird and H. Cohen, "The Directionality Patterns for Acoustic Radiation from a Source on a Rigid Cylinder," *J. Acoust. Soc. Am.* 24:46–49 (1952).

16. J. E. Manning and G. Maidanik, "Radiation Properties of Cylindrical Shells," *J. Acoust. Soc. Am.* 36:1691–1698 (1964).

17. Abramowitz and Stegun, *Handbook*, p. 438, formula (10.1.13).

18. B. O. Peirce, *A Short Table of Integrals*, 3rd ed. (Boston: Ginn, 1929), p. 62, formula (483).

19. B. L. Sandman, "Fluid Loading Influence Coefficients for a Finite Cylindrical Shell," *J. Acoust. Soc. Am.* 60:1256–1264 (1976).

20. G. Chertock, "Sound Radiation from Prolate Spheroids," *J. Acoust. Soc. Am.* 33:871–880 (1961).

21. E. Skudrzyk, *The Foundations of Acoustics* (New York: Springer, 1971), pp. 455–485.

22. Morse and Feshbach, *Methods of Theoretical Physics*, pp. 1502–1513.

23. J. E. Cole III an J. M. Garrelick, "Diffraction of Sound by an Impedance Paraboloid," *J. Acoust. Soc. Am.* 68:1193–1198 (1980).

24. M. Strasberg, "Sound Radiation from Slender Bodies in Axisymmetric Vibration," Paper O–28, 4th Intern. Cong. Acoustics, Copenhagen (Aug. 1962).

25. M. Strasberg, "Radiation from Unbaffled Bodies of Arbitrary Shape at Low Frequencies," *J. Acoust. Soc. Am.* 34:520–521 (1962).

26. G. Chertock, "Sound Radiated by Low-frequency Vibrations of Slender Bodies," *J. Acoust. Soc. Am.* 57:1007–1016 (1975).

27. B. O. Peirce, *A Short Table of Integrals*, p. 46, formula (341) and the definition of j_1 in equation (2.45).

28. H. Lamb, *Hydrodynamics*, 6th ed. (New York: Dover, 1945), p. 155, table where k_1 is our γ.

29. Ibid., where $k_2 = \gamma$; this result can also be obtained from the entrained mass per unit area of the cylinder, equation (6.44b) for $n = 1$.

7.1 Introduction

Chapters 2–6 dealt exclusively with classical acoustic problems, that is, the response of a fluid medium to prescribed velocity sources placed within or on its boundaries. The main objective of this book is to study problems of interaction between vibrating structures and the acoustic fluids in which such structures are immersed. To deal with this subject, we require not only the acoustics background, but also the in vacuo response of the structural elements that form the boundaries of the acoustic medium. In a completely rigorous analysis, such elements would be considered as three-dimensional elastic continua, and therefore capable of supporting both dilatational and distortional (shear) wave motions. However, for the purposes of this book, we consider structural configurations for which one or two characteristic dimensions are much smaller than the remaining dimensions. In such cases, assumptions are made that drastically simplify the differential equations governing the boundary motions; such equations are herein referred to as structural theories, as differentiated from elasticity or continuum theories of structural response.

The structures to be treated fall into two categories. The first set are those that before deformation have no appreciable curvature. These include rods or beams in which the dependent variables, e.g., deflection and stresses, depend on a single coordinate, as well as plates where the variation is over two spatial coordinates. Such structures will be the subject of sections 7.2–7.11. The remainder of this chapter deals with structures that are initially curved, specifically cylindrical and spherical shells.

7.2 Longitudinal Vibrations of an Elastic Bar

The simplest structural vibration problem to be considered is that of longitudinal vibrations of an elastic bar whose length is much greater than any transverse dimension. The longitudinal vibrations of such bars are not usually of themselves an important source of sound radiation, but can excite flexural waves in plates and shells that might be attached to them. These flexural waves can be efficient sound radiators. In the case of underwater transducers, the longitudinal vibrations of piezoelectric elements are themselves used as sources of sound radiation.

Consider a long thin elastic bar, of cross-sectional area S, oriented so that the x axis measures distance along the length of the bar. If, during the motion of the bar, the cross section remains plane and the stress over it is uniform, the equation of motion can be obtained by using a force balance between the elastic and inertial forces acting on a small portion of the bar.

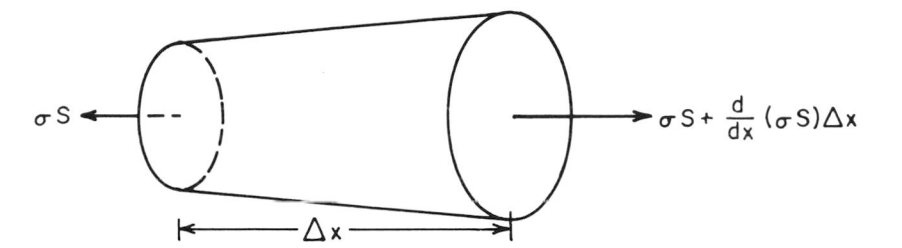

Figure 7.1
Element of a bar of length Δx showing resultant forces applied to it.

Consider a volume element of the bar of length Δx and cross-sectional area S (see figure 7.1). The tensile force that acts on the left-hand side of the bar is σS, where σ is the longitudinal stress, while the force acting on the right-hand face of the elements is $\sigma S + [\partial(\sigma S)/\partial x]\,\Delta x + \cdots$. By Newton's second law of motion, the net force acting on the element is balanced by the inertia of the element:

$$\sigma S + \frac{\partial}{\partial x}(\sigma S)\,\Delta x - \sigma S = \rho_s S \Delta x \frac{\partial^2 u}{\partial t^2}, \tag{7.1}$$

where $u(x, t)$ is the longitudinal displacement of the volume element, and ρ_s is the volume density of the bar material. Since the bar is assumed to be perfectly elastic, we can relate the stress σ to the displacement u by Hooke's law,

$$\sigma = E\frac{\partial u}{\partial x}, \tag{7.2}$$

where E is the Young's modulus of the bar material. For a general description of the stress-strain and strain-displacement relationships, the reader is referred to the book by Fung.[1] Combining equations (7.1) and (7.2) and dividing through by Δx, we obtain

$$\frac{\partial}{\partial x}\left(ES\frac{\partial u}{\partial x}\right) = \rho_s S \frac{\partial^2 u}{\partial t^2}. \tag{7.3}$$

If E and S are assumed constant throughout the length of the bar, the equation of motion governing the displacement u is of exactly the same form

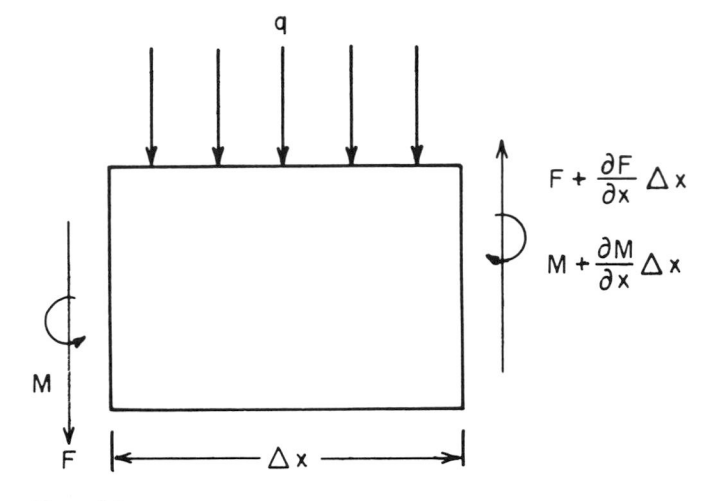

Figure 7.2
Element of a bar of length Δx with resultant forces and moments applied to it.

as that governing the propagation of one-dimensional sound waves given by equation (2.6). The velocity of propagation in this case is known as the bar velocity and is given by

$$c_b = \left(\frac{E}{\rho_s}\right)^{1/2}. \tag{7.4}$$

In order to solve vibration problems for finite length elastic bars, boundary conditions must be prescribed. In the case of a bar fixed at an end, the displacement must vanish, i.e., $u = 0$, while if the end is free, the stress must vanish, and therefore, using equation (7.2), $\partial u / \partial x = 0$.

7.3 Flexural Vibrations in an Elastic Bar

An elastic beam, in addition to its longitudinal motion, can also respond by transverse vibration, that is, motion in a direction perpendicular to its longitudinal axes. Consider a straight bar of uniform cross section whose longitudinal axis is oriented along the x axis. In deriving the equation of motion governing such motion, we shall restrict ourselves to beams whose cross sections are symmetrical about a vertical plane through the centerline (see figure 7.2). Furthermore, the loads causing the motion of the beam act in that vertical plane of symmetry. The neutral surface, that is, the surface

that during bending of the beam undergoes no extension, is perpendicular to the vertical plane of symmetry and passes through the center of gravity of the cross section. The perpendicular distance of a particle from this surface shall be denoted by z and the corresponding displacement component by w. In the elementary, Bernoulli-Euler theory of bending, cross sections initially perpendicular to the neutral surface remain so after deformation. Using this assumption, the longitudinal displacement u is related to w by the geometric relation

$$u = -z\frac{\partial w}{\partial x}. \tag{7.5}$$

The longitudinal strain $\varepsilon = \partial u/\partial x$ is then given by

$$\varepsilon = -z\frac{\partial^2 w}{\partial x^2}. \tag{7.6}$$

Using the one-dimensional Hooke's law, equation (7.2), the stress σ can be written in terms of w:

$$\sigma = -zE\frac{\partial^2 w}{\partial x^2}. \tag{7.7}$$

In this case we see that the longitudinal stress σ varies linearly across the section, whereas in the case of longitudinal vibration the corresponding stress, as given by equation (7.2), was uniform across the section.

Now consider an element of the beam of length Δx and cross-sectional area S with the resultant moments M and shear forces F acting on it as shown in figure 7.2. Neglecting the rotatory inertia of the element, we have the moment equilibrium relation

$$\frac{\partial M}{\partial x}\Delta x = F\Delta x. \tag{7.8}$$

Dividing through by Δx, the equation of motion of the element in the vertical direction becomes

$$-q + \frac{\partial F}{\partial x} = \rho_{\text{s}}S\frac{\partial^2 w}{\partial t^2}, \tag{7.9}$$

where q is the force per unit length applied to the beam. Combining equation (7.8) and equation (7.9), we have

$$\frac{\partial^2 M}{\partial x^2} = \rho_s S \frac{\partial^2 w}{\partial t^2} + q. \qquad (7.10)$$

The resultant moment M can be related to w by using equation (7.7):

$$M = \iint_S \sigma z \, dS = -E \frac{\partial^2 w}{\partial x^2} \iint_S z^2 \, dS \qquad (7.11)$$

$$= -EI \frac{\partial^2 w}{\partial x^2},$$

where $\iint z^2 \, dS = I$ is the moment of inertia of the section. Equation (7.10) then becomes

$$\frac{\partial^2}{\partial x^2} \left[EI \frac{\partial^2 w}{\partial x^2} \right] + \rho_s S \frac{\partial^2 w}{\partial t^2} = -q(x, t). \qquad (7.12a)$$

If we assume that EI, the flexural stiffness, is constant throughout the length of the beam, equation (7.12a) becomes

$$EI \frac{\partial^4 w}{\partial x^4} + \rho_s S \frac{\partial^2 w}{\partial t^2} = -q(x, t). \qquad (7.12b)$$

Equation (7.12b) can be used to find the phase velocity of waves traveling along the beam. In order to do this we seek a solution of equation (7.12b) in the absence of an applied force, $q(x, t) = 0$, in the form

$$w = W e^{i(kx - \omega t)}, \qquad (7.13)$$

where

$$\frac{\omega}{k} = V \qquad (7.14)$$

is the phase velocity. Equation (7.13) is found to be a solution of equation

(7.12) if k and ω are related to each other by the condition

$$EIk^4 - \rho_s S\omega^2 = 0. \tag{7.15}$$

Using the subscript f to denote flexural, equation (7.15) yields

$$k_f = \left(\frac{\rho_s S}{EI}\right)^{1/4} \omega^{1/2}. \tag{7.16a}$$

The corresponding phase velocity is obtained from equation (7.14):

$$V_f = \left(\frac{EI}{\rho_s S}\right)^{1/4} \omega^{1/2}. \tag{7.16b}$$

As opposed to compressional waves, the phase velocity in this case is frequency dependent. The frequency or wavelength dependence of the phase velocity is known as *dispersion*. An arbitrary disturbance can, in principle, always be written as a continuous superposition of components of sinusoidal waves each with a different wave number. Because of the frequency dependence of the phase velocity, each component travels with a different velocity. This leads to a gradually increasing distortion of the disturbance shape as it travels along the length of the beam.

7.4 Group Velocity
We can now introduce the concept of *group velocity*. Instead of assuming a solution of the form given in equation (7.13), let us assume a continuous superposition of simple harmonic waves of the form

$$w(x, t) = \int_{-\infty}^{\infty} f(k) e^{i[kx - \omega(k)t]} dk. \tag{7.17}$$

In this case, $f(k)$ represents the Fourier transform of the initial deflection. In order that this be a solution of equation (7.12), $\omega(k)$ must satisfy the relation equation (7.15). We now seek a solution that will be valid for large t. Using the method of stationary phase (see section 5.3), the solution valid for large time can be shown to be

$$w(x, t) \sim \left[\frac{2\pi}{t|\omega''(k_0)|}\right]^{1/2} f(k_0) e^{i[k_0 x - \omega(k_0)t \pm \pi/4]}, \tag{7.18}$$

where the $+$ or $-$ sign is chosen for $\omega''(k_0) > 0$ or < 0, respectively, and k_0 is that value of k for which

$$\frac{d}{dk}[kx - \omega(k)t] = x - \frac{d\omega}{dk}t = 0, \tag{7.19}$$

or

$$\frac{x}{t} = \frac{d\omega}{dk}\bigg|_{k=k_0}. \tag{7.20}$$

Therefore, for a given value of x and t, the major contribution to the response comes from a group of waves whose wave numbers are close to the value of k that satisfies equation (7.20) and have therefore traveled with a velocity

$$U = \frac{d\omega}{dk}. \tag{7.21}$$

This is the *group velocity*. Using equation (7.14), we can find a relation between the group velocity U and the phase velocity V:

$$U = V + k\frac{dV}{dk}. \tag{7.22}$$

We see from equation (7.22) that, when there is no dispersion, that is, when V is independent of the wave number k, the group and phase velocities are equal.

7.5 Rotatory Inertia and Transverse Shear Effects: Timoshenko Beam Equation

In the particular case of flexural waves, equation (7.22) yields a group velocity that is exactly twice the phase velocity V, or in terms of the flexural wavelength $\lambda_f = 2\pi/k_f$,

$$U = \frac{4\pi}{\lambda_f}\left(\frac{EI}{\rho_s S}\right)^{1/2}. \tag{7.23}$$

Equation (7.23) tells us that the extremely short-wavelength components of a flexural disturbance travel with a very high velocity; in fact, as λ_f ap-

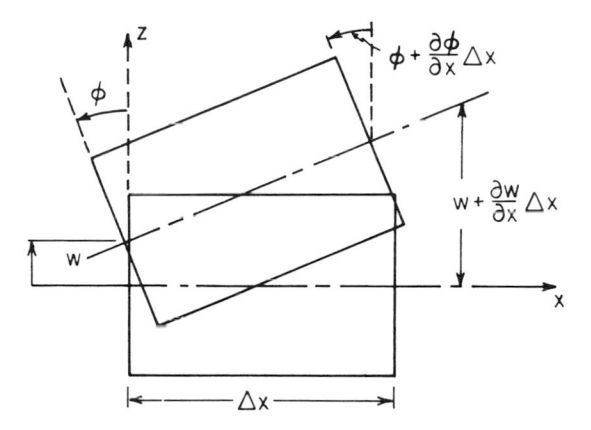

Figure 7.3
Notation used in the derivation of the Timoshenko beam equation.

proaches zero, the group velocity becomes infinite. This is physically un-
reasonable and is a result of the approximate nature of the equation of
motion, equation (7.12). Thus, in addition to the assumption made earlier
that transverse dimensions are small in terms of the beam length, we must
assume that the transverse dimensions are small in terms of a characteristic
wavelength. For cases where this assumption is not satisfied, we can correct
the difficulty as exemplified in equation (7.23) by including the effects of
rotatory inertia and transverse shear.[?] The equation of motion that includes
these effects is the Timoshenko beam equation. To do this we introduce a
new variable, ϕ, which is the angle of rotation of the section; the angular
acceleration of the section is then given by $\partial^2\phi/\partial t^2$ (see figure 7.3). Equation
(7.8) assumes equilibrium of moments, but if we include the effects of
rotatory inertia, the sum of the moments acting on the element must be
balanced by the product of the rotational inertia of the element and its
angular acceleration. Thus modified, equation (7.8) becomes, after dividing
through by Δx,

$$F - \frac{\partial M}{\partial x} = \mathcal{J}\frac{\partial^2\phi}{\partial t^2}, \tag{7.24}$$

where $\mathcal{J} = \rho_s I$ is the rotatory inertia per unit length of the beam.

The inclusion of rotatory inertia, by using equation (7.24) instead of

equation (7.8), would by itself remove the difficulty of having infinite phase velocities for infinitesimally small wavelengths. However, the finite limiting value of the group velocity obtained by including only the rotatory inertia effect is the bar velocity $c_b = (E/\rho_s)^{1/2}$, whereas the three-dimensional theory predicts the limiting value to be that of Rayleigh surface waves. This further deficiency is removed by including the effects of transverse shear. If we relax the condition that plane sections initially perpendicular to the central surface remain perpendicular to the deformed central surface after deformation, the generalization of equation (7.5) becomes

$$u = -z\phi, \tag{7.25}$$

where ϕ is no longer equal to $\partial w/\partial x$ (see figure 7.3).

Using equation (7.25), we see that the longitudinal strain $\varepsilon = \partial u/\partial x$ becomes

$$\varepsilon = -z\frac{\partial \varphi}{\partial x}, \tag{7.26}$$

while, instead of equation (7.11), we now have

$$M = -EI\frac{\partial \varphi}{\partial x}. \tag{7.27}$$

Equation (7.25) also leads to a nonvanishing shear strain

$$\gamma = \frac{\partial u}{\partial z} + \frac{\partial w}{\partial x} = -\varphi + \frac{\partial w}{\partial x}, \tag{7.28}$$

which by Hooke's law yields a shear stress

$$\tau = G\left(\frac{\partial w}{\partial x} - \varphi\right). \tag{7.29}$$

The shear modulus G is related to the Young's modulus E as $E/2(1+v)$, where v is Poisson's ratio. The shearing force F is obtained by integrating equation (7.29) over the area of the cross section:

$$F = G\kappa^2 S\left(\frac{\partial w}{\partial x} - \varphi\right), \tag{7.30}$$

where κ^2 is a factor (close to unity) introduced to account for the fact that the shear stress is not truly independent of z. Using equations (7.27) and (7.30), we can rewrite equations (7.9) and (7.24) as

$$G\kappa^2 S\left(\frac{\partial^2 w}{\partial x^2} - \frac{\partial \varphi}{\partial x}\right) = \rho_s S\frac{\partial^2 w}{\partial t^2} + q, \tag{7.31}$$

$$G\kappa^2 S\left(\frac{\partial w}{\partial x} - \varphi\right) + EI\frac{\partial^2 \varphi}{\partial x^2} = \rho_s I\frac{\partial^2 \varphi}{\partial t^2}. \tag{7.32}$$

where we have replaced $\mathcal{J}$ by $\rho_s I$. Eliminating ϕ from equations (7.31) and (7.32), we finally arrive at an equation for w, which includes the effects of rotatory inertia and transverse shear:

$$EI\frac{\partial^4 w}{\partial x^4} + \rho_s S\frac{\partial^2 w}{\partial t^2} - \left(\rho_s I + \frac{EI\rho_s}{\kappa^2 G}\right)\frac{\partial^4 w}{\partial x^2 \partial t^2} + \rho_s I\frac{\rho_s}{G\kappa^2}\frac{\partial^4 w}{\partial t^4}$$
$$= -q + \frac{EI}{G\kappa^2 S}\frac{\partial^2 q}{\partial x^2} - \frac{\rho_s I}{G\kappa^2 S}\frac{\partial^2 q}{\partial t^2}. \tag{7.33}$$

In order that equation (7.33) reduce to that of the classical beam equation given by equation (7.12b), we must assume that $\mathcal{J} = \rho_s I$ vanishes and that the quantity $G\kappa^2 S$ becomes increasingly large so that the section shear force F in equation (7.30) is finite even though $\phi = \partial w/\partial x$; that is, we neglect both the rotatory inertia and shear deformation of each beam element.

To find an expression for the phase velocity, we again insert a solution of the form $w = We^{i(kx-\omega t)}$, assume $q(x, t) = 0$, and obtain an equation for $V = \omega/k$:

$$\frac{I}{S}\frac{\omega^2}{V^2}\left(\frac{V^2}{\kappa^2 c_s^2} - 1\right)\left(1 - \frac{c_b^2}{V^2}\right) = 1, \tag{7.34}$$

where $c_s = (G/\rho_s)^{1/2}$ and $c_b = (E/\rho_s)^{1/2}$. As equation (7.34) is quadratic in V^2, there are two roots. If we consider the smaller of the two, and take the limiting case of $\omega \to 0$, we arrive at equation (7.16), while the high-frequency

limit of equation (7.34) is

$$V = \kappa c_s. \tag{7.35}$$

According to the exact theory, this should be the velocity of Rayleigh waves c_R. Therefore if we take $\kappa = c_R/c_s$ for the particular material being considered, the equation of motion, equation (7.33), predicts the proper limiting phase velocity. For steel, this ratio is 0.925.

7.6 Forced Vibrations of an Infinite Elastic Beam

Let us now use equation (7.12b) to derive the flexural response of an infinite beam subjected to a point force $Fe^{-i\omega t}$. Without loss of generality, we can assume that the force is applied at the origin. For harmonic time variation, equation (7.12b) becomes

$$EI\frac{d^4w}{dx^4} - \rho_s S\omega^2 w = F\delta(x), \tag{7.36}$$

where $w(x)$ now represents the spatial dependence of the displacement $w(x)e^{-i\omega t}$. Since the beam is assumed infinite, an appropriate method of solution is to transform equation (7.36) by use of a complex Fourier transform as defined by the one-dimensional analogue of equation (5.28a). We then find $\tilde{w}(\gamma)$, the Fourier transform of $w(x)$:

$$\tilde{w}(\gamma) = \frac{F}{EI(\gamma^4 - k_f^4)}, \tag{7.37}$$

where the flexural wave number k_f is given in equation (7.16a). The response $w(x)$ is obtained by using the inversion formula

$$
\begin{aligned}
w(x) &= \frac{1}{2\pi}\int_{-\infty}^{\infty} e^{i\gamma x}\tilde{w}(\gamma)\,d\gamma \\
&= \frac{F}{2\pi EI}\int_{-\infty}^{\infty}\frac{e^{i\gamma x}}{\gamma^4 - k_f^4}\,d\gamma.
\end{aligned}
\tag{7.38}
$$

The simplest way of evaluating equation (7.38) is by integration in the complex plane. The integrand has poles at $\gamma = \pm k_f$, $\pm ik_f$ and can be evaluated using the theory of residues. We then find for $x > 0$

$$w(x) = \frac{iF}{4EIk_f^3}\left(e^{ik_f x} + ie^{-k_f x}\right). \tag{7.39}$$

When equation (7.39) is combined with the time dependence $e^{-i\omega t}$, we see that the response is made up of two parts, a nonpropagating near-field deflection and a propagating wave that travels with the phase velocity

$$V_f = \frac{\omega}{k_f} = \left(\frac{I}{S}\right)^{1/4}(\omega c_b)^{1/2}, \qquad C_b = \sqrt{E/\rho_s} \tag{7.40}$$

which is the same as equation (7.16). As discussed earlier, the frequency dependence of the phase velocity gives rise to dispersion when dealing with transient wave propagation problems and becomes extremely important when one considers the coupling of the structural vibration to the sound field in the ambient medium. Using equation (7.40), one can easily derive an expression for the drive-point impedance, defined as the ratio of force to the resulting velocity at the point of application:

$$Z_0 = \frac{F}{-i\omega w(0)} = 2(1-i)\rho_s S V_f. \tag{7.41}$$

If we look back at section 1.3, where the concept of drive point mobility was defined, we see that Z_0^{-1} corresponds to $Y(0|0)$ for an infinite beam.

7.7 Vibrations of a Finite Elastic Beam
In order to analyze the forced response of a finite beam, we must first study the free vibrations of such a beam, found by setting the forcing function equal to zero and then seeking a solution of the resulting equation that varies as a harmonic function of time:

$$w(x, t) = w(x)e^{-i\omega t}, \tag{7.42}$$

where ω is a real quantity and denotes the frequency of the time variation which for free vibrations is yet to be determined. Equation (7.12b) then becomes, in the absence of an externally applied force,

$$EI\frac{d^4 w}{dx^4} - \omega^2 \rho_s S w = 0, \tag{7.43}$$

whose general solution can be written as

$$w = A \cos k_f x + B \sin k_f x + C \cosh k_f x + D \sinh k_f x, \tag{7.44}$$

where k_f is given in (7.16a). The general solution, equation (7.44), must now be made to satisfy the boundary conditions of the problem. For purposes of illustration, let us assume that the beam is simply supported at its end points, $x = 0$ and $x = L$. The function $w(x)$ must then satisfy the boundary conditions[3]

$$w(0) = w(L) = 0, \tag{7.45a}$$

$$w''(0) = w''(L) = 0. \tag{7.45b}$$

The first set of boundary conditions, equation (7.45a), is a statement of the fact that the deflection is zero at the boundaries, while the second set, equation (7.45b), states that there are no moments applied to the beam at the boundaries. These four boundary conditions lead to a set of four linear homogeneous equations in the four unknown constants A, B, C, D of equation (7.44). For a nontrivial solution to exist, the determinant formed by the coefficients A, B, C, D must vanish, which leads to the characteristic equation. The boundary condition $w(0) = 0$ requires that $A + C = 0$, while the condition $w''(0) = 0$ requires $-A + C = 0$. These two equations can be satisfied simultaneously only if $A = C = 0$. The boundary conditions at $x = L$ are satisfied if

$$B \sin(k_f L) + D \sinh(k_f L) = 0, \tag{7.46}$$

$$-Bk_f^2 \sin(k_f L) + Dk_f^2 \sinh(k_f L) = 0. \tag{7.47}$$

In order that a nontrivial solution exist,

$$\begin{vmatrix} \sin(k_f L) & \sinh(k_f L) \\ -k_f^2 \sin(k_f L) & k_f^2 \sinh(k_f L) \end{vmatrix} = 0. \tag{(7.48)}$$

An evaluation of the determinant in equation (7.48) yields the characteristic equation,

$$2k_f^2 \sin k_f L \sinh k_f L = 0, \tag{7.49}$$

which is satisfied if

$$k_f = \frac{n\pi}{L}, \qquad n = 1, 2, \ldots . \tag{7.50}$$

The solution $k_f = 0$ is discarded since it leads to the trivial result $w(x) = 0$.

Substituting equation (7.50) back into either equation (7.46) or equation (7.47), we find that D must vanish, and consequently

$$w_n(x) = W_n \sin\left(\frac{n\pi x}{L}\right), \qquad n = 1, 2, \ldots, \tag{7.51}$$

where the subscript n is used to denote the fact that for each value of n we have a different function of x that satisfies the differential equation and the particular boundary conditions. These functions are known as the *eigenfunctions* or *characteristic functions* of the problem. Furthermore, using the definition of k_f and equation (7.50), we see that the free-vibration problem admits solutions for only certain discrete frequencies

$$\omega_n = c_b \left(\frac{I}{S}\right)^{1/2} \left(\frac{n\pi}{L}\right)^2. \tag{7.52}$$

These are the natural or resonance frequencies of the system. The natural frequencies as well as eigenfunctions differ for different boundary conditions.

With the solution of the free-vibration problem available, the response due to any prescribed excitation can be readily found. As an example, we consider a concentrated force of harmonic time variation applied at the point $x = x_0$, where $0 < x_0 < L$. Assuming a solution of the form $w(x)e^{-i\omega t}$, where ω is now the excitation frequency, the equation governing $w(x)$ is

$$\frac{d^4 w}{dx^4} - k_f^4 w = \frac{F\delta(x - x_0)}{EI}, \tag{7.53}$$

and the boundary conditions on w are the same as those given by equation (7.45). In order to solve equation (7.53), consider $w(x)$ to be a superposition of the eigenfunctions found from the free-vibration analysis:

$$w(x) = \sum_{n=1}^{\infty} W_n \sin\frac{n\pi x}{L}, \tag{7.54}$$

where the W_n are as yet unknown. Each of the eigenfunctions satisfies the boundary conditions, so that all that remains is to determine the coefficients W_n. If we substitute equation (7.54) into equation (7.53), we find that

$$\sum_{n=1}^{\infty} \left[\left(\frac{n\pi}{L}\right)^4 - k_f^4 \right] W_n \sin \frac{n\pi x}{L} = \frac{F\delta(x-x_0)}{EI}. \tag{7.55}$$

Multiplying both sides of equation (7.55) by $\sin m\pi x/L$, integrating over the interval $(0, L)$, and utilizing the orthogonality condition[3]

$$\int_0^L \sin \frac{n\pi x}{L} \sin \frac{m\pi x}{L} \, dx = \frac{L}{2}\delta_{mn}, \tag{7.56}$$

we find an expression for W_n:

$$W_n = \frac{2F}{LEI} \frac{\sin(n\pi x_0/L)}{(n\pi/L)^4 - k_f^4}. \qquad 7.40; \; K_f^4 \propto \omega^2 \tag{7.57}$$

With ω_n given by equation (7.52), the response can be written as

$$w(x) = \frac{-2F}{M} \sum_{n=1}^{\infty} \frac{\sin(n\pi x_0/L)\sin(n\pi x/L)}{\omega^2 - \omega_n^2}, \tag{7.58}$$

where $M = \rho_s SL$ is the total mass of the beam.

A resonance occurs whenever the excitation frequency ω equals one of the corresponding natural frequencies, unless the location of the point of excitation is such that the corresponding $\sin n\pi x_0/L$ also vanishes. Equation (7.58) represents the response of a finite beam at the location x due to a force applied at x_0. We can therefore use equation (7.58) as an example of the transfer mobility $Y(x|x_0)$ defined by equation (1.3):

$$Y(x|x_0) = \frac{2i\omega}{M} \sum_{n=1}^{\infty} \frac{\sin(n\pi x_0/L)\sin(n\pi x/L)}{\omega^2 - \omega_n^2}.$$

7.8 Flexural Vibrations of Thin Elastic Plates
The flexural motion of plates, unlike that of beams or rods, is a two- rather than a one-dimensional problem. The equation of motion governing the flexural vibrations of plates is the two-dimensional extension of the flexural

equation (7.12) derived for a beam. For a plate of uniform thickness h, oriented so that its undeformed middle surface contains the x and y axes of a rectangular coordinate system, the equation of motion governing the transverse displacement w is[4]

$$D\left(\frac{\partial^4 w}{\partial x^4} + 2\frac{\partial^4 w}{\partial x^2 \partial y^2} + \frac{\partial^4 w}{\partial y^4}\right) + \rho_s h \frac{\partial^2 w}{\partial t^2} = -p_a(x,y,t) \tag{7.59}$$

where $D = Eh^3/12(1 - v^2)$ is the bending stiffness of the plate, v is Poisson's ratio, and $p_a(x,y,t)$ is the applied pressure acting on the plate.

As in the case of a beam, we first derive the response of an infinite plate to a point-load excitation. We then investigate the response of finite plates, which is, however, not as straightforward as the problem of finite beams. Although the eigenfunctions and natural frequencies were derived only for the case of a beam simply supported at both ends, we could easily have done the same for other sets of boundary conditions, as, for example, one end simply supported and the other end clamped, or both ends clamped, and so on. This is true because in the one-dimensional case a general solution equation (7.44) from which all other solutions can be constructed is available. This does not generally hold for the two-dimensional elastic-plate equation. Although we shall present a solution to the problem of a point-excited rectangular plate simply supported along its entire periphery, the reader should not be misled into thinking that the solution to the problem of rectangular plates with other types of boundary conditions is so readily obtainable.

7.9 Point Excitation of an Infinite Plate

The spatial dependence of the response of an infinite plate to a concentrated force that varies harmonically in time, and is applied at the origin, is determined by

$$D\left(\frac{\partial^4 w}{\partial x^4} + 2\frac{\partial^4 w}{\partial x^2 \partial y^2} + \frac{\partial^4 w}{\partial y^4}\right) - \rho_s h \omega^2 w = F\delta(x)\,\delta(y). \tag{7.60}$$

The force acts in the same direction as the positive displacement.

The response of the plate in this case is axisymmetric; that is, in the plane of the plate, the response depends only on the distance from the drive point. For such a problem, it is convenient to introduce cylindrical coordinates whose origin is taken at the drive point. Introducing the

transformation

$$x = r \cos \theta,$$

$$y = r \sin \theta,$$

the equation of motion becomes

$$D\left(\frac{d^2}{dr^2} + \frac{1}{r}\frac{d}{dr}\right)^2 w - \rho_s h \omega^2 w = \frac{F \delta(r)}{2\pi r}. \tag{7.61}$$

Such problems are conveniently solved by use of Hankel transforms, as defined by equations (5.3a) and (5.3b). If we transform equation (7.61), the transform of the displacement becomes

$$\tilde{w}(\gamma) = \frac{F}{2\pi D(\gamma^4 - k_f^4)},$$

where γ is the transform parameter and k_f is now the plate flexural wave number which takes the place of equation (7.16a):

$$k_f = (\rho_s h \omega^2 / D)^{1/4} \qquad D \equiv Eh^3/12(1-\nu^2)$$

$$= \left(\frac{12^{1/2}\omega}{h c_p}\right)^{1/2}. \qquad c_p \equiv [E/(1-\nu^2)\rho_s]^{1/2} \tag{7.62}$$

where c_p is the plate velocity defined in (2.53). The corresponding flexural velocity is

$$c_f = \frac{(h c_p \omega)^{1/2}}{12^{1/4}}. \tag{7.63}$$

The displacement is the inverse transform

$$w(r) = \frac{F}{2\pi D} \int_0^\infty \frac{\mathcal{J}_0(\gamma r) \gamma \, d\gamma}{\gamma^4 - k_f^4}. \tag{7.64}$$

Using the identities in equation (5.19a), we can state equation (7.64) as

$$w(r) = \frac{F}{4\pi D} \int_{-\infty}^{\infty} \frac{H_0^{(1)}(\gamma r)}{\gamma^4 - k_f^4} \gamma \, d\gamma.$$

(7.65)

The integral, being in a form analogous to equation (7.38), is evaluated in the same way:

$$w(r) = \frac{iF}{8\omega\sqrt{\rho_s h D}} \left\{ H_0^{(1)}(k_f r) + \frac{2i}{\pi} K_0(k_f r) \right\},$$

(7.66)

where $K_0(k_f r)$ is the modified Hankel function defined by equation (6.43). For large values of the argument $k_f r \gg 1$ we obtain the asymptotic form

$$w(r) \sim \frac{iF}{8\omega} \sqrt{\frac{2}{\pi \rho_s h D k_f r}} e^{i(k_f r - \pi/4)}.$$

(7.67)

This expression, when combined with the harmonic time variation $e^{-i\omega t}$, represents a wave propagating outward from the origin with a frequency-dependent phase velocity exactly analogous to the response of an infinite beam to a point excitation. The drive-point impedance of the flat plate is

$$Z_p = \lim_{r \to 0} \frac{F}{-i\omega w(0)} = \frac{4}{\sqrt{3}} \rho_s c_p h^2.$$

(7.68)

Thus we obtain the interesting result that the infinite plate drive-point impedance is purely real and frequency independent, unlike (7.41).

Before proceeding to finite plates, it is of interest to quote Powell's observation[5] on the validity of the infinite-plate mathematical model: "If it is the power spectrum of displacement that is required, then the infinite approach can be used only when the sheet is large enough for there to be only negligible reflections from its edges. However, if one is concerned only with the average mean square value of displacement (or related quantities), and if the detailed shape of the spectrum is of no consequence, then the infinite solution may be used, regardless of the magnitude of the damping, provided that the gravest modes are not excited." In the next chapter, we formulate a criterion of the validity of the infinite-plate mathematical model for the purpose of evaluating acoustic power. That criterion reflects exclusively the supersonic portion of the wave-number spectrum, which alone is

responsible for sound radiation. In contrast, Powell's considerations, being concerned with plate response, encompass the entire spectrum.

7.10 Flexural Vibrations of Finite Elastic Plates

We now consider the forced response of a finite rectangular plate simply supported along its entire periphery, of dimensions L_x and L_y, in the x and y directions, respectively. The boundary conditions require that the displacement w and the bending moment vanish along the boundaries. For the rectangular plate, the bending moments M_x and M_y are related to the displacement by the formulas[6]

$$M_x = -D\left(\frac{\partial^2 w}{\partial x^2} + v\frac{\partial^2 w}{\partial y^2}\right), \tag{7.69}$$

$$M_y = -D\left(\frac{\partial^2 w}{\partial y^2} + v\frac{\partial^2 w}{\partial x^2}\right). \tag{7.70}$$

The vanishing of the displacement along an edge—for example, along the edge $x = L_x$, for all values of y—guarantees that $\partial^2 w/\partial y^2$ also vanishes along that same edge. The boundary conditions therefore become

$$w(x,y) = \frac{\partial^2 w(x,y)}{\partial x^2} = 0, \qquad x = 0, L_x, \tag{7.71}$$

$$w(x,y) = \frac{\partial^2 w(x,y)}{\partial y^2} = 0, \qquad y = 0, L_y. \tag{7.72}$$

The functions

$$w_{mn}(x,y) = W_{mn}\sin\frac{m\pi x}{L_x}\sin\frac{n\pi y}{L_y}, \qquad n = 1, 2, 3\ldots, \quad m = 1, 2, 3\ldots, \tag{7.73}$$

satisfy the boundary conditions given by equations (7.71) and (7.72). The equation determining the free vibrations of a plate is equation (7.60) with $p_a(x, y, t) = 0$:

$$D\left(\frac{\partial^4 w}{\partial x^4} + 2\frac{\partial^4 w}{\partial x^2 \partial y^2} + \frac{\partial^4 w}{\partial y^4}\right) - \rho_s h \omega^2 w = 0. \tag{7.74}$$

Each of the functions w_{mn} will satisfy equation (7.74) for the frequency

$$\omega_{mn} = \left(\frac{D}{\rho_s h}\right)^{1/2}\left[\left(\frac{m\pi}{L_x}\right)^2 + \left(\frac{n\pi}{L_y}\right)^2\right]. \tag{7.75}$$

The subscript mn indicates that for each value of m, n in equation (7.73) a different frequency is required. The functions in equation (7.73) are the eigenfunctions for the simply supported plate, and the frequencies in equation (7.75) are the natural frequencies. For a concentrated load acting on the plate at the point $x = x_0, y = y_0$, equation (7.60) becomes

$$D\left(\frac{\partial^4 w}{\partial x^4} + 2\frac{\partial^4 w}{\partial x^2 \partial y^2} + \frac{\partial^4 w}{\partial y^4}\right) - \rho_s h\omega^2 w = F\delta(x - x_0)\delta(y - y_0). \tag{7.76}$$

Following the same procedure as for the beam, we now expand $w(x,y)$ in the eigenfunctions of the system. Using the condition given by equation (7.56), we can show that

$$w(x,y) = -\frac{4F}{M}\sum_{m=1}^{\infty}\sum_{n=1}^{\infty}\frac{\sin(m\pi x_0/L_x)\sin(n\pi y_0/L_y)\sin(m\pi x/L_x)\sin(n\pi y/L_y)}{\omega^2 - \omega_{mn}^2}, \tag{7.77}$$

where $M = \rho_s h L_x L_y$ is the total mass of the plate. Equation (7.77) can be put into the form of a transfer mobility:

$$\begin{aligned}
Y(x,y|x_0,y_0) &= \frac{-i\omega w(x,y)}{F} \\
&= \frac{4i\omega}{M}\sum_{m=1}^{\infty}\sum_{n=1}^{\infty}\frac{\sin(m\pi x_0/L_x)\sin(m\pi x/L_x)\sin(n\pi y_0/L_y)\sin(n\pi y/L_y)}{\omega^2 - \omega_{mn}^2}.
\end{aligned} \tag{7.78}$$

The response given by equation (7.77) has resonances at each of the frequencies $\omega = \omega_{mn}$.

7.11 Thick-Plate Theory; Timoshenko-Mindlin Plate Theory

The equation of motion, equation (7.59), governing the normal deflection of the plate w yields a phase velocity for straight crested waves traveling in the

x direction that is inversely proportional to the wavelength. This is entirely analogous to the situation presented by the classical Bernoulli-Euler beam theory, equation (7.12). In order to remedy the situation in that case, we included the effects of rotatory inertia and transverse shear and thereby derived the Timoshenko beam equation, which gave better results in the short-wavelength limit. By an entirely analogous procedure, which takes into account rotatory inertia and transverse shear, we analyze "thick" plates i.e., plates whose thickness exceeds $\lambda_s/20$, where λ_s is the shear wavelength. As discussed earlier, Timoshenko was the first to include these effects for the case of a one-dimensional bar; Mindlin[7] has shown how the two-dimensional improved flexural equations of motion can be derived directly from the three-dimensional equations of elasticity; we therefore call the governing equation the Timoshenko-Mindlin plate equation.

In the first (1972) edition of this book, a detailed derivation of the Timoshenko-Mindlin plate equation, which followed that of Mindlin,[7] was presented. Rather than repeating it here once again, we present only the final result for completeness. Thus, the two-dimensional analogue of equation (7.33) is

$$
\left[D\nabla^2 - \frac{\rho_s h^3}{12} \frac{\partial^2}{\partial t^2} \right] \left[\nabla^2 - \frac{\rho_s}{\kappa^2 G} \frac{\partial^2}{\partial t^2} \right] w + \rho_s h \frac{\partial^2 w}{\partial t^2}
$$
$$
= -\left(1 - \frac{D}{\kappa^2 Gh} \nabla^2 + \frac{\rho_s h^2}{12\kappa^2 G} \frac{\partial^2}{\partial t^2} \right) p_a(x,y,t). \tag{7.79}
$$

In chapter 8 this equation is used in the determination of the high-frequency sound radiation characteristics of a force-excited elastic plate.

7.12 Introduction to the in Vacuo Vibration of Shells

The previous sections in this chapter have dealt with the vibrations of rods, beams, and plates. In the remaining sections we discuss structures that in their undeformed state have a significant amount of curvature. As mentioned previously, we shall restrict ourselves to consideration of cylindrical and spherical shells.

In the first edition of this book the equations of motion for cylindrical and spherical shells were derived. In this version we shall not rederive the equations, since the major interest here is the study of the acoustic-structure interaction problem rather than a detailed discussion of the structural response.

The latter is subject to the minor differences in the many different sets of equations that have been derived over the years. For those readers interested in such matters, the subject has been reviewed extensively by Leissa.[8] One can generally say that the discrepancies in the differential equations arise from small differences in the strain-displacement relationships, which give rise to unimportant differences in the numerical results.

It is important, however, to restate the basic assumptions, introduced by Love, in deriving the equations of motion of thin elastic shells:

1. The thickness of a shell is small compared with the smallest radius of curvature of the shell.

2. The displacement is small in comparison with the shell thickness.

3. The transverse normal stress acting on planes parallel to the shell middle surface is negligible.

4. Fibers of the shell normal to the middle surface remain so after deformation and are themselves not subject to elongation.

From these assumptions we can assert that the displacements are a linear function of the coordinate normal to the middle surface. The problem can then be shown to reduce to one of a determination of the shell's middle surface deflection.

7.13 Equations of Motion for Cylindrical Shells

Consider a cylindrical shell of thickness h and radius a referred to an r, ϕ, z coordinate system (see figure 7.4), where e_r is the unit vector directed outward along the normal to the shell surface, e_ϕ the tangential unit vector in the direction of increasing ϕ, and e_z the unit vector in the direction of the cylindrical shell generator. The corresponding displacements of the shell midsurface are w, v, and u. We assume that the only external loading acts normally to the cylindrical surface of the shell and is denoted by $p_a(\phi, z)$. Since the acoustic medium is assumed to be inviscid, this loading can represent the fluid loading as well as any externally applied normal load. External loads that act tangentially or longitudinally are for the present assumed to vanish, although they can be easily accommodated by simply adding loading terms to the appropriate equation of motion.

The simplest equations of motion are those that are derived as a dynamic counterpart of Donnell's formulation and are taken from Kraus.[9] In addition to the assumptions already formulated, Kraus states that "...

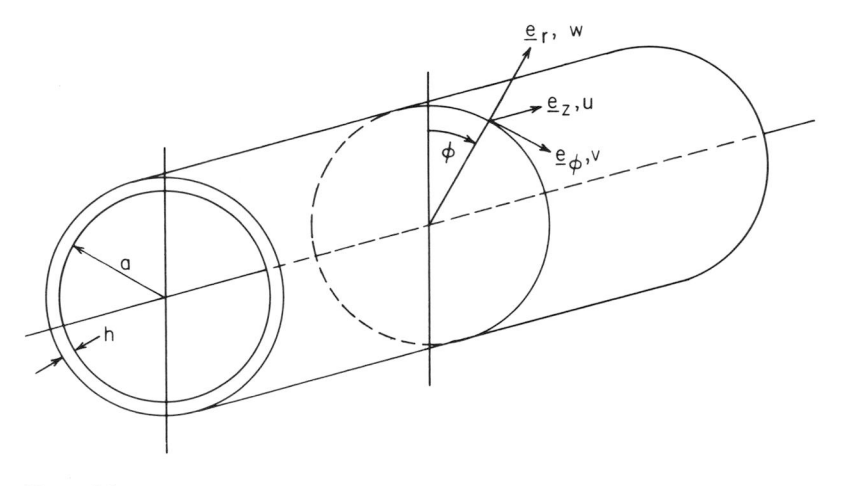

Figure 7.4
Geometry of cylindrical shell showing direction of displacement components.

Donnell's formulation is based upon the assumptions that the expressions for the changes in curvature and twist of the cylinder are the same as those of a flat plate and that the effect of the transverse shearing-stress resultant Q_s on the equilibrium of forces in the circumferential direction is negligible." The equations so derived can be written as

$$\frac{\partial^2 u}{\partial z^2} + \frac{1-v}{2a^2}\frac{\partial^2 u}{\partial \phi^2} + \frac{1+v}{2a}\frac{\partial^2 v}{\partial z \partial \phi} + \frac{v}{a}\frac{\partial w}{\partial z} - \frac{\ddot{u}}{c_p^2} = 0, \tag{7.80a}$$

$$\frac{1+v}{2a}\frac{\partial^2 u}{\partial z \partial \phi} + \frac{1-v}{2}\frac{\partial^2 v}{\partial z^2} + \frac{1}{a^2}\frac{\partial^2 v}{\partial \phi^2} + \frac{1}{a^2}\frac{\partial w}{\partial \phi} - \frac{\ddot{v}}{c_p^2} = 0, \tag{7.80b}$$

$$\frac{v}{a}\frac{\partial u}{\partial z} + \frac{1}{a^2}\frac{\partial v}{\partial \phi} + \frac{w}{a^2} + \beta^2\left(a^2\frac{\partial^4 w}{\partial z^4} + 2\frac{\partial^4 w}{\partial z^2 \partial \phi^2} + \frac{1}{a^2}\frac{\partial^4 w}{\partial \phi^4}\right)$$

$$+ \frac{\ddot{w}}{c_p^2} - \frac{p_a(1-v^2)}{Eh} = 0, \tag{7.80c}$$

where $\beta^2 = h^2/12a^2$. Terms proportional to β^2 represent the contributions of bending stresses. The first two equations, being independent of β^2, are identical with the membrane equations of motion. The third equation differs from the corresponding membrane equation only by inclusion of a term that

can be written as $(h^2/12) \nabla^4 w$, where $\nabla^4 = (\partial^4/\partial z^4) + [(2/a^2)(\partial^4/\partial z^2 \partial \phi^2)] + [(1/a^4)(\partial^4/\partial \phi^4)]$. This is exactly the same type of term that occurs in the equation of motion for thin plates.

From equation (7.80c) we see that the normal displacement of the shell w is coupled to u and v, the two in-plane displacements of the shell. For flat plates, the normal displacement is independent of the in-plane displacement components.

For a cylindrical shell of finite length, the specification of boundary conditions is required. The simplest set, for which most results are available, is that of a simply supported cylindrical shell. If its ends are located at $z = \pm L$, the displacement boundary conditions are

$$w = \frac{\partial^2 w}{\partial z^2} = v = \frac{\partial u}{\partial z} = 0 \qquad \text{at } z = \pm L. \tag{7.81}$$

The boundary conditions for a shell that is clamped at its ends are

$$u = v = w = \frac{\partial w}{\partial z} = 0 \qquad \text{at } z = \pm L. \tag{7.82}$$

7.14 Planar Vibrations of a Thin Cylindrical Shell

Now that the equations of motion governing the vibrations of cylindrical shells have been stated, it is instructive to consider some examples. One of the simplest cases to deal with is that of plane strain in which we assume that the axial component of displacement u vanishes and the circumferential and radial components v and w, respectively, are independent of the axial coordinate z. To study the free vibrations of this system, we set the applied force p_a equal to zero and assume a solution of the form

$$u = 0,$$
$$v = V_n \sin n\phi\, e^{-i\omega t}, \tag{7.83}$$
$$w = W_n \cos n\phi\, e^{-i\omega t}.$$

For such a dynamic configuration, equation (7.80a) is trivially satisfied, while (7.80b) and (7.80c) require

$$(\Omega^2 - n^2) V_n - nW_n = 0, \tag{7.84a}$$

$$\beta = \frac{h}{\sqrt{12}\, a}$$

$$nV_n - (\Omega^2 - 1 - \beta^2 n^4)\, W_n = 0, \qquad\qquad (7.84b)$$

where $\Omega = \omega a / c_p$ is a nondimensional frequency parameter.

The set of equation (7.84) is a homogeneous system of two linear algebraic equations in the unknown modal amplitudes V_n and W_n. The determinant formed by the coefficients of V_n and W_n must vanish for a nontrivial solution to exist. This results in a frequency equation

$$\Omega^4 - \Omega^2(1 + n^2 + \beta^2 n^4) + \beta^2 n^6 = 0, \qquad\qquad (7.85)$$

which, for $n > 0$, yields two real nonnegative roots for Ω^2, denoted $\Omega_n^{(1)}$ and $\Omega_n^{(2)}$:

$$\Omega_n^2 = \tfrac{1}{2}[1 + n^2 + \beta^2 n^4 \mp \sqrt{(1 + n^2 + \beta^2 n^4)^2 - 4\beta^2 n^6}\,], \qquad (7.86)$$

where the $\mp$ sign corresponds to $\Omega_n^{(1)}$, $\Omega_n^{(2)}$, respectively. These two natural frequencies associated with each value of $n > 0$ reflect the fact that the two nonvanishing components of displacement give the system two degrees of freedom. In the more familiar case of beam vibrations, for example, the longitudinal and transverse components of motion are uncoupled, which leads to two entirely independent frequency equations. Figure 7.5 shows how the two resonance frequencies $\Omega_n^{(1)}$ and $\Omega_n^{(2)}$ vary with n. Calculations performed for two values of h/a indicate that $\Omega_n^{(1)}$ is dependent on h/a, while $\Omega_n^{(2)}$ is essentially independent of h/a. In fact, for large n, $\Omega_n^{(1)} \sim \beta n^2$ while $\Omega_n^{(2)} \sim n$. These results are similar to those obtained by Rayleigh wherein he considered inextensional and extensional types of modes, respectively. Once the frequency equation (7.85) is solved, we can use (7.84) to determine the ratio $|W_n/V_n|$ corresponding, respectively, to $\Omega_n^{(1)}$ and $\Omega_n^{(2)}$. In the former case, we find that for $n > 1$, $|W_n/V_n| = n$, the motion is primarily radial. For $n = 0$, there is no tangential motion, so that the motion is all radial, therefore displaying a single natural frequency, $\Omega_0 = 1$. For $n = 1$, the smaller of the two natural frequencies is zero, being associated with rigid body translation for which $|W_1/V_1| = 1$.

7.15 Forced Planar Vibrations of a Thin Cylindrical Shell

The results of the preceding section can now be used to determine the response of a cylindrical shell to a radial exciting force, which is independent of the axial coordinate and symmetric with respect to the $\phi = 0$ axis.

The applied pressure $f(\phi)e^{-i\omega t}$ is expressed as a Fourier expansion

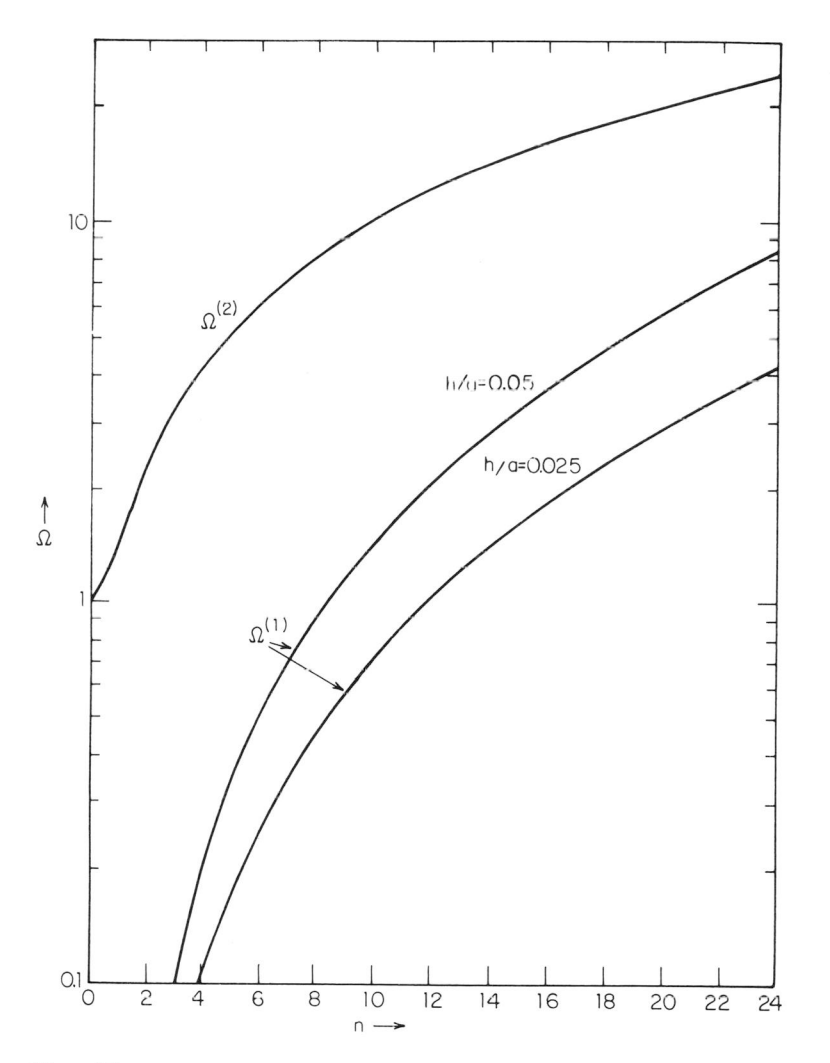

Figure 7.5
Resonance frequencies for an infinite cylindrical shell (no axial dependence) as a function of the circumferential mode number (upper branch does not vary significantly with h/a; $v = 0.3$; curves shown as a continuous function for convenience only).

$\sum_{n=0}^{\infty} f_n \cos n\phi\, e^{-i\omega t}$, where

$$f_n = \frac{\varepsilon_n}{\pi} \int_0^{\pi} f(\phi') \cos n\phi'\, d\phi', \qquad \varepsilon_0 = 1, \quad \varepsilon_n = 2 \text{ for } n \geq 1. \tag{7.87}$$

The response displacements are then assumed to be

$$u = 0,$$

$$v = \sum_{n=1}^{\infty} V_n \sin n\phi\, e^{-i\omega t}, \tag{7.88}$$

$$w = \sum_{n=0}^{\infty} W_n \cos n\phi\, e^{-i\omega t}.$$

Substituting equations (7.87) and (7.88) into the equation of motion, equation (7.80), and utilizing the orthogonality conditions

$$\frac{\varepsilon_n}{2\pi} \int_{-\pi}^{\pi} \cos m\phi \cos n\phi\, d\phi = \delta_{mn}, \qquad \frac{1}{\pi} \int_{-\pi}^{\pi} \sin m\phi \sin n\phi\, d\phi = \delta_{mn}, \tag{7.89}$$

we obtain two nonhomogeneous, linear algebraic equations for the modal amplitudes V_n and W_n:

$$(\Omega^2 - n^2) V_n - nW_n = 0,$$

$$-nV_n + [\Omega^2 - (1 + \beta^2 n^4)] W_n = -\frac{f_n a^2 (1 - v^2)}{Eh}. \tag{7.90}$$

Solving for V_n and W_n, we obtain expressions for the nonvanishing deflections v and w:

$$v = -\frac{a^2 (1 - v^2)}{Eh} \sum_{n=1}^{\infty} \frac{n f_n \sin n\phi\, e^{-i\omega t}}{[\Omega^2 - (\Omega_n^{(1)})^2][\Omega^2 - (\Omega_n^{(2)})^2]},$$

$$w = -\frac{a^2 (1 - v^2)}{Eh} \sum_{n=0}^{\infty} \frac{f_n (\Omega^2 - n^2) \cos n\phi\, e^{-i\omega t}}{[\Omega^2 - (\Omega_n^{(1)})^2][\Omega^2 - (\Omega_n^{(2)})^2]}. \tag{7.91}$$

where $\Omega_n^{(1)}$ and $\Omega_n^{(2)}$ were discussed in section 7.14. In later chapters it will be found convenient to express the response in terms of the modal mechanical impedance

$$\mathcal{Z}_n = -\frac{i\rho_s c_p}{\Omega} \frac{h}{a} \frac{[\Omega^2 - (\Omega_n^{(1)})^2][\Omega^2 - (\Omega_n^{(2)})^2]}{(\Omega^2 - n^2)}. \tag{7.92}$$

Equation (7.91) can now be written

$$\dot{v}(\phi) = \sum_{n=1}^{\infty} \frac{f_n n \sin n\phi\, e^{-i\omega t}}{\mathcal{Z}_n (\Omega^2 - n^2)}, \qquad \dot{w}(\phi) = \sum_{n=0}^{\infty} \frac{f_n \cos n\phi\, e^{-i\omega t}}{\mathcal{Z}_n}. \tag{7.93}$$

7.16 Nonplanar Vibrations of a Cylindrical Shell

In the preceding section we have considered the special case of a cylindrical shell of infinite length responding to a force excitation that is independent of the axial coordinate. When the response displays axial dependence, the motion of an infinite or simply supported shell is described by the displacement components

$$u = \sum_{m,n} U_{mn} \cos n\phi \sin k_m z,$$

$$v = \sum_{m,n} V_{mn} \sin n\phi \cos k_m z, \tag{7.94}$$

$$w = \sum_{m,n} W_{mn} \cos n\phi \cos k_m z.$$

The motion so described consists of standing waves in both the circumferential and axial directions. The radial displacement nodal lines in the circumferential direction are separated by the distances $\pi a/n$, and the axial wavelength is given by $2\pi/k_m$. If we substitute the foregoing displacement representation in the simplified equations of motion for the cylinder, equation (7.80), we obtain a homogeneous set of three linear algebraic equations for the displacement amplitudes U_{mn}, V_{mn}, and W_{mn}. In order that a nontrivial solution of this set exist, the determinant of the coefficients must vanish:

$$\begin{vmatrix} -\Omega^2 + k_m^2 a^2 & \frac{1}{2}(1+v)nk_m a & vk_m a \\ \quad + \frac{1}{2}(1-v)n^2 & & \\ \frac{1}{2}(1+v)nk_m a & -\Omega^2 + \frac{1}{2}(1-v)k_m^2 a^2 & n \\ & \quad + n^2 & \\ vk_m a & n & -\Omega^2 + 1 \\ & & \quad + \beta^2 (k_m^2 a^2 + n^2)^2 \end{vmatrix} = 0.$$

$$\tag{7.95}$$

This is in the form of a cubic equation in Ω^2:

$$(\Omega^2)^3 - A_2(\Omega^2)^2 + A_1(\Omega^2) - A_0 = 0, \tag{7.96}$$

where the coefficients A_0, A_1, and A_2 are

$$A_0 = \left(\frac{1-v}{2}\right)[(1-v^2)k_m^4 a^4 + \beta^2(k_m^2 a^2 + n^2)^4], \tag{7.97a}$$

$$A_1 = \left(\frac{1-v}{2}\right)[(k_m^2 a^2 + n^2)^2 + n^2 + (3+2v)k_m^2 a^2]$$
$$\qquad + \beta^2\left(\frac{3-v}{2}\right)(k_m^2 a^2 + n^2)^3, \tag{7.97b}$$

$$A_2 = 1 + \left(\frac{3-v}{2}\right)(k_m^2 a^2 + n^2) + \beta^2(k_m^2 a^2 + n^2)^2. \tag{7.97c}$$

For $n > 0$, the solution of equation (7.96) thus gives us a set of three natural frequencies corresponding to each modal configuration. In section 7.14, where we considered the planar vibrations of cylindrical shells, we obtained two frequencies corresponding to each modal configuration since in that case we assumed zero axial displacement and hence one degree of freedom less. Specializing the determinant, (7.95), to ($n = 0$) modes, the v term is uncoupled from the u and w terms. Axisymmetric v motion represents *torsional* vibrations whose natural frequency is

$$\Omega_v = k_m a[(1-v)/2]^{1/2}. \tag{7.98a}$$

Here and below the subscript identifies the predominant displacement component of the resonant mode. The remaining terms in (7.98) yield the characteristic equation

$$(-\Omega^2 + k_m^2 a^2)(-\Omega^2 + 1 + \beta^2 k_m^4 a^4) - v^2 k_m^2 a^2 = 0.$$

An approximate solution applicable to thin shells of large aspect ratio is constructed by assuming $\Omega^2 \ll 1$ and neglecting the β term's contribution to the potential energy of longitudinal modes. This yields

$$\Omega_u \simeq k_m a(1-v^2)^{1/2}, \tag{7.98b}$$

$$\Omega_w \simeq (1 + \beta^2 k_m^4 a^4)^{1/2}. \tag{7.98c}$$

Longitudinal resonances occur when the shell length measures m half compressional bar wavelengths c_b/f. At lower order *radial* resonances the shell circumference measures one compressional plate wavelength, c_p/f. Radial modes are effectively degenerate until, with increasing m, the β term in (7.95c) becomes significant, viz., of order 10^{-1}. For $h/a = O(10^{-2})$, i.e., $\beta^2 = O(10^{-5})$, this requires $k_m a = O(10)$. We shall see in section 9.6 that radiation loading eliminates the degeneracy of axisymmetric radial modes.

Consider a cylindrical shell of length L that is simply supported at its ends, $z = \pm L$. For the modes whose radial displacement w is symmetric with respect to z and ϕ, the modal configuration is given by equation (7.94) with

$$k_m = (2m + 1)\frac{\pi}{2L}, \qquad m = 0, 1, 2, \ldots .$$

The simply supported boundary conditions, equation (7.81), are seen to be satisfied. We can then use equation (7.96) to solve for the natural frequencies of such a shell. A finite-length shell may vibrate with any of three distinct frequencies for a modal pattern having the same number of circumferential and longitudinal nodal lines.

Typical results are shown in figure 7.6, where for a fixed $k_m a$, the three natural frequencies are plotted as a function of n. The two upper-branch frequencies increase monotonically with n, while the lower branch, which corresponds to predominantly radial modes, shows the somewhat unexpected result of the frequency initially decreasing with n and then increasing. Arnold and Warburton[10] offered an explanation of the phenomenon by considering the relative strain energies of the bending and stretching motions of the shell's middle surface for a cylinder simply supported at both ends. For low circumferential-mode numbers the stretching energy far exceeds that of bending energy. The situation reverses for higher circumferential-mode numbers, thereby decreasing the potential energy and hence the natural frequencies. It is only when the circumferential flexural term, $\beta^2 n^4$ in equation (7.95), becomes sufficiently large to compensate for the decrease in membrane stiffness that the potential energy, and hence the natural frequencies, again increases.

The location of the natural-frequency minimum can be identified analytically by means of an approximate expression constructed by Heckl,[11]

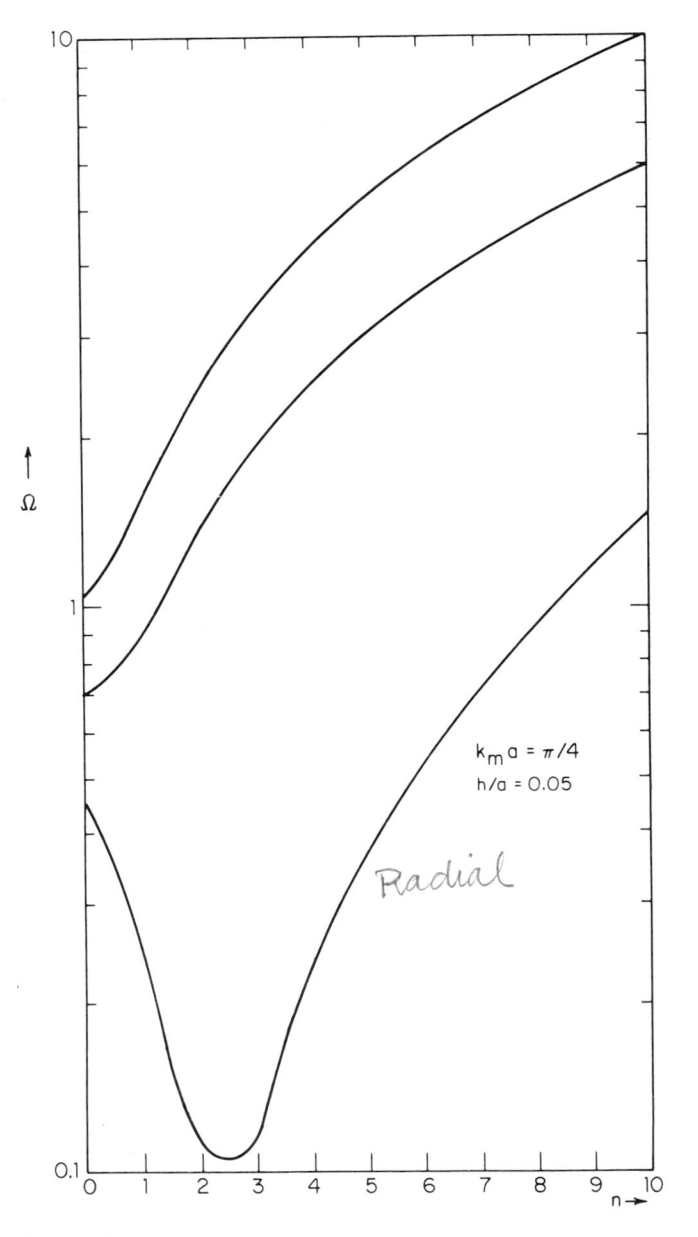

Figure 7.6
Resonance frequencies of a finite cylindrical shell with simply supported ends as a function of circumferential mode number, ($v = 0.3$; curves shown as a continuous function for convenience only).

Table 7.1
Natural frequencies Ω_{mn} of the predominantly radial modes of cylindrical shells computed from the rigorous theory (7.95) and from the approximate expression (7.99)

$k_m a$	π/1		π/2		π/4	
h/a	0.05		0.05		0.01	
n	(7.95)	(7.99)	(7.95)	(7.99)	(7.95)	(7.99)
1	0.257	0.258	0.574	0.481	0.257	0.258
2	0.127	0.128	0.338	0.336	0.113	0.115
3	0.143	0.144	0.249	0.250	0.0635	0.638
4		0.235	0.286	0.287	0.0578	0.579
5		0.363	0.397	0.398		0.0759
6		0.521	0.550	0.551	0.105	
7		0.709	0.736	0.737	0.142	
8		0.925		0.952	0.185	
9		1.17		1.20	0.234	
10		1.44		1.47	0.289	

which ignores tangential inertial forces. Here we account for the contribution of circumferential motion to the kinetic energy by multiplying the kinetic energy associated with the predominantly radial motion by $[1 + (V_{mn}/W_{mn})^2]$, where the ratio $|V_{mn}/W_{mn}|$ of circumferential to radial motion is approximated by the n^{-1}, viz., by the corresponding ratio for inextensional planar modes [see the discussion of (7.86)].

The approximate expression for the natural frequencies of predominantly radial modes becomes

$$\Omega_{mn} \simeq \left[(1 - \nu^2)\frac{k_m^4 a^4}{k_s^4 a^4} + \beta^2 k_s^4 a^4 \right]^{1/2} (1 + n^{-2})^{-1/2}, \qquad n > 0, \tag{7.99}$$

where k_s is the helical wave number:

$$k_s = \left(k_m^2 + \frac{n^2}{a^2} \right)^{1/2}. \tag{7.100}$$

The first term in (7.99) embodies membrane stresses, the second term flexural stresses. A comparison of the resonance frequencies constructed with this expression and rigorous theory is found in table 7.1. The approximate

in vacuo natural frequencies are compared with those computed from the exact theory and with the natural frequencies of the submerged shell in figure 9.3. Equation (7.99) is seen to be adequate for practical purposes, except for $n = 1$ modes of short axial wavelength, i.e., large $k_m a$.

Setting the derivative of (7.99) with respect to n equal to zero, one can solve for

$$n_{\min} \simeq \left(\frac{k_m a}{\beta^{1/2}} - k_m^2 a^2\right)^{1/2}. \tag{7.100a}$$

For the parameters of figure 7.6 this yields $n_{\min} = 2.4$, a result consistent with the rigorous calculations. The integer value of n closest to (7.100a) identifies the mode with the smallest resonance frequency.

To conclude the discussion of cylindrical shells, we note that the membrane stresses of the lower modes result in a low-frequency impedance that is higher than that of the infinite plate of the same thickness and material. At high frequencies, where both structures are controlled by flexural stresses, the impedances are comparable. This is illustrated in figure 7.7 for a simply supported shell of aspect ratio 2 excited by a concentrated force at midspan. The rigorous shell response is computed from equation (7.93). Also shown is an expression for the radial admittance embodying approximations similar to those underlying (7.99). Except in accounting for the contribution of circumferential motion to kinetic energy, coupling between radial and tangential motions has been neglected, thereby eliminating the effect of resonances of predominantly longitudinal and circumferential modes. Radial mobility, formulated exclusively in terms of the natural frequencies of predominantly radial modes, is therefore approximated as follows:

$$
\begin{aligned}
Y_c &\equiv \frac{\dot{W}}{F} \\
&= \frac{i}{\pi \rho_s c_p h L} \sum_{m=1,3,\dots} \left(\left[\Omega - \frac{\Omega_{m0}^2}{\Omega}(1 - i\eta_s)\right]^{-1} \right. \\
&\quad \left. + 2 \sum_{n=1,2,\dots} \left[\Omega - \frac{\Omega_{mn}^2}{\Omega}(1 - i\eta_s)\right]^{-1} (1 + n^{-2})^{-1}\right).
\end{aligned}
\tag{7.101}
$$

The breathing-mode natural frequency Ω_{m0} is given as Ω_w in equation (7.95c) and Ω_{mn} for $n > 0$ in equation (7.99).

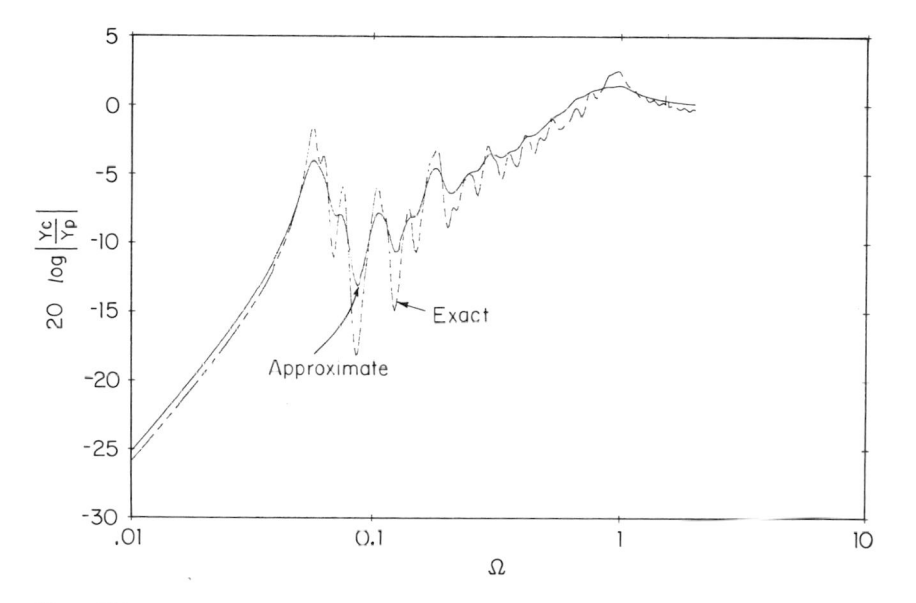

Figure 7.7
Drive-point admittance $Y_c = \dot{w}/F$ of a simply supported cylindrical shell excited by a concentrated radial force applied at midspan, normalized to the admittance $Y_p = \mathcal{Z}_p^{-1}$ of a point-excited infinite plate (7.68). The aspect ratio of the shell is 2; $h/a = 10^{-2}$; $\eta_s = 0.1$. The exact curve is computed from equation (7.80), the approximate curve from equation (7.101).

7.17 Spherical Shells; Equations of Motion

As our final example of structures with curvature, we consider the vibrations of a closed spherical shell in which both membrane (extensional) and flexural (inextensional) effects are included. A detailed derivation of the equations of motion using Hamilton's variational principle is found in the first edition of this book.

The shell is oriented with its center at the origin of a spherical coordinate system as shown in figure 7.8. The middle surface of the shell is given by $R = a$. As a result of the assumptions discussed in section 7.12, the general displacements of the spherical shell can be expressed in terms of the shell's midsurface deflections. Since we shall only consider the nontorsional axisymmetric motions, the only nonvanishing components of motion are w and u, the midsurface radial and tangential motions in the direction of increasing θ, respectively. The tangential component of motion in the ϕ direction vanishes, while both w and u are independent of ϕ.

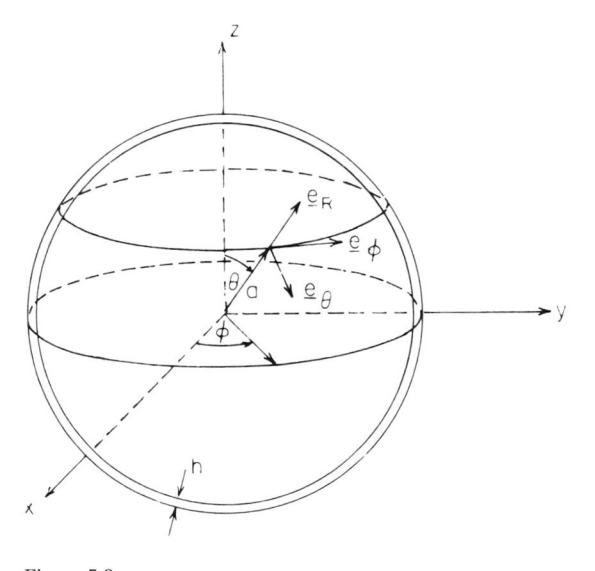

Figure 7.8
Spherical shell showing spherical coordinate system used.

The spherical-shell equations of motion in terms of the displacement functions u and w are

$$(1 + \beta^2)\left[\frac{\partial^2 u}{\partial \theta^2} + \cot \theta \frac{\partial u}{\partial \theta} - (v + \cot^2 \theta) u\right] - \beta^2 \frac{\partial^3 w}{\partial \theta^3}$$

$$- \beta^2 \cot \theta \frac{\partial^2 w}{\partial \theta^2} + [(1 + v) + \beta^2 (v + \cot^2 \theta)] \frac{\partial w}{\partial \theta} - \frac{a^2 \ddot{u}}{c_p^2} = 0,$$

(7.102)

$$\beta^2 \frac{\partial^3 u}{\partial \theta^3} + 2\beta^2 \cot \theta \frac{\partial^2 u}{\partial \theta^2} - [(1 + v)(1 + \beta^2) + \beta^2 \cot^2 \theta] \frac{\partial u}{\partial \theta}$$

$$+ \cot \theta [(2 - v + \cot^2 \theta) \beta^2 - (1 + v)] u - \beta^2 \frac{\partial^4 w}{\partial \theta^4} - 2\beta^2 \cot \theta \frac{\partial^3 w}{\partial \theta^3}$$

$$+ \beta^2 (1 + v + \cot^2 \theta) \frac{\partial^2 w}{\partial \theta^2} - \beta^2 \cot \theta (2 - v + \cot^2 \theta) \frac{\partial w}{\partial \theta}$$

$$- 2(1 + v) w - \frac{a^2 \ddot{w}}{c_p^2} = -p_a \frac{(1 - v^2) a^2}{Eh}.$$

(7.103)

It will be more convenient to work in terms of a new independent variable $\eta = \cos\theta$. If we make this transformation, the foregoing set of equations, assuming harmonic time variation, can be written as

$$L_{uu}u + L_{uw}w + \Omega^2 u = 0, \tag{7.104}$$

$$L_{wu}u + L_{ww}w + \Omega^2 w = -p_a \frac{a^2(1-v^2)}{Eh}, \tag{7.105}$$

where the operators L_{uu}, L_{uw}, L_{wu}, and L_{ww} are given by

$$L_{uu} = (1+\beta^2)\left\{(1-\eta^2)^{1/2}\frac{d^2}{d\eta^2}(1-\eta^2)^{1/2} + (1-v)\right\}, \tag{7.106}$$

$$L_{uw} = (1-\eta^2)^{1/2}\left\{[\beta^2(1-v) - (1+v)]\frac{d}{d\eta} + \beta^2\frac{d}{d\eta}\nabla_\eta^2\right\}, \tag{7.107}$$

$$L_{wu} = -\left\{[\beta^2(1-v) - (1+v)]\frac{d}{d\eta}(1-\eta^2)^{1/2} + \beta^2\nabla_\eta^2\frac{d}{d\eta}(1-\eta^2)^{1/2}\right\}, \tag{7.108}$$

$$L_{ww} = -\beta^2\nabla_\eta^4 - \beta^2(1-v)\nabla_\eta^2 - 2(1+v), \tag{7.109}$$

and

$$\nabla_\eta^2 = \frac{d}{d\eta}(1-\eta^2)\frac{d}{d\eta}.$$

7.18 Free Axisymmetric Nontorsional Vibrations of a Spherical Shell

Before treating force-excited vibrations, we consider the free-vibration problem in order to determine the resonant frequencies and eigenfunctions appropriate to the problem. For this purpose, consider the set of equations (7.104) and (7.105) with the forcing function p_a set equal to zero. The solutions can be expressed in terms of $P_n(\eta)$, the Legendre polynomial of the first kind of order n:

$$v(\eta) = \sum_{n=0}^{\infty} V_n(1-\eta^2)^{1/2}\frac{dP_n}{d\eta}, \tag{7.110}$$

$$w(\eta) = \sum_{n=0}^{\infty} W_n P_n(\eta).$$

(7.111)

Equations (7.104) and (7.105) are satisfied if the expansion coefficients V_n and W_n satisfy the equations

$$[\Omega^2 - (1 + \beta^2)(v + \lambda_n - 1)]V_n$$
$$- [\beta^2(v + \lambda_n - 1) + (1 + v)]W_n = 0,$$

(7.112)

$$-\lambda_n[\beta^2(v + \lambda_n - 1) + (1 + v)]V_n$$
$$+ [\Omega^2 - 2(1 + v) - \beta^2\lambda_n(v + \lambda_n - 1)]W_n = 0,$$

(7.113)

where $\lambda_n = n(n + 1)$.

The foregoing set of equations is a homogeneous system of two linear algebraic equations in the unknowns V_n and W_n, so that the determinant formed by the coefficients of the unknowns V_n and W_n must vanish for a nontrivial solution to exist. This leads to the frequency equation that is a quadratic in Ω^2:

$$\Omega^4 - [1 + 3v + \lambda_n - \beta^2(1 - v - \lambda_n^2 - v\lambda_n)]\Omega^2 + (\lambda_n - 2)(1 - v^2)$$
$$+ \beta^2[\lambda_n^3 - 4\lambda_n^2 + \lambda_n(5 - v^2) - 2(1 - v^2)] = 0.$$

(7.114)

Therefore for each mode characterized by a specific value of $n > 0$ there are two distinct resonant frequencies. For $n = 0$ corresponding to what is sometimes called the "breathing mode," there is only one real frequency. The larger of the two nondimensional resonant frequencies is denoted by $\Omega_n^{(2)}$ and the lower by $\Omega_n^{(1)}$, so that equation (7.14) can be rewritten in the form

$$[\Omega^2 - (\Omega_n^{(1)})^2][\Omega^2 - (\Omega_n^{(2)})^2] = 0.$$

(7.115)

Figure 7.9 shows the variation of the $\Omega_n^{(1)}$ and $\Omega_n^{(2)}$ as a function of n.

7.19 Forced Vibrations of a Spherical Shell
We can now consider an excitation pressure of the form $p_a = f(\eta)e^{-i\omega t}$. Expanding $f(\eta)e^{-i\omega t}$,

$$f(\eta) = \sum_{n=0}^{\infty} f_n P_n(\eta),$$

(7.116)

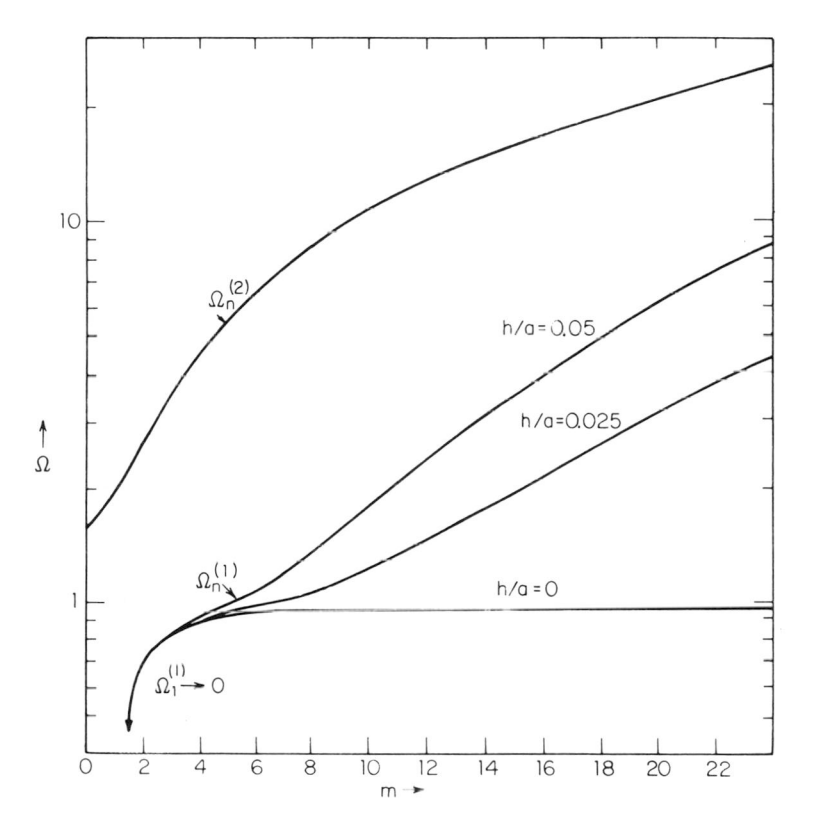

Figure 7.9
Resonance frequencies for axisymmetric vibrations of a spherical shell $v = 0.29$ (upper branch does not vary significantly with h/a).

where

$$f_n = \left(\frac{2n+1}{2}\right) \int_{-1}^{1} f(\eta) P_n(\eta)\, d\eta. \tag{7.117}$$

As in the case of free vibration, we also express v and w in terms of an infinite series of Legendre polynomials, by equations (7.110) and (7.111). Substituting the foregoing expressions into the equations of motion, we find that equations (7.104) and (7.105) can be satisfied if for each value of n the coefficients V_n and W_n satisfy

$$[\Omega^2 - (1 + \beta^2)(v + \lambda_n - 1)] V_n$$
$$- [\beta^2 (v + \lambda_n - 1) + (1 + v)] W_n = 0,$$
$$-\lambda_n [\beta^2 (v + \lambda_n - 1) + (1 + v)] V_n \tag{7.118}$$
$$+ [\Omega^2 - 2(1 + v) - \beta^2 \lambda_n (v + \lambda_n - 1)] W_n = -\frac{a^2 (1 - v^2)}{Eh} f_n.$$

We can then solve these equations for V_n and W_n to obtain

$$V_n = -\frac{a^2 (1 - v^2)}{Eh} f_n \frac{[\beta^2 (v + \lambda_n - 1) + (1 + v)]}{(\Omega_n^{(1)})^2 - (\Omega_n^{(2)})^2}$$
$$\cdot \left\{ \frac{1}{\Omega^2 - (\Omega_n^{(1)})^2} - \frac{1}{\Omega^2 - (\Omega_n^{(2)})^2} \right\}, \tag{7.119}$$

$$W_n = -\frac{a^2 (1 - v^2)}{Eh} f_n \frac{[\Omega^2 - (1 + \beta^2)(v + \lambda_n - 1)]}{(\Omega_n^{(1)})^2 - (\Omega_n^{(2)})^2}$$
$$\cdot \left\{ \frac{1}{\Omega^2 - (\Omega_n^{(1)})^2} - \frac{1}{\Omega^2 - (\Omega_n^{(2)})^2} \right\}. \tag{7.120}$$

The actual displacements are then found using equations (7.110) and (7.111). As was the case for the cylindrical shell, it will be found useful to define a modal mechanical impedance given by $Z_n = f_n/(-i\omega W_n)$, given by

$$Z_n = -\frac{i \rho_s c_p}{\Omega} \frac{h}{a} \frac{[\Omega^2 - (\Omega_n^{(1)})^2][\Omega^2 - (\Omega_n^{(2)})^2]}{[\Omega^2 - (1 + \beta^2)(v + \lambda_n - 1)]}. \tag{7.121}$$

References

1. Y. C. Fung, *Foundations of Solid Mechanics* (Englewood Cliffs, N.J.: Prentice-Hall, 1965), pp. 127–131.

2. S. Timoshenko, *Vibration Problems in Engineering*, 3rd ed. (Princeton, N.J.: Van Nostrand, 1955), pp. 329–331.

3. Ibid., p. 326.

4. Ibid., pp. 441–452.

5. A Powell, "On the Approximation to the 'Infinite' Solution by the Method of Normal Modes for Random Vibrations," *J. Acoust. Soc. Am.* 30:1136–1139 (1958).

6. S. Timoshenko and S. Woinowsky-Krieger, *Theory of Plates and Shells*, 2nd ed. (New York: McGraw-Hill, 1959), pp. 81–83.

7. R. D. Mindlin, "Influence of Rotary Inertia and Shear on Flexural Motion of Isotropic Elastic Plates," *J. Appl. Mech.* 18:31–38 (1951). M. C. Junger and D. Feit, *Sound, Structures, and Their Interaction*, 1st ed. (Cambridge, MA: MIT Press, 1972), pp. 152–156.

8. A. W. Leissa, "Vibration of Shells," NASA SP-288, National Aeronautics and Space Administration, Washington, D.C. (1973).

9. H. Kraus, *Thin Elastic Shells* (New York: John Wiley, 1967), p. 297.

10. R. N. Arnold and G. B. Warburton, "Flexural Vibrations of the Walls of Thin Cylindrical Shells Having Freely Supported Ends," *Proc. Royal Soc. Lond.* A197:238–256 (1949).

11. M. Heckl, "Vibrations of Point-driven Cylindrical Shells," *J. Acoust. Soc. Am.* 34:1553–1557 (1962).

The preceding chapters dealt with either the purely acoustic problem, that is, the sound field given a prescribed velocity distribution, or the purely structural problem, that is, the response of an elastic structure to prescribed loading. In this chapter we initiate the study of interaction problems by considering the response of elastic plates in contact with relatively dense fluids such as water. We first direct our attention to the concept of coincidence frequency (section 8.1) and the effect of radiation loading on the flexural wave phase velocity (section 8.2). The next few sections deal with plates sufficiently extended and damped to respond effectively as infinite structures. In section 8.3 we examine the acoustic far-field and the effect of damping on the pressure radiated by point-excited plates. Section 8.4 introduces the reader to the effect of a frame stiffener on the plate response. We then examine the far-field (section 8.5) and power (section 8.6) radiated by uniform plates subjected to distributed or multipolar exciting forces. Sound radiation by plates sufficiently small and undamped to display a predominantly standing-wave pattern is the subject of section 8.7. Finally, the results in sections 8.3 and 3.10 are combined to examine the alteration of the sound field by a laminar bubble swarm interposed between the plate and the ambient liquid medium.

8.1 Coincidence Frequency

It was shown in section 5.11 that an infinite train of straight-crested waves with a *subsonic wave number* $k_s > k$ does not radiate acoustic energy, the pressure field decaying exponentially with distance from the boundary. For *supersonic wave numbers*, $k_s < k$, plane sound waves are radiated in a direction intersecting the normal to the plate at an angle

$$\theta_c = \sin^{-1}(k_s/k). \tag{8.1}$$

From equation (8.1) it is obvious that a real angle θ exists only when $k_s < k$. This is consistent with equations (5.60)–(5.62) which indicate that straight crested waves in a planar boundary radiate sound pressure only when $k_s < k$. If that boundary is a plate, and if for the time being we assume that fluid loading is negligible, as for example in the atmosphere, k_s can be taken equal to the flexural wave number k_f, equation (7.62). Because k_f is frequency dependent, true sound pressure is radiated only for frequencies such that

$$\frac{k_f}{k} = \left(\frac{\rho_s h}{D}\right)^{1/4} \frac{c}{\sqrt{\omega}} < 1. \tag{8.2}$$

for $1/16''$ of Al $f_c = 16 \times 9.3\,kHz = 148.8\,kHz$

$\omega \geq \omega_c$ for radiation

Solving this equation for frequency ω, true radiated pressure waves exist only for $\omega \geq c^2(\rho_s h/D)^{1/2}$. The lower bound of this frequency range is the *coincidence* or *critical* frequency ω_c.

This is compactly expressed in terms of the compressional wave velocity introduced in the latter of equations (7.62):

$$\omega_c = c^2\left(\frac{\rho_s h}{D}\right)^{1/2} = \frac{\sqrt{12}c^2}{hc_p}.$$

$f_c(steel) = \dfrac{\sqrt{12}*2.25\times10^{10}}{5.24\times10^5 * 2\pi * 2.54}$ for $2.54\,cm$ of steel

$= 9.32\,kHz$

$$(8.3)$$

$= 23.7\,kHz/cm$

In an alternative interpretation the coincidence frequency is defined as the frequency at which the phase speed of flexural waves in the absence of fluid loading coincides with the sound speed of the ambient fluid. Equation (8.3) indicates that this frequency is smaller for thick plates and large c_p or, in other words, the stiffer the plate the lower the frequency. To provide a reference point, it is noted that for one-inch aluminum or steel plates the coincidence frequency is about 9,300 Hz. In deriving equation (8.3), we assumed that fluid loading is negligible and that the wave train extends over the entirety of the infinite plane $z = 0$. It was shown in section 5.12 that, because boundaries give rise to supersonic wave numbers, a displacement distribution confined to a finite portion of the planar surface radiates sound even if $k_s > k$.

As stated in section 7.11 the validity of thin plate theory is limited to plates (or frequency ranges) where the plate thickness is less than one-twentieth of the shear wavelength. For steel or aluminum plates immersed in water, this criterion is satisfied when $\omega/\omega_c < 0.7$. The Timoshenko-Mindlin theory,[1] developed in section 7.11, is therefore required when analyzing coincidence effects.

In water: $c = 1.5E5\,cm/sec$

upsidedown or $1/\text{fty factor}$ of 2.?

8.2 Phase Velocity of Flexural Waves in a Submerged Plate

From the discussion of the coincidence frequency relevant to flexural waves in the absence of fluid loading, we now move on to consider the phase velocity of flexural waves in a submerged plate. For this purpose, we use the equation of motion governing flexural waves modified to account for the presence of fluid loading. Equation (7.59) then becomes

Thin, non-radiating plates; $\omega \leq \omega_c$

$$D\nabla^4 w + \rho_s h\frac{\partial^2 w}{\partial t^2} = -p|_{z=0},\qquad (8.4)$$

where $p|_{z=0}$, the pressure exerted by the fluid on the vibrating plate, is a

$\dfrac{h}{\lambda_s} < \dfrac{1}{20} \Rightarrow \dfrac{fh}{c_s} < \dfrac{1}{20} \Rightarrow \dfrac{\omega h}{\omega_c h} < \dfrac{2\pi c_s}{20\omega_c h} = \dfrac{\pi c_s c_p}{10\sqrt{12}c^2} = 1.$

$c_s(Al) = 6.42E5,\ c_p(Al) = 5.45E5$

solution of the wave equation (2.31). The pressure on the surface is related to the surface displacement w by the boundary condition, equation (2.1) or (2.24), which is merely a statement of conservation of momentum for fluid particles at the water-plate interface.

We again seek the condition under which a solution of the form

$$w = W \exp[i(\gamma x - \omega t)] \tag{8.5}$$

can be used to solve the system of equations (2.24), (2.32), and (8.4). Equation (8.5) is a solution provided the dispersion relation

$$K = \sqrt{K_x^2 + K_z^2} \; ; \; K_x = \gamma$$

$$F(\omega, k) = \frac{\rho \omega}{(\gamma^2 - k^2)^{1/2}} + \omega \rho_s h \left(1 - \frac{D\gamma^4}{\rho_s h \omega^2}\right) = 0 \tag{8.6}$$

is satisfied. The phase velocity of a fluid-loaded flexural wave is therefore the value of $c_f = \omega/\gamma$ that is a solution of this equation. Writing equation (8.6) in terms of c_f, we obtain

$$\frac{\rho c}{\omega \rho_s h} + \left(\frac{c^2}{c_f^2} - 1\right)^{1/2} \left(1 - \frac{\omega^2 c^4}{\omega_c^2 c_f^4}\right) = 0, \tag{8.7}$$

where ω_c is the coincidence frequency defined in equation (8.3).

To find the frequency-dependent normalized roots c_f/c of equation (8.7) it is convenient to introduce two nondimensional parameters following Crighton and Innes.[2] One is $\Omega = \omega/\omega_c$, the frequency normalized to the coincidence frequency, and the other, $\varepsilon = \rho c/\omega_c \rho_s h$, is the "fluid loading at coincidence" parameter. The latter definition comes from the notion that for a plate, fluid loaded on one side, of mass per unit area $\rho_s h$, the appropriate nondimensional measure of fluid loading at any frequency ω is $\rho c/\omega \rho_s h$.

Now setting $(c^2/c_f^2) - 1 = \tau^2$, equation (8.7) can be rewritten in the form

$$\tau^5 + 2\tau^3 + (1 - \Omega^{-2})\tau - \varepsilon \Omega^{-3} = 0. \tag{8.8}$$

Equation (8.8) is a fifth-order polynomial with real coefficients that has at least one real positive root at all frequencies. A numerical analysis reveals that this root corresponds to a subsonic wave, $c_f/c < 1$. For a steel plate submerged in water, figure 8.1 illustrates the frequency dependence of this

$$\Rightarrow \frac{\omega_c}{\omega} > 0.71 \Rightarrow \theta_c = \sin^{-1}\left(\frac{\omega_c}{\omega}\right)^{1/2}$$

i.e. $0.71 < \left(\frac{\omega_c}{\omega}\right) < 1; \quad 45° < \theta_c < 90°$

phase velocity. The other, complex roots of equation (8.8) are at this juncture of no consequence to our analysis.

One can obtain an approximation to the real root by rewriting equation (8.6) in the form

$$\gamma^4 = k_f^4 \left[1 + \frac{\rho}{\rho_s h (\gamma^2 - k^2)^{1/2}} \right], \qquad k_f = \left(\frac{12^{1/2}\omega}{h\,C_p} \right)^{1/2} \qquad (8.9)$$

where we have used equation (7.62) to define k_f. As a first approximation we can substitute $\gamma = k_f$ in the right-hand side of equation (8.9). Introducing the parameters ε and Ω, equation (8.9) yields

$$\gamma = k_f \left[1 + \frac{\varepsilon}{\Omega^{1/2}(1 - \Omega)^{1/2}} \right]^{1/4}. \qquad (8.10)$$

This approximation is valid for $\Omega < 1$, where it is extremely accurate (figure 8.1).

Examination of equation (8.10) reveals that fluid loading increases the fluid-loaded real wave number γ over its in vacuo value k_f. Since phase velocity is inversely proportional to γ, the fluid decreases the flexural wave velocity. The effect of other, complex roots of equation (8.8) on the response variables depends on which of a number of possible branch cuts one uses in the transform plane[3] when dealing with the forced vibration problems (section 8.3). As a result, no particular physical significance can be attributed to them.

Another very low-frequency approximation ignores the inertia of the plate and the compressibility of the fluid. Equation (8.6) now simplifies to

$$(\rho\omega/\gamma) - (D\gamma^4/\omega) = 0, \qquad \omega/\omega_c \ll 1. \qquad (8.11)$$

This equation can be solved explicitly. Its real root,

$$\gamma = (\rho\omega^2/D)^{1/5}, \qquad \omega/\omega_c \ll 1. \qquad (8.12)$$

is the dotted line in figure 8.1. It is seen to be fairly accurate only at extremely low frequencies. Since this result is obtained by ignoring the compressibility of the liquid and the inertia of the plate, flexural waves thus modeled recall Biot's interface waves described in connection with equation (2.58b).

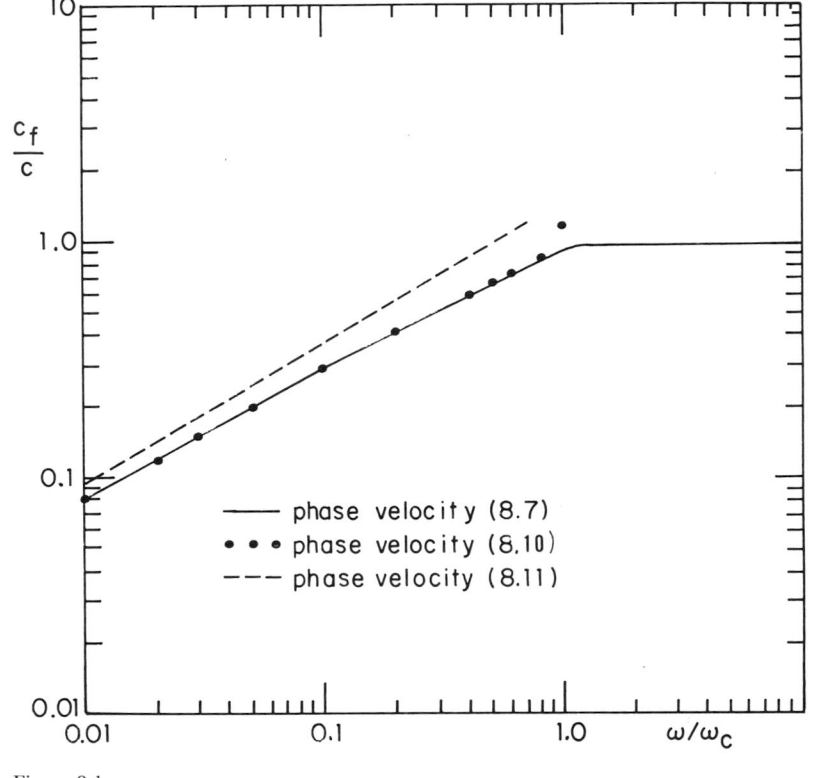

Figure 8.1
Phase velocity of flexural wave on an elastic plate that is loaded by an acoustic fluid ($\rho_s/\rho = 7.8$; $c_p/c = 3.5$).

$$C_p = \sqrt{E/(1-\nu^2)\rho_s}$$

We can now proceed to study the problem of sound radiation by locally excited plates.

8.3 Effectively Infinite Locally Excited Plates

8.3.1 THE PLATE RESPONSE
Even though literally such plates, of course, do not exist, they provide a valid mathematical model for numerous practical situations where the plate exhibits sufficient structural as well as radiation damping and is of sufficient area so that the boundary-reflected waves are negligible compared with the waves propagating from the excitation area. In fact, even when these

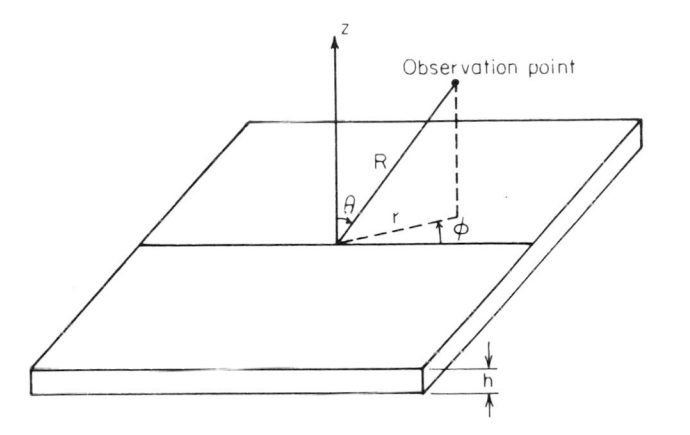

Figure 8.2
Geometry of the plate and coordinates defining observation point.

requirements are only met marginally, the infinite plate model provides insight into the sound radiation process, because differences in the subsonic $(k_s > k)$ portion of the wave number spectra of the finite and infinite plate responses have no effect on the far-field. A criterion for the validity of the infinite plate model is formulated in section 8.7 in terms of the loss factor, the plate parameters, and frequency.

The analysis will first be carried through with thin-plate theory, whereby forced motion of the plate is governed by equation (7.59). The distributed load p_a in that equation now becomes the sum of a concentrated harmonic transverse driving force $Fe^{-i\omega t}$ and of the, as yet unknown, acoustic surface pressure. Because of the cylindrical symmetry of the problem, the analysis is formulated in terms of the cylindrical coordinates r and z. The $z = 0$ plane coincides with the midplane of the plate. The drive point is at $r = 0$ (see figure 8.2). Factoring out the flexural rigidity D in equation (7.59), the inertia term becomes proportional to $\rho_s h \omega^2 / D$, which equals k_f^4. The equation of motion then becomes

$$D(\nabla_\sigma^4 - k_f^4)w(r) = -p(r,0) + F\delta(r)/2\pi r. \qquad (8.13)$$

The surface Laplace operator ∇_σ^2 was defined in equation (2.61). Because of symmetry, derivatives with respect to ϕ vanish. The problem therefore lends itself to formulation in terms of the Hankel transform techniques

illustrated in section 5.3. Applying the Hankel transform to equation (8.13), one obtains a relation between the transforms $\tilde{w}$ and $\tilde{p}$:

$$D(\gamma^4 - k_f^4)\tilde{w}(\gamma) = -\tilde{p}(\gamma;0) + (F/2\pi), \tag{8.14}$$

where γ is the transform parameter.

Using the definition

$$\tilde{Z}_p = -i\omega\rho_s h(1 - \gamma^4/k_f^4), \tag{8.15}$$

equation (8.14) becomes

$$-i\omega\tilde{Z}_p\tilde{w}(\gamma) = -\tilde{p}(\gamma;0) + (F/2\pi). \tag{8.16}$$

The impedance in equation (8.15) is based on classical plate theory used in equation (7.59). The Timoshenko-Mindlin theory, equation (7.79), required to determine the plate response in the coincidence frequency range and higher yields a more complicated expression for $\tilde{Z}_p$:

$$\tilde{Z}_p = -i\omega\rho_s h \frac{[1 - (1/k_f^4)(\gamma^2 - k_p^2)(\gamma^2 - k_R^2)]}{1 + (h^2/12)(c_p^2/c_R^2)(\gamma^2 - k_p^2)}, \tag{8.17}$$

where $k_p = \omega/c_p$, $k_R = \omega/c_R$, and $c_R = \kappa(G/\rho_s)^{1/2}$ is the Rayleigh wave speed.

Since both $\tilde{w}$ and $\tilde{p}$ are unknown, a second relation between these two transforms must be introduced. This is provided by the transform solution of the Helmholtz equation, equation (5.14a), subject to the boundary condition in equation (5.14b). The resulting relation between $\tilde{w}(\gamma)$ and $\tilde{p}(\gamma;z)$, equation (5.15), is formulated here in terms of the specific acoustic impedance transform $\tilde{Z}_a$:

$$\tilde{p}(\gamma;z) = -i\omega\tilde{Z}_a(\gamma)\tilde{w}(\gamma)e^{iz\sqrt{k^2-\gamma^2}}, \tag{8.18}$$

where

$$\tilde{Z}_a(\gamma) = \frac{\rho\omega}{\sqrt{k^2-\gamma^2}}. \tag{8.19}$$

When this relation is substituted in equation (8.16), an equation that can be solved for $\tilde{w}$ is obtained:

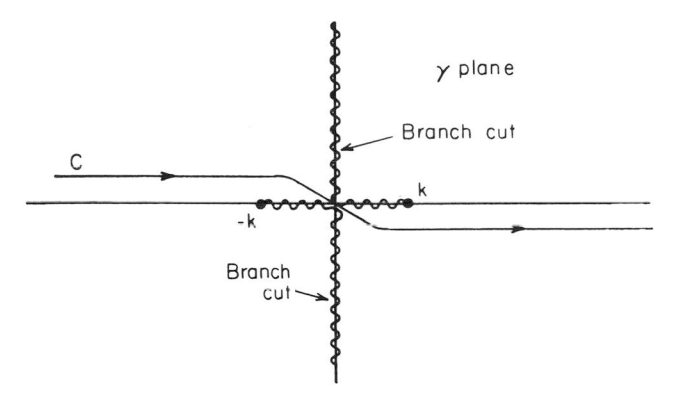

Figure 8.3
Integration path in γ plane for integrals defined in equations (8.21) and (8.28).

$$\tilde{w}(\gamma) = \frac{iF}{2\pi\omega}(\tilde{Z}_a + \tilde{Z}_p)^{-1}. \tag{8.20}$$

The inverse transform of this expression is the plate response, which can be expressed as a contour integral in the complex γ plane. Using equations (5.3b) and (5.19a), the plate response is given by

$$w(r) = \frac{iF}{4\pi\omega}\int_C \frac{H_0(\gamma r)\gamma\,d\gamma}{\tilde{Z}_a(\gamma) + \tilde{Z}_p(\gamma)}. \tag{8.21}$$

The contour C in figure 8.3 is along the real axis, being taken slightly above the negative portion of the real axis and slightly below its positive portion. The branch cut shown was chosen so that $\mathrm{Re}(\sqrt{\gamma^2 - k^2}) > 0$ along the contour of integration.

Using Cauchy's theorem, we can write the integral defined in equation (8.21) as a sum of two parts, one an integral along the branch cut and the other a sum of residue terms associated with the singularities of the integrand in the upper half of the γ plane, $\mathrm{Im}(\gamma) \geq 0$. In order to examine the residue term contributions we must first determine the location of the poles of the integrand. Substituting $\zeta = (\gamma^2 - k^2)^{1/2}$ in the denominator of the integrand of equation (8.21), we obtain the characteristic equation whose roots identify the poles of the integrand:

$$\text{from}\{\tilde{Z}_a(\gamma) + \tilde{Z}_p(\delta)\} = 0$$

$$\gamma \equiv [(c/c_f)^2 - 1]^{1/2}$$

$$\zeta^5 + 2k^2\zeta^3 + (k^4 - k_f^4)\zeta - \frac{\rho}{\rho_s k_f h}k_f^5 = 0. \tag{8.22}$$

Recognizing that $\zeta = k\tau$, equation (8.22) is equivalent to equation (8.8).

This characteristic equation has five roots, three of which satisfy the desired condition $\mathrm{Re}\,\zeta = \mathrm{Re}\,(\gamma^2 - k^2)^{1/2} > 0$. In general, it is necessary to solve equation (8.22) numerically. In the low-frequency limit where the equation reduces to

$$\zeta^5 - \frac{\rho}{\rho_s k_f h}k_f^5 = 0, \qquad \frac{\omega}{\omega_c} \ll 1, \tag{8.23}$$

these roots can be constructed analytically:

$$\zeta_n = k_f\left(\frac{\rho}{\rho_s k_f h}\right)^{1/5} \exp\left(\frac{i2\pi n}{5}\right), \qquad n = 0, 1, 2, 3, 4. \tag{8.24}$$

Since the real part of ζ must be greater than zero, the relevant pole roots correspond to $n = 0, 1, 4$. The corresponding poles in the γ plane are then given approximately by

$$\gamma_1 = \delta,$$
$$\gamma_2 = \delta e^{i2\pi/5}, \tag{8.25}$$
$$\gamma_3 = \delta e^{i4\pi/5},$$

where $\delta = k_f(\rho/\rho_s k_f h)^{1/5}$. The approximations made here are equivalent to the approximations made in obtaining equation (8.12).

The contributions of γ_2 and γ_3 to the field response $k_f r \gg 1$ are exponentially small since they have a large imaginary part. The residue contribution of the pole at γ_1 is evaluated by means of equation (8.21), using the asymptotic form of $H_0(\gamma r)$, equation (5.19b):

$$w_1(r) \simeq (F/10)(2/\pi D\rho r)^{1/2} \exp(i\gamma_1 r - i\pi/4), \qquad k_f r \gg 1, \qquad \omega/\omega_c \ll 1. \tag{8.26}$$

As in the discussion preceding equation (8.10), we note that the amplitude of the above far-field contribution depends only on the stiffness of the plate

and the density of the fluid, and is independent of the fluid compressibility and plate inertia.

An asymptotic evaluation for large r of the branch cut integral yields a disturbance with spatial dependence of the form $(1/r)^2$ and a phase speed characteristic of acoustic waves. Therefore, in the far-field of the drive point, this latter wave is negligible as compared with the fluid modified flexural wave in equation (8.26).

The above discussion is restricted to the extreme low-frequency limit $\Omega = \omega/\omega_c \ll 1$. If we are not as restrictive and merely assume that the approximation derived in equation (8.10) can be used for γ_1, the residue contribution at γ_1 to the response is

$$w_1(r) = \frac{(iF/\omega\mathcal{Z}_p) H_0(\gamma_1 r)}{(\gamma/k_f)^2 \{1 + \varepsilon/[4\Omega(\gamma_1^2/k_f^2)(\gamma_1^2/k_f^2 - \Omega)]^{3/2}\}}, \qquad \Omega < 1, \qquad (8.27)$$

where $\mathcal{Z}_p$ is the in vacuo impedance of a point driven plate, equation (7.68). Once again, the far-field contributions of γ_2 and γ_3 can be neglected. The plate response is therefore associated with the flexural wave traveling at a fluid modified flexural wave speed whose amplitude displays cylindrical spreading of the form $(1/r)^{1/2}$.

In order to determine the response $w(r)$ at any arbitrary frequency and position, one must resort to a numerical evaluation of the integral in equation (8.21). The integral can again be evaluated by integration, over the contour C (figure 8.3). This equals the sum of the integral along the branch cuts and the residue contributions in the upper half-plane, $\mathrm{Im}\,\gamma > 0$. The other procedure is to evaluate the integral over C as an integral over real values of γ.

This first procedure is the "contour integral method" and the second the "direct integration" method. In this section we shall discuss the numerical results obtained by the latter method.[4] At all frequencies the integrand displays a real pole that must be removed from the axis before the integral can be evaluated directly. This is accomplished by introducing structural damping embodied in the complex Young's modulus, $E^* \equiv E(1 - i\eta_s)$, where η_s is the structural loss factor for the plating.

Figure 8.4 illustrates the response of a fluid-loaded plate as a function of $k_f r$ for two frequencies below coincidence. These calculations were normalized to the drive-point impedance of an in vacuo plate performed for a 1-cm-thick steel plate in water so that the coincidence frequency is 23.6 kHz. It is recalled that the in vacuo drive point velocity calculated from

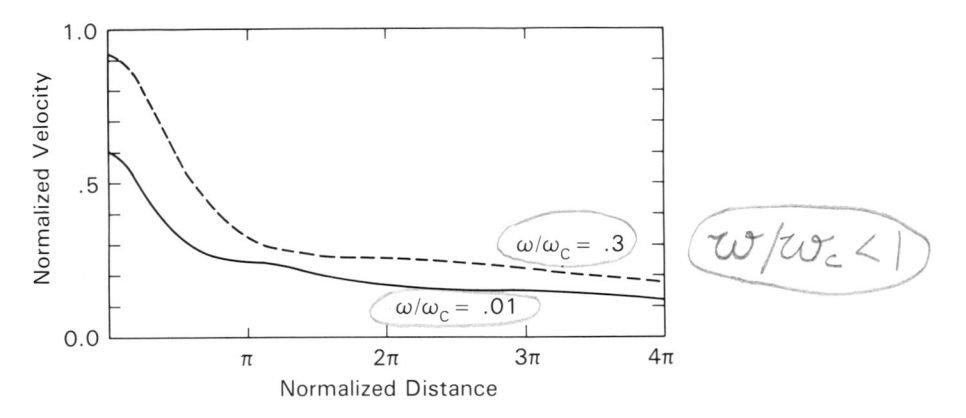

Figure 8.4
Velocity normalized to the drive-point velocity of a point-driven plate (in vacuo) as a function of distance normalized to the flexural wavelength.

equation (7.66) is frequently independent. The drive point admittance $\dot{w}(0)/F = -i\omega w(0)/F$, whose real and imaginary parts are normalized to the purely real in vacuo admittance, the inverse equation (7.68), is shown in figure 8.5. As the frequency approaches coincidence, the real part of the admittance approaches the in vacuo value while the imaginary part approaches zero.

8.3.2 THE PRESSURE FIELD IN RESPONSE TO A POINT FORCE
The pressure also is obtained as an inverse Hankel transform by substituting $\tilde{w}(\gamma)$ from equation (8.20) in equation (8.18). This inverse transform can be expressed as a contour integral similar to equation (8.21):

$$p(r,z) = \frac{F}{4\pi} \int_C \frac{\tilde{Z}_a(\gamma)\, e^{iz(k^2-\gamma^2)^{1/2}}}{\tilde{Z}_a(\gamma) + \tilde{Z}_p(\gamma)} H_0(\gamma r)\, \gamma\, d\gamma. \tag{8.28}$$

This integral can be evaluated in the far-field by the method of stationary phase illustrated in section 5.3. Combining equations (8.18), (8.20) and (5.22) yields

$$p(R,\theta) = \frac{-i\omega\rho F e^{ikR}}{2\pi R} [\tilde{Z}_a(k\sin\theta) + \tilde{Z}_p(k\sin\theta)]^{-1}. \tag{8.29}$$

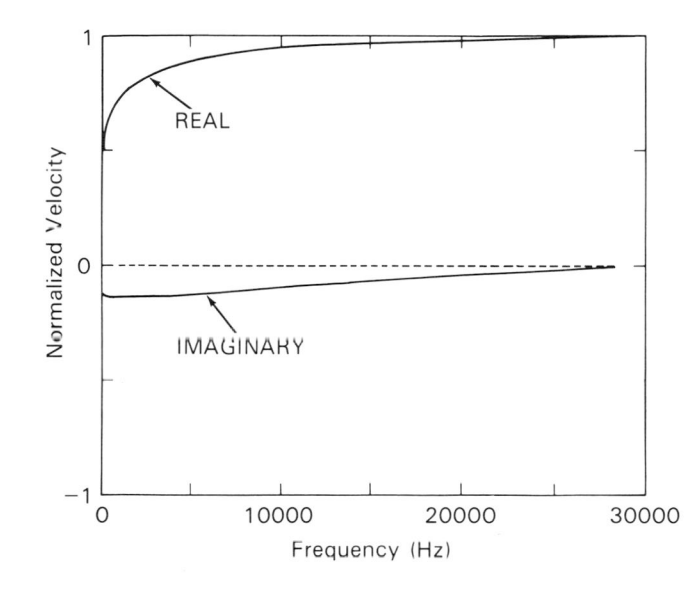

Figure 8.5
Drive-point normalized velocity of a point-driven fluid loaded plate as a function of frequency (coincidence frequency for this plate is 23.6 kHz).

Here we have transformed to spherical coordinates, $z = R \cos \theta$, $r = R \sin \theta$ (see figure 8.2). Using the definition of $\tilde{Z}_a$ and $\tilde{Z}_p$ to evaluate equation (8.29) explicitly, the far-field pressure becomes

$$p(R, \theta) = \frac{-ikFe^{ikR}}{2\pi R} \frac{\cos \theta}{1 - ikh(\rho_s/\rho) \cos \theta [1 - (\omega^2/\omega_c^2) \sin^4 \theta]}. \tag{8.30}$$

In the foregoing expression we have rewritten k^4/k_f^4 as ω^2/ω_c^2.

As discussed earlier, in the high-frequency range the thin-plate equations of motion must be replaced by the Timoshenko-Mindlin equation (7.79). Using equation (8.17) for $\tilde{Z}_p(\gamma)$, the expression for the far-field pressure becomes

$$p(R, \theta) = -ikF[\exp(ikR)/2\pi R] \cos \theta$$
$$\cdot \left\{ 1 - ikh(\rho_s/\rho) \cos \theta \frac{[1 - (\omega/\omega_c)^2[\sin^2 \theta - (c/c_p)^2][\sin^2 \theta - (c/c_R)^2]]}{1 + (k^2 h^2/12)(c_p/c_R)^2[\sin^2 \theta - (c/c_p)^2]} \right\}^{-1}. \tag{8.31}$$

Set to zero for θ_c!

Steel & Aluminum

This is of the form $\left(\dfrac{A^2}{A - iB}\right)$; $B = 0 \Rightarrow$ max.!?!

8.3.3 PRESSURE MAXIMUM; EFFECT OF STRUCTURAL DAMPING

The pressure along the normal to the plate $\theta = 0$ coinciding with the axis of the driving force reduces to

$$p(R,0) = -\frac{iFk}{2\pi R}\frac{\exp(ikR)}{1 - ikh(\rho_s/\rho)} \qquad \text{from Eq. 8.30}$$

$$\simeq -\frac{iFk}{2\pi R}\exp(ikR), \qquad \left(kh\frac{\rho_s}{\rho}\right)^2 \ll 1. \qquad (8.32)$$

The notable feature of this result is that it depends on only one plate parameter, that is, the weight per unit area $h\rho_s$. Specifically, the flexural rigidity D has no effect on the pressure. Along the normal through the drive point a plate-radiated field is therefore indistinguishable from the field radiated by a membrane of comparable weight. At low frequencies, where the kh factor in equations (8.30) and (8.31) is negligible compared with unity, the pressure field tends to $F/\lambda R$, i.e., twice the field radiated by a force applied directly to the liquid medium, equation (6.34a). If the plate were exposed to the fluid medium on both its surfaces, thereby doubling the radiation loading, the two expressions would be identical.

In the frequency range below coincidence, the maximum pressure is given by $p(R,0)$. Above coincidence, a larger maximum appears near the coincidence angle θ_c, equation (8.1), where the trace velocity of an acoustic wave $c/\sin\theta_c$ matches the flexural velocity in the plate c_f. For thin-plate theory, using equations (8.1)–(8.3),

$$\theta_c = \sin^{-1}\left(\frac{\omega_c}{\omega}\right)^{1/2}. \qquad \text{i.e., when (Eq. 8.1)} \quad \frac{c}{c_f} = \sin\theta_c \qquad (8.33)$$

The imaginary term in the denominator of equation (8.30) vanishes for $\theta = \theta_c$. When the corresponding value of $\cos\theta_c$ computed from equation (8.33), viz., $[1 - (\omega_c/\omega)]^{1/2}$, is substituted in equation (8.30), the pressure at the coincidence angle is found to be

$$p(R,\theta_c) = -\frac{iFke^{ikR}}{2\pi R}(1 - \omega_c/\omega)^{1/2}. \qquad \text{Here the acoustic fluid is not a liquid.} \qquad (8.34)$$

This maximum pressure is thus located in the proximity of a conical surface defined by the coincidence angle, equation (8.33).

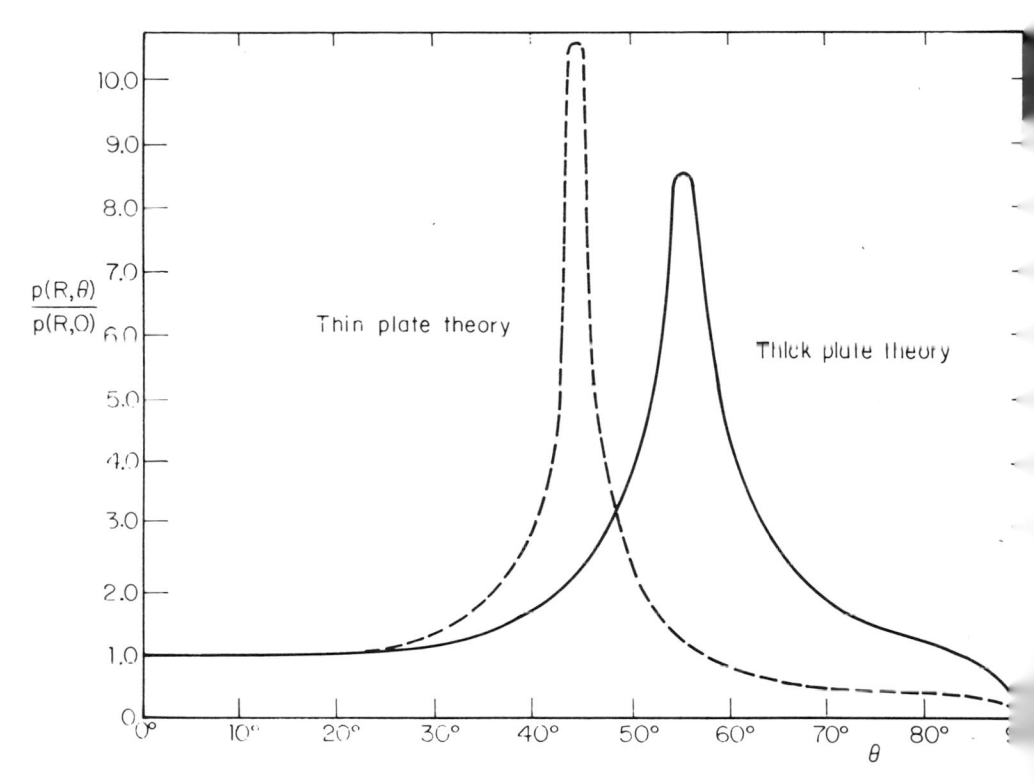

Figure 8.6
Far-field radiated by a point-excited effectively infinite plate computed from thin-plate theory [equation (8.30)] and thick-plate theory [equation (8.31)].

TM Theory

When the acoustic fluid is a liquid, the Timoshenko-Mindlin theory is mandatory if meaningful results are to be obtained in the coincidence range. Thick-plate theory yields larger values of θ_c and smaller values of peak pressure (see figure 8.6). The phase velocity of the structural wave predicted by this theory (section 7.11) tends to the Rayleigh wave velocity c_R, as is the case for the Timoshenko beam [equation (7.34)]. The coincidence angle thus tends to

$$\lim_{\omega \to \infty} \theta_c = \sin^{-1}\left(\frac{c}{c_R}\right). \tag{8.35}$$

In lieu of equation (8.34), the pressure at this coincidence angle is given by

For TM Theory have from

Eq. 8.31 : $\sin^2\theta_c = \frac{1}{2}\{(x+y) \pm [(x-y)^2 + 4\frac{\omega_c^2}{\omega^2}]^{1/2}\}$

θ_c = "Coincidence" angle. At $\theta_c = 90°$ $\omega = \bar{\omega} = $ Coincidence frequency :

$$\lim_{\omega \to \infty} p(R, \theta_c) = -\frac{iFk}{2\pi R}\left[1 - \left(\frac{c}{c_R}\right)^2\right]^{1/2} \exp(ikR). \tag{8.36}$$

Eq. 8.31 neglecting the second term. ⟹ Scattering angle; see Eq. 8.31

This pressure differs only by the factor $[1 - (c/c_R)^2]^{1/2}$ from the low-frequency limit of the pressure obtained by neglecting the ρ_s term in equation (8.30). In contrast, thin-plate theory, equation (8.34), predicts a high-frequency limit that approximately equals the low-frequency limit of equation (8.32).

The above results neglect structural damping. Small loss factors are accounted for by multiplying c_p by $(1 - i\eta_s)^{1/2} \approx (1 - \frac{1}{2}i\eta_s)$, or ω_c by $(1 + \frac{1}{2}i\eta_s)$. The imaginary component of this coincidence frequency keeps the kh term in the denominator of the expression for the pressure field of a point-excited plate, equation (8.30), from vanishing at the coincidence angle. The coincidence pressure, which takes the place of equation (8.34), becomes

$$p(R, \theta_c) = -\frac{iFk[1 - (\omega_c/\omega)]^{1/2}}{2\pi R\{1 + \eta_s kh(\rho_s/\rho)[1 - \omega_c/\omega]^{1/2}\}}. \tag{8.37}$$

In the high-frequency range above coincidence, the maximum far-field pressure is therefore markedly reduced by structural damping (see figure 8.7).

Below coincidence, the maximum pressure displays no significant dependence on damping, because equation (8.32) is independent of the plate stiffness and hence of both E and η_s. Below coincidence, the importance of structural damping derives from the fact that, in its absence, the infinite plate would never be a suitable mathematical model for a real, necessarily finite plate.

The above results have been summarized in table 8.1, where the factor $(k/2\pi R)$ is written as $(\lambda R)^{-1}$. Furthermore, a characteristic frequency ω_m is defined that marks the lower limit of the frequency range where the inertial impedance of the plate predominates over radiation loading, i.e., where $\omega\rho_s h > \rho c$. The ratio of this frequency to the coincidence frequency, equation (8.3) is independent of the plate thickness. It equals the parameter ε defined in connection with (8.8):

$X \& Y > 1$ or $X \& Y < 1$ for $\bar{\omega} = $ real.

$$\frac{\omega_m}{\omega_c} = \varepsilon = \frac{\rho c_p}{\sqrt{12\rho_s c}}. \tag{8.38}$$

$(1-X)(1-Y) \Rightarrow X \& Y > 1$

$\to \bar{\omega}^2 = \omega_c^2 / [1 - X - Y + XY]$; $X \equiv c^2/c_p^2$

$Y \equiv c^2/c_R^2$

Note $XY - X - Y = 0 \Rightarrow$
$c^2 = c_p^2 + c_R^2$

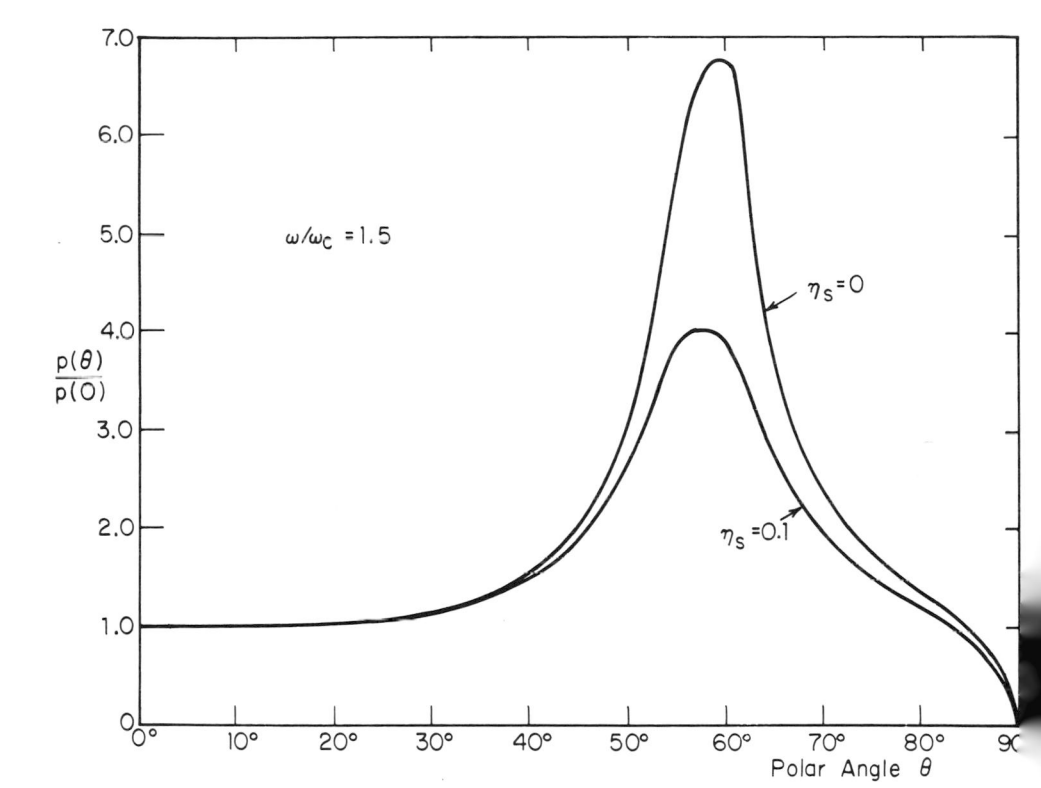

Figure 8.7
Effect of structural damping on the far-field of a point-excited efffectively infinite plate.
[Reproduced from Feit[5]]

For a steel plate in water, this ratio is 0.13. The acoustic wavelength at frequency ω_m is defined as λ_m.

8.4 Infinite Line-Driven Elastic Plate

Section 8.3 dealt with point-excited elastic plates of infinite extent. A plate excited by a line force or moment is a prerequisite to the analysis of the response and radiation characteristics of frame-stiffened plates. In such problems the action of the frame on the plate is accounted for by replacing it by a distribution of force and moment acting on the intersection line of the frame with the plate. The strength of the force and moment distribution is obtained from the dynamical equations governing the frame motion (see section 8.4.3).

Table 8.1
Asymptotic expressions for maximum far-field pressure radiated by a large plate excited by a concentrated force

Frequency range	Plate parameter governing peak pressure	Location of pressure maximum	Maximum pressure $p(R)/F$
Low $\omega < \omega_m$	Independent of plate parameters	Beam $\theta = 0$	$(\lambda R)^{-1}$
Middle $\omega_m < \omega < \omega_c$	Mass (per unit area)	Beam $\theta = 0$	$(2\pi h R \rho_s / \rho)^{-1} = (\lambda_m R)^{-1}$ Note: valid at high frequencies at $\theta = 0$
High $\omega_c < \omega$	Plate stiffness, structural damping	Coincidence $\theta = \theta_c$ $\sec \theta_c = [1 - (\omega_c/\omega)]^{-1/2}$	$\left[\lambda R \left(\sec \theta_c + \dfrac{\omega}{\omega_m}\eta_s\right)\right]^{-1}$
Ultra-high (thick-plate theory) $\omega_c \ll \omega$		Coincidence $\theta = \theta_c$ $\sec \theta_c = [1 - (c/c_R)^2]^{-1/2}$	
1. Lossless material	Plate stiffness		$\dfrac{[1 - (c/c_R)^2]^{1/2}}{\lambda R}$
2. Damped material $(\omega_m/\eta_s) \ll \omega$	Structural damping		$(2\pi \eta_s R \rho_s / \rho)^{-1} = (\lambda_m R \eta_s)^{-1}$

8.4.1 THE PLATE RESPONSE TO A LINE FORCE

We again use thin plate theory and assume that the plate lies in the (x, y) plane with the acoustic fluid occupying the half space $z > 0$. The line force excitation acts on a line $x = x_0$, which is parallel to the y axis. In a fashion exactly analogous to the treatment in section 8.3.1, the transform of the displacement is stated as

$$\tilde{w}(\gamma) = (iF'/\omega) \exp(-i\gamma x_0)/(\tilde{Z}_a + \tilde{Z}_p), \tag{8.39a}$$

where F' is the force per unit length acting in the positive z direction [see (8.20) for comparison]. The discussion following equation (8.21) also applies to the problem at hand, the only difference being the form of the inverse transform:

$$w(x) = (iF'/2\pi\omega) \int_C (\tilde{Z}_a + \tilde{Z}_p)^{-1} \exp[i\gamma(x - x_0)] \, d\gamma, \tag{8.39b}$$

where C is again the contour shown in figure 8.3.

The dispersion equation $\tilde{Z}_a + \tilde{Z}_p = 0$ cannot be solved explicitly in the γ plane because it is equivalent to a fifth-order polynomial [equation (8.22)]. This leads to difficulties in exactly representing the residue contributions when evaluating the contour integral. In addition, the branch cut integral contributions are not easily recognized in terms of well-known functions. Nevertheless, certain attributes of the response are worth presenting.

As discussed earlier, there is always one pole on the real axis. For $\Omega < 1$, where a good approximation to this pole is given by equation (8.10), its contribution to the plate response is (where $m = \rho_s h$)

$$w_1(x) = \frac{(iF'k_f/4m\omega^2) \exp(i\gamma_1 x)}{(\gamma_1/k_f)^3 \{[1 + \varepsilon/[4\Omega^{1/2}(\gamma_1/k_f)^2(\gamma_1^2/k_f^2 - \Omega)^{3/2}]\}}. \tag{8.40}$$

In the absence of damping, the magnitude of this contribution is independent of x. From our earlier discussion we have noted that the contributions of the remaining poles are not significant. The remaining important contribution to the response comes from the integral along the L-shaped branch cut shown in figure 8.3. The integral on the lower portion of the L, i.e., $0 < \gamma < k$, represents wavelets with supersonic trace wave speeds capable of radiating to the far-field. The vertical portion of the branch cut produces a near-field distortion component that exhibits an exponential decay away from the drive point.

If we deform the L-shaped branch cut to the vertical line emanating from $\gamma = k$, the branch cut integral becomes

$$w_b(x) = (iF'/\pi\omega)\,(\rho/m)\,k_f^4 \exp ik(x - x_0)\,I(x),\tag{8.41}$$

where

$$I(x) = \int_0^\infty \frac{\tau e^{-ux}\,du}{[(k + iu)^4 - k^4]^2\tau^2 - (\rho/m)^2 k_f^8},\tag{8.42}$$

$$\tau = [(k + iu)^2 - k^2]^{1/2} = (2iku - u^2)^{1/2}.$$

$I(x)$ is evaluated asymptotically by expanding the factor of e^{-ux} in the integral about $u = 0$:

$$
\begin{aligned}
I(x) &\cong \frac{(2k)^{1/2} e^{i\pi/4} \int_0^\infty u^{1/2} e^{-ux}\,du}{-(\rho/m)^2 k_f^8}\\[2mm]
&= \frac{(2k)^{1/2} e^{i\pi/4} (\sqrt{\pi}/2x^{3/2})}{-(\rho/m)^2 k_f^8}.
\end{aligned}\tag{8.42a}
$$

Combining the above equations, we arrive at an asymptotic result for the branch cut contribution:

$$w_b(x) \simeq (F'/\rho c^2 \sqrt{2\pi})\,\frac{\exp\{i[k(x - x_0) - \pi/4]\}}{(kx)^{3/2}}.\tag{8.43}$$

$w_b(x)$ is completely independent of the plate parameters, and travels with a phase speed characteristic of sound waves in the acoustic medium. Its magnitude shows a $(kx)^{-3/2}$ spatial dependence as compared with $w_1(x)$, whose magnitude is independent of x. The structural response of a line-driven fluid-loaded plate can be thought of as the sum of two principal contributions, one a fluid-modified flexural wave with no spatial decay, in the absence of structural damping, and an acoustic wave launched at the drive point that decays as $(kx)^{-3/2}$.

8.4.2 RESPONSE GREEN'S FUNCTION FOR A LINE-LOADED PLATE

In section 8.3.1 we briefly described the two possible methods by which inversion integrals of the form given by equation (8.21) or (8.28) could be

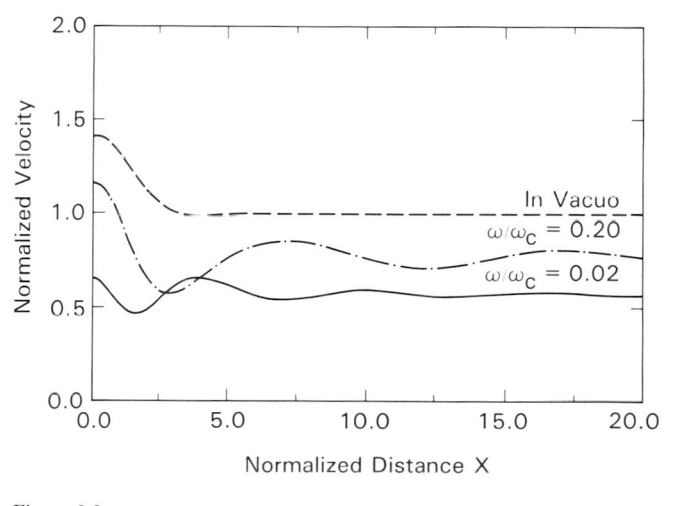

Figure 8.8
Velocity of a line-driven plate normalized to $k_f F''/4\omega m$ plotted as a function of the normalized distance $X = k_f x$ for two frequencies below coincidence. [Reproduced from Feit and Liu[7]]

evaluated numerically in the absence of explicit procedures. In the present case of a line-loaded plate, Nayak[6] was one of the first to use the "contour integral" to calculate the drive-line admittance of a fluid-loaded plate. This approach was used by Feit and Liu[7] to construct the transfer admittance derived below.

The two essential parameters governing the response and radiated pressure field of fluid-loaded plates were seen to be the nondimensional frequency $\Omega = \omega/\omega_c$ and the fluid-loading parameter at coincidence defined in connection with equation (8.8), which for a steel plate in water is $\varepsilon = 0.129$. Figure 8.8 illustrates response as a function of $k_f x$ for a steel plate in water for two different values of frequency below coincidence, as well as the normalized response in vacuo, which is independent of frequency. The response curves display a spatially decaying near-field with a modulation in space where modulation arises from the interference between the flexural wave and the acoustical wave launched at the drive point. Because the magnitude of the acoustic component varies as $(kx)^{-3/2}$, the modulation diminishes with increasing kx. These functions can be thought of as the response Green's function, since they represent the deflection at x due to a unit force applied at $x = x_0$. Similar curves can be generated for the

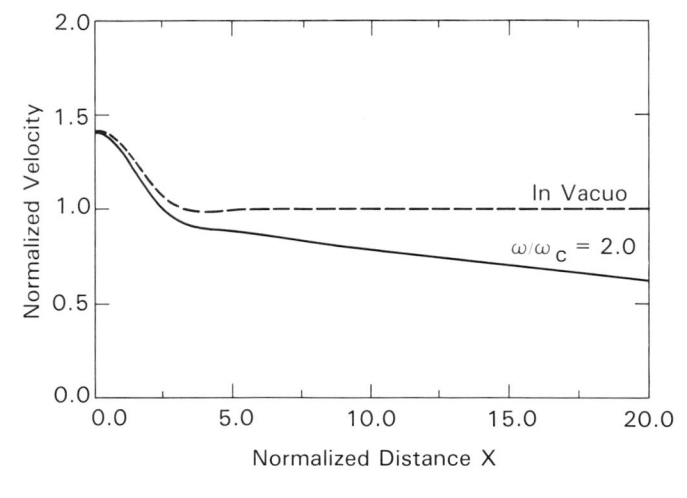

Figure 8.9
Velocity of a line-driven plate as a function of distance, both ordinate and abscissa normalized as in figure 8.8, for a frequency above coincidence. [Reproduced from Feit and Liu[7]]

other Green's functions of the problem, i.e., the plate slope at x due to a unit load at x_0, the deflection at x due to unit moment at x_0, etc.

Figure 8.9 shows the same type of results above coincidence. The spatial decay with increasing $k_f x$ is due to the leakage of energy away from the plate into the fluid in the form of significant acoustic radiation. As frequency increases the effect of fluid loading diminishes and the decay due to acoustic leakage decreases.

8.4.3 SCATTERING OF A FLEXURAL WAVE BY A PLATE DISCONTINUITY
Using the above results, we now outline the methods by which the response and radiation characteristics of frame-stiffened plates can be obtained. For purposes of illustration we assume that the frames are attached to the plate in a symmetric fashion, thereby avoiding the generation of in-plane reactions. The frames do, however, generate both translational (transverse) and rotational forces on the plate as a result of their inertial characteristics.

The frame reaction on the plate is a force F_1 and a moment M_1 of unknown magnitudes. For a line force of magnitude F_0 at $x = x_0$, and a frame located at $x = x_1$, the equation of motion of the plate becomes

or TM Theory

$$\underbrace{D\frac{d^4 w}{dx^4}} - \rho_s h\omega^2 w = F_0\,\delta(x - x_0) + F_1\,\delta(x - x_1) + M_1\,\delta'(x - x_1) - p(x, 0).$$
(8.44)

If M_F and $\mathcal{J}_F$ are the mass and rotary inertia of the frame per unit length, respectively,

$$F_1 = -M_F \ddot{w}_F,$$
$$M_1 = -\mathcal{J}_F \ddot{\theta}_F,$$
(8.45)

where w_F and θ_F are the translational and angular rotations of the frame. The Fourier transform of equation (8.44) is

$$(\tilde{Z}_p + \tilde{Z}_a)\tilde{w} = F_0 e^{-i\gamma x_0} + (F_1 + i\gamma M_1)e^{-i\gamma x_1}.$$ *See Eq. 8.20* (8.46)

Using the results of section 8.4.2, the appropriate Green's functions are constructed:

$$G_{11}(x|x_0) = \frac{1}{2\pi}\int_{-\infty}^{\infty}\frac{e^{i\gamma(x-x_0)}\,d\gamma}{(\tilde{Z}_a + \tilde{Z}_p)}$$
(8.47)

and

$$G_{21}(x|x_0) = \frac{i}{2\pi}\int_{-\infty}^{\infty}\frac{\gamma e^{i\gamma(x-x_0)}\,d\gamma}{(\tilde{Z}_a + \tilde{Z}_p)}$$
(8.48)

We can immediately invert equation (8.46):

$$w(x) = F_0 G_{11}(x|x_0) + F_1 G_{11}(x|x_1) + M_1 G_{21}(x|x_1),$$
$$\theta(x) = \frac{dw}{dx} = F_0 G'_{11}(x|x_0) + F_1 G'_{11}(x|x_1) + M_1 G'_{21}(x|x_1),$$
(8.49)

where G'_{11} and G'_{21} denote the derivatives of these functions with respect to x. Using equation (8.47) and (8.48), we note that $G'_{11} = G_{21}$. Because the functions $\tilde{Z}_a$ and $\tilde{Z}_p$ are even functions of γ, we also find that $G'_{11}(x|x) = G_{21}(x|x) \equiv 0$.

F_1 and M_1 are related to w_F and θ_F through equations (8.45) and to $w(x_1)$ and $\theta(x_1)$ by compatibility of displacement and slope between the

plate and frame. Equations (8.49), evaluated at $x = x_1$, yield two equations in the two unknowns $\dot{w}_F$ and $\dot{\theta}_F$. These equations are

$$\dot{w}_F[1 - M_F\omega^2 G_{11}(x_1|x_1)] = F_0 G_{11}(x_1|x_0),$$

$$\dot{\theta}_F[1 - J_F\omega^2 G'_{21}(x_1|x_1)] = F_0 G'_{11}(x_1|x_0). \qquad (8.50)$$

With $\dot{w}_F$ and $\dot{\theta}_F$ determined, we can find F_1 and M_1 and determine the response of the plate as *Take Fourier transform and substitute into Eq 5.36 for the scattered pressure.*

$$\dot{w}(x) = F_0 G_{11}(x|x_0) - M_F\omega^2\dot{w}_F G_{11}(x|x_1) - J_F\omega^2\dot{\theta}_F G_{21}(x|x_1) \qquad (8.51)$$

This same technique can be extended to an arbitrary number of frames or line discontinuities attached to the plate. For each discontinuity there are two equations, and in general all the equations are coupled to each other by transfer admittance functions, which, as already mentioned, are usually not explicitly known, but can be evaluated numerically once the two essential parameters Ω and ε are specified.

8.5 Pressure and Power Radiated by an Infinite Plate Driven by Distributed Loads

8.5.1 TRANSFORM SOLUTIONS

The results of section 8.3 allowed us to calculate the pressure field radiated by a concentrated force acting on a fluid-loaded elastic plate of effectively infinite extent. The same technique yields the far-field radiated pressures of plates excited by loads arbitrarily distributed over specified areas. For a plate in the (x, y) plane, we can write the external load distribution as $p_e(x, y)$ and its two-dimensional transform as $\tilde{p}_e(\gamma_x, \gamma_y)$, where γ_x and γ_y are the transform parameters over the x and y coordinates, respectively [equation (5.28)].

If such a force distribution is substituted into the plate equation that accounts for radiation loading as well as the applied external load, we can solve for the transform of the displacement: *$\tilde{Z}_a = Eq. 8.19$ $\tilde{Z}_p \Rightarrow$ Thin Plates: Eq 8.15.*

$$\tilde{\tilde{w}}(\gamma_x, \gamma_y) = (-i\omega)\tilde{p}_a(\gamma_x, \gamma_y)/[\tilde{Z}_a(\gamma_x, \gamma_y) + \tilde{Z}_p(\gamma_x, \gamma_y)]. \qquad (8.52)$$

$\tilde{Z}_a$ and $\tilde{Z}_p$ have the same functional dependence as in equations (8.15) and (8.19) with $\gamma^2 = \gamma_x^2 + \gamma_y^2$. Substituting this result in equation (5.36), we obtain the far-field radiated pressure: *Thick Plates: Eq. 8.17.*

$$p(R, \theta, \phi) = T(R, \phi) \tilde{p}_a (k \sin \theta \cos \phi, k \sin \theta \sin \phi), \tag{8.53}$$

where $T(R, \theta)$ is the pressure radiated per unit concentrated force, equation (8.30) or (8.31), with $F = 1$. This result could have been constructed directly from the convolution theorem.[8] In the direction normal to the plate, θ vanishes and therefore $\bar{\gamma}_x = k \sin \theta \cos \phi$, $\bar{\gamma}_y = k \sin \theta \sin \phi$ also vanish:

$$p(R, 0) = T(R, 0) \tilde{p}_a (0, 0). \tag{8.54}$$

The transform of the applied pressure, $\tilde{p}_a (\gamma_x, \gamma_y)$ evaluated at $\gamma_x = 0, \gamma_y = 0$, reduces to a surface integral equal to the resultant force:

$$F = \int p_a (x, y) \, dx \, dy.$$

Consequently

$$p(R, 0) = FT(R, 0) \tag{8.55}$$

no matter what the force distribution.

For axisymmetric load distributions, the Hankel transform of the acceleration distribution takes the place of (8.52). The corresponding far-field pressure is given by (5.22):

$$p(R, \theta) = 2\pi T(R, \theta) \tilde{p}_a (k \sin \theta). \tag{8.56}$$

On the axis, where

$$\tilde{p}_a (0) = F/2\pi,$$

this reduces to (8.55).

8.5.2 EXAMPLES OF LOAD DISTRIBUTIONS
First consider a uniform load applied over a rectangular area (figure 8.10a):

$$p_a (x, y) = PH(L_x - |x|) H(L_y - |y|),$$

$$p_a (\gamma_x, \gamma_y) = 4P \int_0^{L_x} \int_0^{L_y} \cos \gamma_x x \cos \gamma_y y \, dx \, dy \tag{8.57}$$

$$= (4P/\gamma_x \gamma_y) \sin \gamma_x L_x \sin \gamma_y L_y.$$

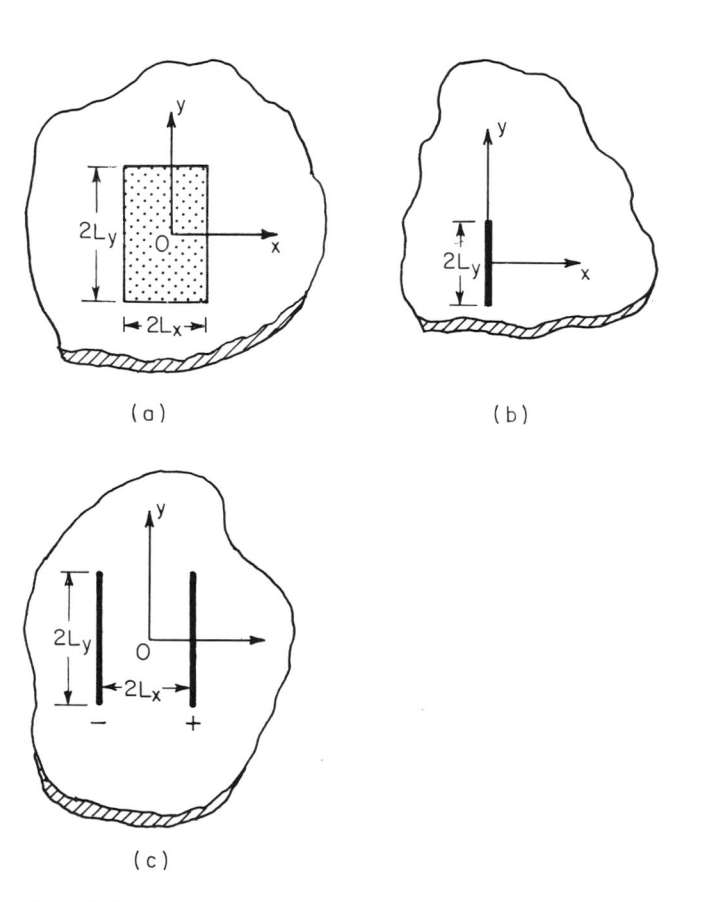

Figure 8.10
Applied loads: (a) distributed load, equation (8.57); (b) line load, equation (8.57) with $L_x \to 0$; (c) Line moment, equation (8.59).

Equation (8.53) yields the pressure

$$p(R, \theta, \phi) = T(\theta) F j_0 (kL_x \sin\theta \cos\phi)$$
$$\cdot j_0 (kL_y \sin\theta \sin\phi), \tag{8.58}$$

where j_0 is defined in equation (2.15). This distribution -in-angle displays nulls if kL_x or $kL_y > \pi$. It is therefore more directive than the field of the concentrated force, (8.30), a conclusion we shall reach for all distributed loads.

The pressure field of a line load extending in the y direction (figure 8.10b) is constructed simply by setting $j_0 (kL_x \sin\theta \cos\phi)$ equal to unity. The pressure radiated by the moment M generated by two out-of-phase line forces located at $x = \pm L_x$ can be similarly constructed (figure 8.10c):

$$p_a(x, y) = (M/4L_xL_y) H(L_y - |y|) [\delta(L_x - x) - \delta(L_x + x)],$$
$$\tilde{p}_a(\gamma_x, \gamma_y) = -i(M/L_x) j_0 (\gamma_y L_y) \sin\gamma_x L_x. \tag{8.59}$$

The pressure field is

$$p(R, \theta, \phi) = -i(M/L_x) T(\theta) \sin(kL_x \sin\theta \cos\phi) \cdot j_0 (kL_y \sin\theta \sin\phi)$$
$$\cong -iMT(\theta) j_0 (kL_y \sin\theta \sin\phi) k \sin\theta \cos\phi, \qquad k^2 L_x^2 \ll 1. \tag{8.60}$$

In the low-frequency limit, the pressure displays a quadrupolelike k^2 dependence on wave number.

Turning to axisymmetric loads, we first consider a pressure applied uniformly over a circular surface:

$$p_a(r) = PH(a - r),$$
$$\tilde{p}_a(\gamma) = P \int_0^a \mathcal{J}_0 (\gamma r) r \, dr = F \mathcal{J}_1 (\gamma a)/\gamma a \pi. \tag{8.61}$$

The corresponding pressure field is

$$p(R, \theta) = 2FT(\theta) \mathcal{J}_1 (ka \sin\theta)/ka \sin\theta$$
$$\simeq 2FT(\theta) (1 - 0.125k^2 a^2 \sin^2\theta); \qquad k^2 a^2 \sin^2\theta \ll 1. \tag{8.62}$$

Nulls occur if ka exceeds the first zero of $\mathcal{J}_1$, viz., 3.83. A circular line force

$$p_a(r) = (F/2\pi a)\,\delta(a-r),$$
$$\tilde{p}_a(\gamma) = F\mathcal{J}_0(\gamma a)/2\pi,$$

$$(8.63)$$

radiates a pressure field

$$p(R,\theta) = FT(\theta)\mathcal{J}_0(ka\sin\theta)$$
$$\simeq FT(\theta)(1 - 0.25k^2a^2\sin^2\theta) \qquad k^2a^2\sin^2\theta \ll 1.$$

$$(8.64)$$

The main lobe is narrower than that of the distributed force, which in turn is more directive than the concentrated force.

Finally, consider a multipole force distribution consisting of a concentrated force concentric and out-of-phase with a circular line force:

$$p_a(r) = \frac{F}{2\pi}\left[\frac{\delta(r)}{r} - \frac{1}{a}\delta(a-r)\right],$$

$$\left.\begin{array}{l}
\tilde{p}_a(\bar{\gamma}) = \dfrac{F}{2\pi}[1 - \mathcal{J}_0(ka\sin\theta)] \\[2ex]
\qquad \simeq \dfrac{F}{8\pi}k^2a^2\sin^2\theta, \qquad k^2a^2 \ll 1.
\end{array}\right\} \qquad \bar{\gamma} = k\sin\theta$$

$$(8.65)$$

This axisymmetric couple generates a far-field

$$|p(R,\theta)| = \frac{Fk^3a^2\sin^2\theta\cos\theta}{8\pi R}, \qquad k^2a^2 \ll 1. \tag{8.66}$$

This result is interesting in that even though the pressure field is in phase, it varies as a higher order pole, as k^3.

8.6 Power Radiated by Plates

We noted two facts: (i) the on-axis pressure $p(R,0)$ normalized to the resultant exciting force is independent of the load distribution; (ii) loads whose extent is commensurate with the acoustic wavelength are associated with directivity patterns characterized by nulls. The combination of these two facts indicates that distributed loads radiate less power when normalized to unit force than do point excitations. For this purpose, the power integral, (3.27), will be evaluated for an adjustable e.g., an exponentially decaying

load distribution

$$p_a(r) = P \exp(-mr),$$

$$\tilde{p}_a(\gamma) = (F/2\pi)\,[\,(\gamma/m)^2 + 1]^{-3/2},$$

(8.67)

whose far-field is

$$p(R, \theta) = FT(\theta)\,[\,(k/m)^2 \sin^2\theta + 1]^{-3/2}$$

$$\cong FT(\theta)\,[1 - (3k^2 \sin^2\theta/2m^2)\,], \qquad (k/m)^2 \sin^2\theta \ll 1.$$

(8.68)

For $(m/k)^2 \to \infty$, this tends to the concentrated force. The force can be distributed over an arbitrarily wide area by letting $(m/k) \ll 1$. We cannot, however, set $(m/k) = 0$ because the far-field conditions embodied in the stationary-phase integration of the inverse Hankel transform, and hence in equation (8.56), are not satisfied.

The acoustic power is obtained by integrating the far-field intensity, as indicated in equation (3.27). Since the pressure is ϕ independent, the ϕ integration reduces to multiplication by 2π. The region of the acoustic field being restricted to the half-space $0 \le \theta \le \pi/2$, the limits of the θ integration must be modified accordingly:

$$\Pi = \frac{\pi R^2}{\rho c} \int_0^{\pi/2} |p(R, \theta)|^2 \sin\theta \, d\theta. \qquad \text{thin plates in air!}$$

(8.69)

This integral can be evaluated analytically in the low-frequency limit when the $(\omega/\omega_c)^2$ term in equation (8.30) is negligible. For the concentrated force, identified here by the subscript c, the integral is readily evaluated:

$$\left.\begin{aligned}
\Pi_c &\approx \frac{1}{\rho c \pi}\left(\frac{F\rho}{2\rho_s h}\right)^2 \left(1 - \frac{\rho}{kh\rho_s}\tan^{-1}\frac{kh\rho_s}{\rho}\right) \\
&\approx \frac{1}{\rho c \pi}\left(\frac{F\rho}{2\rho_s h}\right)^2, \qquad \frac{kh\rho_s}{\rho} \gg 1
\end{aligned}\right\} \qquad \omega^2 \ll \omega_c^2.$$

(8.70)

The latter inequality is compatible with the restriction on ω for thin plates vibrating in air but not for submerged plates. For the latter case, a more useful relation is obtained from the small-argument approximation for the inverse tangent, $\tan^{-1} x \approx x - (x^3/3)$. Equation (8.70) can thus be reduced further:

$$\Pi_c \approx \frac{F^2 k^2}{12\pi\rho c}, \qquad \frac{kh\rho_s}{\rho} \ll 1. \tag{8.71}$$

Like the pressure field, the latter low-frequency limit of the power is seen to be independent of the plate parameters. Similar results were obtained by Heckl[9,10] and Skudrzyk,[11] partly by substantially different analytical procedures.

Turning to the field radiated by the distributed load, equation (8.68), the power, equation (8.69), becomes

$$\frac{\Pi}{F^2} = \frac{R^2 \pi}{\rho c} \int_0^{\pi/2} \left| T(\theta) \frac{\tilde{p}_a(k\sin\theta)}{\tilde{p}_a(0)} \right|^2 \sin\theta \, d\theta, \tag{8.72}$$

where

$$|\tilde{p}_a(k\sin\theta)/\tilde{p}_a(0)|^2 = [(k/m)^2 \sin^2\theta + 1]^{-3}. \quad \textit{Eq. 8.67}$$

This weighting factor decreases monotonically from unity for a concentrated load, $m = \infty$, to zero for a uniformly distributed load, $m = 0$. An analytical expression for the power can be readily constructed when radiation loading predominates over plate inertia: $\Rightarrow \omega < \omega_m \ (Eq. 8.38) \ !^2 (\omega_m/\omega_c) = 0.13$

$$\frac{\Pi}{F^2} = \frac{k^2}{4\pi\rho c} \int_0^{\pi/2} \frac{\cos^2\theta \sin\theta \, d\theta}{[(k/m)^2 \sin^2\theta + 1]^3}, \qquad \left(\frac{kh\rho_s}{\rho}\right)^2 \ll 1. \tag{8.73}$$

for steel in water.

Thick or Thin Plate Theory ???

Normalizing this integral to the power Π_c radiated by the concentrated force, (8.70), we finally have

p. 250

$\Rightarrow \omega/\omega_c < 1$

for no radiation ? thin plates.

$$\frac{\Pi}{\Pi_c} = \frac{3m^6}{k^6} \int_0^1 \frac{x^2 \, dx}{[1 + (m/k)^2 - x^2]^3}$$

$$= \frac{3m^2}{4k^2} - \frac{3(m/k)^2}{8[(k/m)^2 + 1]} \cdot \left\{ 1 + \frac{(m/k)^2}{[(m/k)^2 + 1]^{1/2}} \right.$$

$$\left. \cdot \tanh^{-1}\left[\left(\frac{m}{k}\right)^2 + 1\right]^{1/2} \right\} \cong 1, \qquad m^2/k^2 \gg 1 \tag{8.74}$$

$$\cong 0.75 m^2/k^2, \qquad m^2/k^2 \ll 1.$$

This is plotted in figure 8.11.

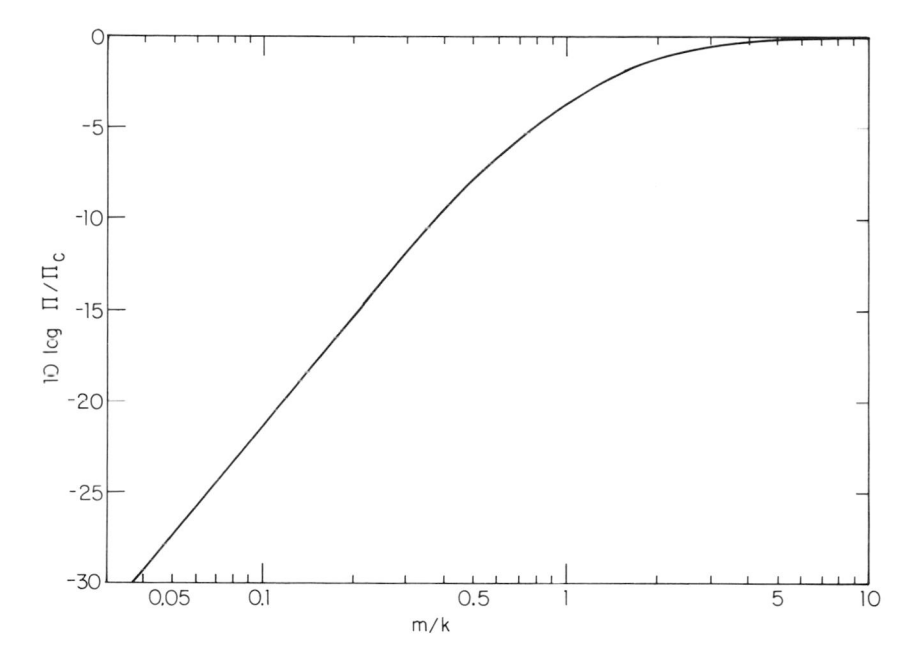

Figure 8.11
Power Π radiated by an exponentially shaded load distribution, $p_a(r) = P\exp(-mr)$, normalized to the power Π_c radiated by a concentrated force, (8.70). [Reproduced from Junger[12]]

8.7 Sound Radiation from Rectangular Plates

8.7.1 PLATE RESPONSE

The normal modes of a plate simply supported at its boundaries and vibrating in vacuo were considered in section 7.10. For our present purposes, we select a somewhat different notation to facilitate the introduction of the acoustic solution generated in chapter 5. We take our origin of coordinates at the center of the plate and consider the plate to be of length $2L_x$ in the x direction and $2L_y$ in the y direction. The origin of coordinates will be located at the intersection of the planes of symmetry. For excitations symmetric about the origin, the plate response is in the form of even modes:

$$w(x,y) = \sum_{m=0}^{\infty} \sum_{n=0}^{\infty} W_{mn} \cos k_m x \cos k_n y, \qquad (8.75a)$$

where

$$k_m = \left(\frac{2m+1}{2}\right)\frac{\pi}{L_x},$$

$$k_n = \left(\frac{2n+1}{2}\right)\frac{\pi}{L_y}.$$

(8.75b)

Loads not symmetric about the planes of symmetry will of course excite modes embodying a $\sin k_m x$ or $\sin k_n y$ dependence, where

$$k_m = m\pi/L_x,$$

$$k_n = m\pi/L_y.$$

(8.76)

Sound radiation by antisymmetric modes was studied in chapter 5 (see in particular table 5.2). The modes of a plate vibrating in vacuo are characterized by discrete wave numbers, equations (8.75) and (8.76), so that each modal configuration has attached to it a pair of wave numbers k_m and k_n. These modal configurations are by definition orthogonal to each other, or, in other words, each mode can be excited independently of the other modes by an appropriately distributed load. When such a plate is submerged, each of these modes generates an acoustic pressure in the plane of the plate. The resulting pressure distribution due to a single mode is not orthogonal to the other modes. Therefore, if one attempts to formulate the problem in terms of the in vacuo normal modes, these modes become coupled. Consider the equation of motion of the plate with a concentrated force applied at the origin,

$$\left[D\left(\frac{\partial^2}{\partial x^2} + \frac{\partial^2}{\partial y^2}\right)^2 - \rho_s h\omega^2\right] w(x,y) = F\delta(x)\delta(y) - p(x,y,0),$$

(8.77)

where $p(x,y,0)$ is the acoustic pressure due to the presence of the fluid. Using equation (8.76) and the procedures set forth in section 7.10, we obtain a set of equations for the modal amplitude W_{mn} in the form

$$L_x L_y \rho_s h[\omega_{mn}^2 - \omega^2] W_{mn} = F_{mn},$$

(8.78)

where ω_{mn} is the in vacuo natural frequency of the (m,n) mode,

$$\omega_{mn} = \frac{hc_p}{\sqrt{12}}(k_m^2 + k_n^2) = \left(\frac{D}{\rho_s h}\right)^{1/2}(k_m^2 + k_n^2).$$

(8.79)

If we wish to include the effects of structural damping, c_p is considered as a complex number, as discussed earlier in section 8.3.3. Here F_{mn} is the generalized force acting on the (m, n) mode and is associated with the driving force as well as the radiation loading term $p(x, y, 0)$. For this particular case, the generalized force F_{mn} becomes

$$F_{mn} = F - \int_{-L_x}^{L_x} \int_{-L_y}^{L_y} p(x, y, 0) \cos k_m x \cos k_n y \, dx \, dy. \tag{8.80}$$

The surface pressure is given by equation (5.36) with $z = 0$. Introducing the modal sum for the displacement distribution,

$$p(x, y, 0) = -\frac{i\rho\omega^2}{4\pi^2} \sum_{q,r} \int_{-\infty}^{\infty} \int_{-\infty}^{\infty} \frac{\tilde{w}_{qr}(\gamma_x, \gamma_y) \cos \gamma_x x \cos\gamma_y y \, d\gamma_x \, d\gamma_y}{\sqrt{k^2 - \gamma_x^2 - \gamma_y^2}},$$

where

$$\tilde{w}_{qr}(\gamma_x, \gamma_y) = W_{qr} \int_{-L_x}^{L_x} \int_{-L_y}^{L_y} \cos k_q x \cos k_r y \cos \gamma_x x \cos \gamma_y y \, dx \, dy \tag{8.81}$$

and

$$k_q = (2q + 1)\pi/2L_x, \qquad k_r = (2r + 1)\pi/2L_y.$$

The generalized force in (8.80) can now be constructed:

$$F_{mn} = F - \int_{-L_x}^{L_x} \int_{-L_y}^{L_y} p(x, y, 0) \cos k_m x \cos k_n y \, dx \, dy$$

$$= F + \frac{i\rho\omega^2}{4\pi^2} \sum_{q,r} \int_{-L_x}^{L_x} \int_{-L_y}^{L_y} \left[\int_{-\infty}^{\infty} \int_{-\infty}^{\infty} \frac{\tilde{w}_{qr}(\gamma_x, \gamma_y) \cos \gamma_x x \cos \gamma_y y \, d\gamma_x \, d\gamma_y}{(k^2 - \gamma_x^2 - \gamma_y^2)^{1/2}} \right]$$

$$\cdot \cos k_m x \cos k_n y \, dx \, dy. \tag{8.82}$$

Referring to equation (8.81), we see that the integral over the variables x and y can be identified as the transform $\tilde{w}_{mn}(\gamma_x, \gamma_y)/W_{mn}$. The generalized force F_{mn} thus becomes

$$F_{mn} = F + \frac{i\rho\omega^2}{4\pi^2 W_{mn}} \sum_{q,r} \int_{-\infty}^{\infty} \int_{-\infty}^{\infty} \frac{\tilde{w}_{qr}(\gamma_x, \gamma_y) \tilde{w}_{mn}(\gamma_x, \gamma_y) \, d\gamma_x \, d\gamma_y}{(k^2 - \gamma_x^2 - \gamma_y^2)^{1/2}}. \tag{8.83}$$

If we explicitly evaluate the transform of the displacement functions, we obtain

$$F_{mn} = F + i\omega \sum_{q,r} I_{mnqr} W_{qr}, \tag{8.84}$$

where we used the definition

$$I_{mnqr} = \frac{4\rho\omega k_m k_n (-1)^{m+n}}{\pi^2} \sum_{q,r} (-1)^{q+r} k_q k_r$$

$$\cdot \left\{ \int_{-\infty}^{\infty} \int_{-\infty}^{\infty} \frac{\cos^2 \gamma_x L_x \cos^2 \gamma_y L_y \, d\gamma_x \, d\gamma_y}{(k^2 - \gamma_x^2 - \gamma_y^2)^{1/2} (k_m^2 - \gamma_x^2)(k_n^2 - \gamma_y^2)(k_q^2 - \gamma_x^2)(k_r^2 - \gamma_y^2)} \right\}.$$

We can now construct modal coupling coefficients in the form of mutual specific acoustic resistance r_{mnqr} and reactance $-i\omega m_{mnqr}$:

$$r_{mnqr} = \text{Re} \frac{I_{mnqr}}{4L_x L_y},$$

$$m_{mnqr} = -\frac{1}{\omega} \text{Im} \frac{I_{mnqr}}{4L_x L_y}. \tag{8.85}$$

The modal equations of motion, equation (8.78), now finally become

$$\rho_s h (\omega_{mn}^2 - \omega^2) W_{mn} - 4 \sum_{q,r} (\omega^2 m_{mnqr} + i\omega r_{mnqr}) W_{qr} = \frac{F}{L_x L_y}. \tag{8.86}$$

We have a doubly infinite set of equations coupled by the mutual impedance components in equation (8.86) linking an (m, n) mode to a (q, r) mode. I_{mnqr} must be evaluated numerically as has been done, for example, by Lax,[13] who computed the analogous integral for the case of circular plates, or asymptotically, in restricted frequency domains, as by Davies.[14] For the higher-order modes of a plate, $k_q L_x$ and $k_r L_y \gg 1$, equation (8.81) simplifies to

$$\tilde{w}_{qr}(\gamma_x, \gamma_y) = \lim_{\substack{k_q L_x \to \infty \\ k_r L_y \to \infty}} \left\{ W_{qr} \int_{-L_x}^{L_x} \cos k_q x \cos \gamma_x x \, dx \int_{-L_y}^{L_y} \cos k_r y \cos \gamma_y y \, dy \right\}$$

$$= 4\pi^2 \, \delta(k_q - \gamma_x) \, \delta(k_r - \gamma_y) \, W_{qr}.$$

The surface pressure in equation (8.80) now becomes

$$p(x, y, 0) = -i\rho\omega^2 \sum_{q,r} W_{qr} \frac{\cos k_q x \cos k_r y}{\sqrt{k^2 - k_q^2 - k_r^2}}, \qquad k_m L_x, k_n L_y \gg 1.$$

The generalized force in equation (8.83) reduces to

$$F_{mn} = F + \frac{i\rho\omega^2 L_x L_y W_{mn}}{\sqrt{k^2 - k_m^2 - k_n^2}}, \qquad k_m L_x, k_n L_y \gg 1.$$

Therefore, we see that in the large $k_m L_x$, $k_n L_y$ limit the coupling impedance terms vanish and the generalized force contribution due to the radiation loading corresponds to the accession to inertia per unit area and specific acoustic resistance given by equations (5.63) and (5.61) for an infinite train of straight-crested parallel waves or of straight crested orthogonal waves. The modal amplitudes can now be solved for

$$W_{mn} = iF \Bigg/ \left\{ L_y L_x \omega \left[\frac{\rho\omega}{\sqrt{k^2 - k_m^2 - k_n^2}} - i\omega\rho_s h \left(1 - \frac{\omega_{mn}^2}{\omega^2}\right) \right] \right\},$$

$$k_m L_x, k_n L_y \gg 0.$$

(8.87)

8.7.2 FAR-FIELD SOUND PRESSURE

We shall now concern ourselves with the far-field pressure radiated by finite plates. We shall restrict ourselves to plates whose dimensions are larger than half of an acoustic wavelength. We shall consider two classes of modes.

Efficiently Radiating Modes These modes radiate well by virtue of their spatial characteristics, their effective structural wavelength being larger than the acoustic wavelength at the given frequency. In terms of wave numbers the modal wave number must be smaller than the acoustic wave number:

$$k_m^2 + k_n^2 < k^2.$$

(8.88)

In all that is to follow, we shall ignore the modes whose wavelength in one of the coordinate directions is larger than the acoustic wavelength while the other wavelength is smaller. These modes are characterized, for example, by $k_m < k$, while $k_m^2 + k_n^2 > k^2$, and are known as edge modes. For a dis-

cussion of these see section 5.9. The modal wave number can be rewritten in terms of ω_{mn}, the modal resonant frequency, and ω_c, the coincidence frequency [equations (8.79) and (8.3)]:

$$k_m^2 + k_n^2 = \frac{\omega_{mn}\omega_c}{c^2}. \tag{8.89}$$

For a plate submerged in water, ω_c is relatively large (as mentioned earlier, $\omega_c/2\pi = 9,300$ Hz for a one-inch steel or aluminium plate in water). Using equation (8.88) we can say that in this situation efficiently radiating modes have resonant frequencies well below the excitation frequency, $\omega_{mn}^2 \ll \omega^2$. Consequently, such modes are characterized by a comparatively large inertial impedance. Using equation (8.87), we can approximate W_{mn}, the modal amplitude, by the relation

$$W_{mn} = \frac{iF}{L_y L_x \omega}\left(\frac{\rho\omega}{\sqrt{k^2 - k_m^2 - k_n^2}} - i\omega\rho_s h\right)^{-1}, \qquad \omega^2 \gg \omega_{mn}^2. \tag{8.90}$$

The previous remarks do not apply to plates radiating into the atmosphere, because the sound velocity in air is approximately five times smaller, thereby making the coincidence frequency smaller by a factor of 25. As far as sound radiation is concerned, the relatively small amplitude of the modes as given by equation (8.90) is compensated for by the large ratio of peak pressure to acceleration, equation (5.50).

Resonant Modes   Resonant modes are characterized by a large modal amplitude, which comes about because of good frequency matching, or, in other words, these are modes whose resonant frequency is close to the excitation frequency. If the excitation frequency is below the coincidence frequency, then these modes are associated with the relatively inefficient sound radiation characteristic of the range $k_m^2 + k_n^2 > k^2$. With regard to the far-field sound radiation, the corresponding small ratio of peak pressure to acceleration, equation (5.48), is compensated for by the large modal amplitude.

Now that we have classified the two types of modes that make up the plate motion, we can combine the results of the present chapter with those of chapter 5, and obtain the peak pressures radiated by these modes. First we consider the pressure maxima of efficiently radiating modes whose modal wave numbers satisfy equation (8.88). Each of these modes has a pressure maximum given by equation (5.50) that occurs at the angle (θ_0, ϕ_0) asso-

ciated with the coincidence cone:

$$k_m = k \sin \theta_0 \cos \phi_0,$$

$$k_n = k \sin \theta_0 \sin \phi_0.$$

Here W_{mn} is determined by equation (8.90) and is calculated at the wave numbers given by the preceding equation. When this is done, the pressure maximum corresponding to the modal wave numbers (k_m, k_n) is given by

$$p_{\max} = -\left(\frac{ikFe^{ikR}}{2\pi R}\right)\left\{\cos\theta_0 \Big/ \left[1 - \frac{i\omega\rho_s h}{\rho c}\cos\theta_0\left(1 - \frac{\omega^2}{\omega_c^2}\sin^4\theta_0\right)\right]\right\}. \quad (8.91)$$

If we compare this result to the far-field pressure radiated by a point-excited infinte plate, equation (8.30), we find the two expressions are exactly the same. In this way we see that *the pressure maxima caused by the efficiently radiating modes of a finite plate have as their envelope the pressure radiated by an infinite plate.* As the frequency increases, more modes become efficiently radiating and the locus of the pressure maxima becomes exactly equivalent to the far-field pressure of the infinite plate.

To the preceding results we must now add the effects of the resonant modes. If the excitation frequency is below the coincidence frequency of the plate, there are no resonant modes that are also efficiently radiating modes. The resonant mode contributions to the far-field come about when the real component of the denominator of (8.87) vanishes:

$$\left\{\frac{\rho\omega}{\sqrt{k_m^2 + k_n^2 - k^2}} + \omega\rho_s h\left[1 - \frac{D(k_m^2 + k_n^2)^2}{\rho_s h\omega^2}\right]\right\} = 0, \qquad \omega = \omega_{mn}, \quad (8.92)$$

where we have rewritten ω_{mn} in terms of the wave number using equation (8.79). This equation can be solved approximately:

$$(k_m^2 + k_n^2) \sim k_f^2\left[1 + \frac{\rho c}{\omega\rho_s h}\left(\frac{\omega_c}{\omega} - 1\right)^{-1/2}\right]^{1/2}, \qquad \omega = \omega_{mn}. \quad (8.93)$$

Therefore for the resonant modes

$$\ddot{W}_{mn} = \frac{i\omega F}{L_x L_y r_s}, \qquad \omega = \omega_{mn}, \quad (8.94)$$

where r_s is the structural resistance and is what remains of the denominator of equation (8.87) after we have satisfied equation (8.93). The expression for r_s is given by

$$r_s = \frac{\eta_s D}{\omega} (k_m^2 + k_n^2)^2.$$

Substituting equation (8.93) gives

$$r_s \simeq \eta_s \omega \rho_s h \left[1 + \frac{\rho c}{\omega \rho_s h} \left(\frac{\omega_c}{\omega} - 1 \right)^{-1/2} \right]. \qquad (8.95)$$

In the case that we are considering (excitation frequency less than the coincidence frequency), resonance amplitude is controlled by structural rather than radiational resistance, and the pressure peak occurs on the normal $\theta = 0$. This pressure maximum is obtained by combining (8.94) and (5.48):

$$|p| = 2\rho F \left/ \left\{ \pi R k_m k_n L_x L_y \rho_s h \eta_s \left[1 + \frac{\rho c}{\omega \rho_s h} \left(\frac{\omega_c}{\omega} - 1 \right)^{-1/2} \right] \right\} \right. . \qquad (8.96)$$

If $k_m \simeq k_n$, we can use equation (8.93) to solve for k_n and finally obtain

$$|p| = 4\rho F \left/ \left\{ \pi R k_f^2 L_x L_y \rho_s h \eta_s \left[1 + \frac{\rho c}{\omega \rho_s h} \left(\frac{\omega_c}{\omega} - 1 \right)^{-1/2} \right]^{3/2} \right\} \right. . \qquad (8.97)$$

Having obtained the maximum far-field pressure associated with a modal resonance, we can determine the conditions under which such a resonance is significant. Comparing equation (8.97) with the far-field pressure of the nonresonant efficiently radiating modes, evaluated on axis $\theta = 0$, we note that if

$$\eta_s < 8\rho \left[1 + \left(\frac{\omega \rho_s h}{\rho c} \right)^2 \right]^{1/2} \left/ \left\{ k k_f^2 L_y L_x \rho_s h \left[1 + \frac{\rho c}{\omega \rho_s h} \left(\frac{\omega_c}{\omega} - 1 \right)^{-1/2} \right]^{3/2} \right\} \right. ,$$
$$\qquad (8.98)$$

the resonant mode contribution exceeds the contribution of the efficiently radiating modes. If for the moment we neglect fluid loading $\omega \rho_s h / \rho c \gg 1$, the foregoing criterion becomes

$$\eta_s < \frac{32}{k_f^2 A},$$
(8.99)

where we have rewritten the product $L_x L_y$ in terms of the plate area A. If the inequality is satisfied, the maximum far-field pressure is associated with a resonant mode, and if not, the maximum pressure is due to the efficiently radiating modes. The form of the inequality and the conclusions are precisely the same as those obtained by Heckl[9] on the basis of a comparison of radiated power. An alternative interpretation of equation (8.99) is that if that inequality is not satisfied, the infinite plate theory is adequate.

8.8 Low-Impedance Layers

To conclude the topic of sound radiation by plates, we illustrate the construction of the sound field radiated by a layered plate, specifically by a point-excited elastic plate separated from a semiinfinite body of liquid by a layer of low-impedance material. This layer, whether a plastic foam or a bubble *swarm*, can be analyzed as a fluid. Its low sound velocity requires that standing waves across the layer's thickness be accounted for even at relatively low frequencies. Wave propagation in the low-impedance material introduces two pressure components: the outgoing pressure p_o and the pressure p_r reflected from the layer-liquid interface. Together with the pressure p_t transmitted into the body of liquid and the plate response w, this constitutes four unknown quantities, as opposed to two for the bare-plate radiator. Two boundary conditions at the layer-liquid interface provide the needed additional equations. These require that the three component pressure fields display continuity of pressure and of normal fluid particle acceleration, (2.24). The layer material parameters are identified by the subscript d. If the layer thickness is H, these boundary conditions become

$$\left.\begin{array}{l} p_o + p_r = p_t \\[2mm] \dfrac{1}{\rho_d}\left(\dfrac{\partial p_o}{\partial z} + \dfrac{\partial p_r}{\partial z}\right) = \dfrac{1}{\rho}\dfrac{\partial p_t}{\partial z} \end{array}\right\} \quad z = H.$$
(8.100)

The boundary condition at the plate-layer interface accounts for the presence of two rather than a single pressure component:

$$\ddot{w} = -\frac{1}{\rho_d}\left(\frac{\partial p_o}{\partial z} + \frac{\partial p_r}{\partial z}\right), \quad z = 0.$$
(8.101)

The unknown quantities obey four differential equations—the three Helmholtz equations governing the transmitted pressure p_t in the semiinfinite liquid, and the pressure components p_o and p_r in the fluid layer—and the equation of motion of the plate, which differs from (8.13) in that the single pressure component $-p(r, 0)$ is replaced by the sum $-[p_o(r, 0) + p_r(r, 0)]$. When the four differential equations and three boundary conditions are Hankel transformed, one obtains an algebraic equation governing the plate response transform in the form of (8.20), and three ordinary differential equations similar to (5.14a). For the two outgoing pressure components, p_o and p_t, the transforms are in the form of (8.18), i.e., of exponentials of positive imaginary power indicating propagation in the positive z direction, but the interface-reflected pressure transform is proportional to $\exp[-i(k_d^2 - \gamma^2)^{1/2}z]$. Here, as in other expressions applicable to the region $0 < z < H$, one must be careful to use the parameters associated with the decoupling layer rather than the liquid. Thus, the radiation impedance equation (8.19) becomes

$$\tilde{Z}_a = \rho_d \omega (k_d^2 - \gamma^2)^{1/2}.$$

Applying the familiar procedures for eliminating the unwanted transforms $\tilde{w}$, $\tilde{p}_o$, and $\tilde{p}_r$, the transmitted pressure transform can finally be formulated. For the sake of conciseness, two phase angles are introduced:

$$\phi_d = (k_d^2 - \gamma^2)^{1/2} H,$$

$$\phi = (k^2 - \gamma^2)^{1/2} H.$$

The former accounts for the acoustic path length across the layer, the latter for the phase shift of the transmitted wave originating at $z = H$ rather than $z = 0$. In addition to these phase angles, the layer-liquid interface reflection coefficient transform is introduced:

$$\bar{R}(\gamma) = \frac{1 - (\rho_d \phi / \rho \phi_d)}{1 + (\rho_d \phi / \rho \phi_d)}.$$
(8.102)

The transmitted pressure transform that takes the place of the transform in the integrand of (8.28) finally becomes

$$\tilde{p}_t(\gamma, z) = \frac{F(1 + \bar{R}) \exp[i(k^2 - \gamma^2)^{1/2}z + i(\phi_d - \phi)]}{2\pi\{1 + \bar{R}[1 - (\tilde{Z}_p / \tilde{Z}_a)] \exp(i2\phi_d) + (\tilde{Z}_p / \tilde{Z}_a)\}}.$$
(8.103)

The point of stationary phase in the inverse transform remains unaltered by the low-impedance layer. Asymptotic far-field evaluation of the integral parallels the bare-plate solution, (8.28)–(8.30):

$$p_t(R, \theta) = \frac{-iFk \cos \theta \, (1 + \bar{R}) \exp[i(\phi_d - \phi) + ikR]}{2\pi R\{1 + \bar{R}[1 - (\tilde{Z}_p/\tilde{Z}_a)] \exp(i2\phi_d) + (\tilde{Z}_p/\tilde{Z}_a)\}}. \tag{8.104}$$

Here the various γ-dependent functions are evaluated at $\bar{\gamma} = k \sin \theta$. Specifically, the impedance ratio, phase angles, and reflection coefficient become

$$\tilde{Z}_p/\tilde{Z}_a = -i(\rho_s \underset{\text{plate thickness}}{h}/\rho_d)\,[1 - (k \sin \theta/k_f)^4]\,(k_d^2 - k^2 \sin^2 \theta)^{1/2}, \quad \textit{Eqs. 8.15, 8.19}$$

$$\phi_d = (k_d^2 - k^2 \sin^2 \theta)^{1/2}\underset{\text{coating thickness}}{H,}$$

$$\phi = kH \cos \theta, \tag{8.105}$$

$$\bar{R}(\theta) = \frac{\rho(k_d^2 - k^2 \sin^2 \theta)^{1/2} - \rho_d k \cos \theta}{\rho(k_d^2 - k^2 \sin^2 \theta)^{1/2} + \rho_d k \cos \theta}.$$

The layer insertion loss is defined as $20 \log|p/p_t|$, where p is the pressure radiated by the bare plate, equation (8.30).

The pressure ratio takes a particularly simple form on the normal to the plate through the drive point. Here $\bar{R}$ becomes the plane wave reflection coefficient

$$\bar{R}(0) = (1 - \zeta)/(1 + \zeta), \tag{8.106}$$

where ζ is the impedance ratio:

$$\zeta = \rho_d c_d/\rho c. \tag{8.107}$$

For a bubble swarm, this becomes

$$\zeta = (1 - \alpha)k/k_d. \tag{8.108}$$

The other quantities in (8.105) reduce to

$$\phi_d = k_d H, \qquad \phi = kH,$$

$$\frac{\tilde{Z}_p}{\tilde{Z}_a} = \frac{ik_d h \rho_s}{\rho}.$$

The pressure ratio now becomes

$$\left|\frac{p(R,0)}{p_{\mathrm{t}}(R,0)}\right| = \left|\cos k_{\mathrm{d}}H - \frac{(\rho_{\mathrm{s}}k_{\mathrm{d}}h/\rho) + i\zeta}{1 - i(\rho_{\mathrm{s}}kh/\rho)}\sin k_{\mathrm{d}}H\right|. \tag{8.109}$$

This result will be interpreted without computations in terms of asymptotic limits. For vanishing frequencies, the ratio is of order unity. At somewhat higher frequencies the layer amplifies radiation of certain frequency components corresponding to zeros of the real part of (8.109). These thickness resonances can be approximated by means of an asymptotic form of the pressure ratio:

$$|p(R,0)/p_{\mathrm{t}}(R,\theta)| = |\cos k_{\mathrm{d}}H - (\rho_{\mathrm{s}}k_{\mathrm{d}}h/\rho)\sin k_{\mathrm{d}}H|, \qquad \zeta^4 \ll (kh\rho_{\mathrm{s}}/\rho)^2 \ll 1. \tag{8.110}$$

To verify the range of validity defined by these inequalities, consider a layer in the form of a *bubble swarm* in water. For frequencies far below the bubble breathing mode natural frequency, (3.34), the velocity in the bubble swarm is given by (3.43). Substituting this result in (8.108), one obtains an explicit expression for the bubble swarm impedance ratio:

$$\zeta = (\gamma P_{\mathrm{s}}/\alpha\rho c^2)^{1/2}, \qquad \alpha \ll 1, \qquad f^2 \ll f_0^2$$
$$= 0.79 \times 10^{-2}(P_{\mathrm{s}}/\alpha)^{1/2} \qquad (P_{\mathrm{s}} \text{ in atm}). \tag{8.111}$$

For $P_{\mathrm{s}} = 4$ atm, i.e., a depth of 100 ft, and a fractional air volume $\alpha = 10^{-2}$, this yields $\zeta = 0.16$, $\zeta^4 = 6.2 \times 10^{-4}$. For a steel plate of 2 cm thickness, $(kh\rho_{\mathrm{s}}/\rho)^2 = 0.43 \times 10^{-6}f^2$, with f in Hz. The first inequality in (8.111) is therefore satisfied for $f > 100$ Hz. The upper limit requires $f < 500$ Hz. For air bubbles in water, the breathing mode resonance frequency, (3.34), equals

$$f_0 = 655P_{\mathrm{s}}^{1/2}/2a \qquad (\text{in Hz}; P_{\mathrm{s}} \text{ in atm}; 2a \text{ in cm}).$$

At the assumed 100-ft depth, the low-frequency asymptotic expression for ζ, (8.111), is therefore applicable for bubbles not exceeding 1.5 cm. The insertion loss displays minima at frequencies where the layer amplifies sound radiation, when the two terms in (8.110) cancel. The lowest frequency can be approximated by setting the cosine equal to unity and replacing the sine by its argument:

$$k_d = (\rho/\rho_s hH)^{1/2}, \qquad k_d^2 H^2/3 \ll 1. \tag{8.112}$$

When this is set equal to the low-frequency expression for k_d obtained by combining (8.108) and (8.111),

$$k_d = \omega(\alpha\rho/\gamma P_s)^{1/2}, \qquad f^2 \ll f_0^2, \qquad \alpha \ll 1,$$

one can solve for the fundamental natural frequency of the layer:

$$f_n = \frac{1}{2\pi}\left(\frac{\gamma P_s}{\alpha\rho_s hH}\right)^{1/2}. \tag{8.113}$$

This can be interpreted as the resonance of a simple oscillator whose mass per unit area $\rho_s h$ is that of the plate and whose stiffness $\gamma P_s/\alpha H$ is that of the compliant layer. For the parameters selected earlier, and a layer thickness of $H = 10$ cm, this resonance occurs at 190 Hz. Additional thickness resonances associated with higher roots of equation (8.110) exist if $k_d H$ is sufficiently large. At and above the bubble breathing mode resonance, a very high insertion loss is computed from (8.109), in part because of the impedance mismatch at the layer-water interface, but primarily because of the high attenuation in (3.47). In fact, for any practical parameters, the attenuation is so large as to make the insertion loss meaningless in practice, the sound field in the decade above the bubble resonance frequency being dominated by flanking path transmission.[15] The insertion loss drops precipitously as antiresonance is approached, being down to 4 dB at that frequency for the parameters selected here. Consequently, large insertion losses occur in a frequency range that, for a bubble diameter of 1 cm, $P_s = 4$ atm, $\alpha = 10^{-2}$, falls between the breathing mode resonance at 1,330 Hz and the antiresonance at 52 kHz.

    Closed-cell foam displays less dispersion, the impedance ratio in (8.107) being relatively constant compared with that of bubble swarms. Resonant amplification is predicted at a frequency obtained by replacing $\gamma P_s/\alpha$ in (8.113) by the foam bulk modulus B_d:

$$f_n = \frac{1}{2\pi}\left(\frac{B_d}{\rho_s hH}\right)^{1/2}. \tag{8.114}$$

The mean insertion loss predicted by (8.109) increases with frequency until $k_d H = O(1)$. As this condition is fulfilled at frequencies when $k^2 h^2$ is still

much smaller than unity, the mean insertion loss levels off at a value of $-20 \log \zeta$, where

$$\zeta = 6.7 \times 10^{-6} (\rho_d B_d)^{1/2},$$

with the bulk modulus in μbar and the density in grams. For an ideal foam displaying the very low characteristic impedance of air, $\zeta = 2.8 \times 10^{-4}$, the high-frequency mean insertion loss tends to 71 dB. Even this large value obtained with an ideal, unrealizable coating falls far short of the insertion loss predicted for bubble swarm layers by means of the theory presented in section 3.10 and confirmed by the measurements reproduced in figure 3.10.

References

1. R. D. Mindlin, "Waves and Vibrations in Isotropic Elastic Plates," in J. N. Goodier and N. J. Hoff, eds., *Proceedings of the First Symposium on Naval Structural Mechanics* (New York: Pergamon, 1960), pp. 199–232.

2. D. G. Crighton and D. Innes, "Low-frequency Acoustic Radiation and Vibration Response of Locally Excited Fluid Loaded Structures," *J. Sound Vib.* 91:293–314 (1983).

3. D. G. Crighton, "The Free and Forced Waves on a Fluid Loaded Elastic Plate," *J. Sound Vib.* 63:225–235 (1978).

4. R. V. Waterhouse and F. S. Archibald," Diffraction at a Fixed Point on a Fluid-Loaded Plate," *J. Acoust. Soc. Am.* 74 (Sl):S83 (A) (1983).

5. D. Feit, "Pressure Radiated by a Point-Excited Elastic Plate," *J. Acoust. Soc. A,.* 40:1489–1494 (1966).

6. P. R. Nayak, "Line Admittance of Infinite Isotropic Fluid-Loaded Plates," *J. Acoust. Soc. Am.* 47:191–201 (1970) [and see errata in 47:390 (1971)].

7. D. Feit and Y. N. Liu, "The Nearfield Response of a Line Driven Fluid-Loaded Plate," *J. Acoust. Soc. Am.* 78:763–767 (1985).

8. See, for example, A. Papoulis, *The Fourier Integral and Its Applications* (New York: McGraw-Hill, 1962), p. 25 et seq.

9. M. Heckl, "Radiation from a Point-Excited Infinitely Large Plate under Water," *Acustica* 13:182 (1963) (in German).

10. M. Heckl, "Sound Radiation by Point-Excited Plates," *Acustica* 9:371–380 (1959) (in German).

11. E. Skudrzyk, "Sound Radiation of a System with a Finite or an Infinite Number of Resonances," *J. Acoust. Soc. Am.* 30:1152–1158 (1958).

12. M. C. Junger, "Pressure Radiated by an Infinite Plate Driven by Distributed Load," *J. Acoust. Soc. Am.* 74:649–653 (1983).

13. M. Lax, "The Effect of Radiation on the Vibration of a Circular Diaphragm," *J. Acoust. Soc. Am.* 16:5–13 (1944).

14. H. G. Davies, "Acoustic Radiation from Fluid Loaded Rectangular Plates," MIT, TR71476-1, Dec. 1969; also *J. Acoust. Soc. Am.* 47:56 (A) (1969).

15. For plots of insertion loss versus frequency, and for asymptotic insertion loss expressions applicable to various frequency ranges, see M. G. Junger and J. E. Cole, III, "Bubble Swarm Acoustics: Insertion Loss of a Layer on a Plate," *J. Acoust. Soc. Am.* 68:241–247 (1980).

9 Sound Radiation by Shells at Low and Middle Frequencies

9.1 Introduction

The function of ship hulls, and in fact of any structure designed to exclude the ambient liquid medium from an air-filled space, requires either a boxlike or, more commonly, a shell-like configuration. The theory developed here is of course applicable to similar structures responsible for radiation of airborne sound. To make the dynamic interaction of these shells with an acoustic fluid analytically tractable, they are usually idealized as cylindrical or spherical shells. If we restrict ourselves to spherical shells, to infinitely periodic cylindrical shells, and to finite simply supported, cylindrical shells, that is, to three rather academic configurations whose in vacuo vibrations are analytically tractable, we find that the interaction problem is only tractable for the former two. For the third configuration, the one of greatest practical interest, we can obtain an explicit, rapidly converging series solution at the price of a relatively crude development. An alternative, more rigorous approach to the submerged cylindrical shell requires a numerical formulation. Because the latter does not reveal as readily the relation between the shell parameters and the radiated pressure, the crude analytical development is presented here. The analytical tools prerequistic for studying the vibrations of and sound radiation by submerged cylindrical and spherical shells were obtained in chapter 6, for the pressure field, and in chapter 7, for the dynamic properties of the shell.

The spherical shell and the infinite cylindrical shell will be studied first, because the interaction solution takes an exceptionally simple form for these two geometries. Specifically, the submerged shell retains normal modes not coupled by radiation loading. The spherical shell is studied in detail, both when surrounded by fluid and filled with fluid (sections 9.2–9.5). These analytical solutions provide insight into the effect of the ambient liquid on the dynamics of curved elastic plates in general. Another shell geometry that preserves uncoupled modes when submerged, the infinite cylindrical shell, is then analyzed (sections 9.6 and 9.7). This provides the point of departure for the study of the point-exicted finite cylindrical shell, which is not analytically tractable because its modes are coupled by radiation when it is exposed to an ambient fluid. A heuristic, phenomenological study of the finite shell, its sound field, and the effect of structural damping, is presented in sections 9.8–9.13. The effect of structural damping on peak pressures is examined in section 9.14. Finally, the conditions for the existence of normal modes in a submerged structure are stated, with emphasis on asymptotic situations of practical interest.

This chapter is limited to the low- and middle-frequency range, where

the ratio of circumference to wavelength is of order less than five. The high-frequency range, where the wave-harmonic series formulation converges poorly, is dealt with in chapter 12. Because that chapter is also concerned with high-frequency diffraction, it is placed after the two chapters that deal with scattering.

9.2 Characteristic Equation of the Submerged Spherical Shell

A spherical or other closed shell cannot be deformed without significant stretching of its middle surface. This entails substantial membrane stresses, represented by a linear term in (h/a) in the equations of motion. In contrast, flexural stresses are proportional to $\beta^2 h/a$, that is, $(h/a)^3$. Consequently, for thin-walled shells, and for the lower-order modes to which this anaysis is confined, membrane stresses predominate over flexural stresses. The β^2 terms in the equations of motion derived in chapter 7 will therefore be neglected. As in chapter 7, we confine ourselves to a radial excitation $f(\eta) = f(\cos\theta)$ axisymmetric about the $\theta = 0$ (or $\eta = 1$) axis [equation (7.116)]. The motion is independent of the spherical longitude ϕ and is described by two displacement components, a radial component w and a tangential component v [equations (7.110) and (7.111)].

After deriving the equation of motion of the submerged shell, we shall demonstrate that the submerged shell admits uncoupled modes of this same form, even though the ratios of radial to tangential displacement components are altered by radiation loading. The natural frequencies will be found to be lowered by accession to inertia. Furthermore, a radiation damping term is introduced in the equations of motion.

The loading, in the absence of a driving force, is expressed in terms of the specific acoustic impedance defined in chapter 6. For axisymmetric modes,

$$
\begin{aligned}
p_a(\theta) &= -\sum_{n=0}^{\infty} z_n \dot{W}_n P_n(\cos\theta) \\
&= -\sum_{n=0}^{\infty} (-i\omega W_n r_n - \omega^2 W_n m_n) P_n(\cos\theta).
\end{aligned} \tag{9.1}
$$

The negative sign arises because the applied pressure p_a was defined in Chapter 7 as positive outward, while the radiation loading is positive inward. The impedance components are frequency dependent, as seen by referring to equations (6.30)–(6.32). Since the pressure generated by the mode of order n embodies the same spatial dependence as the modal configuration of the radial displacement component, the pressure, like the mode, is orthogonal to all modes embodying a different θ dependence.

When solving the equations of motion of the submerged shell, it is convenient to combine the accession to inertia with the distributed inertial force of the shell proper, and to display the resistive impedance as a damping term. Equation (7.113) becomes

$$
\left[-\Omega^2 \left(1 + \frac{m_n}{\rho_s h} \right) - i\Omega \frac{a}{h} \frac{r_n}{\rho_s c_p} + 2(1+v) \right] W_n + (1+v)(n^2+n) V_n = 0,
$$

$$
(9.2)
$$

where λ_n has been written explicitly to avoid confusion with the wavelength. The equation of motion for the tangential component is not changed by radial loading and is therefore still given by equation (7.112). Equations (9.2) and (7.112) can be written as a set of homogeneous equations for the modal amplitudes W_n and V_n. The undamped natural frequencies are then determined as the roots of the real part of the determinant. The resulting equation is

$$
\Omega^4 \left(1 + \frac{m_n}{\rho_s h} \right) - \Omega^2 \left[1 + 3v + \left(1 + \frac{m_n}{\rho_s h} \right)(n^2+n) - \frac{m_n}{\rho_s h}(1-v) \right]
$$

$$
+ (1-v^2)(n^2+n-2) = 0.
\qquad (9.3)
$$

This equation takes the place of equation (7.114) derived for the shell in vacuo. It is a transcendental equation, since m_n is frequency dependent.

9.3 Natural Frequencies, Modal Configurations, and Radiation Damping of Submerged Spherical Shells

The natural frequencies of the submerges shell will be overscored, $\bar{\Omega}_n$, to distinguish them from the corresponding in vacuo natural frequencies. For the purely radial $n = 0$ breathing mode, there is a single real root. For $n \geq 1$, equation (9.3) yields two roots, $\bar{\Omega}_n^{(1)}$ and $\bar{\Omega}_n^{(2)}$, each associated with a different ratio V_n/W_n obtained by substituting the corresponding root in equations (7.112) and (9.2).

The natural frequencies and ratios V_n/W_n are tabulated in table 9.1 for a shell in vacuo and when submerged.[1] Radiation loading lowers the undamped natural frequencies, particularly for the predominantly radial modes. *Radiation loading alters the ratio V_n/W_n, but not the location of the nodal lines. Thus, for the spherical shell the identity of the in vacuo normal modes is preserved when the shell is submerged.*

Table 9.1
Effect of submergence on the normal modes of a spherical steel shell[a]
($h/a = 10^{-2}$, $v = 0.30$, $\rho_s/\rho = 7.67$, $c_p/c = 3.53$)

Mode order		Undamped dimensionless natural frequency ($\Omega_n \equiv \omega_n a/c_p$)		Ratio of tangential to radial displacement (V_n/W_n)		Radiation loss factor (η_n)
Branch	n	In vacuo	Submerged	In vacuo	Submerged	Submerged
Lower ($j = 1$)	1	0	0	1.000	1.000	0
	2	0.701	0.318	0.270	0.250	0.0309
	3	0.830	0.392	0.123	0.117	0.0102
	4	0.881	0.461	0.0702	0.0681	0.00202
	5	0.905	0.521	0.0457	0.0447	0.000699
	6	0.920	0.561	0.0321	0.0317	0
	7	0.928	0.585	0.0239	0.0237	0
	8	0.934	0.604	0.0208	0.0183	0
	9	0.938	0.621	0.0147	0.0146	0
Upper ($j = 2$)	0	1.61	1.22	0	0	1.74
	1	1.98	1.82	-0.500	-0.646	0.912
	2	2.72	2.55	-0.616	-1.08	0.179
	3	3.64	3.42	-0.680	-3.28	0.00862
	4	4.60	4.42	-0.713	-5.50	0.000145

a. Reproduced from Greenspon.[1]

Because flexural effects were omitted from equation (9.2), the modes of the lower branch become degenerate with increasing n, thus rapidly approaching the condition $(\omega_n a/c)^2 \equiv (ka)_n^2 \ll (2n - 1)$. In this range, the small-argument asymptotic expression for the accession to inertia [equation (6.32)] can be used. This reduces equation (9.3) to a nontranscendental, ordinary algebraic equation. The corresponding natural frequencies can be further simplified by ignoring the weak inertial coupling between the two families of modes:

$$\overline{\Omega}_n^{(1)} \approx (1 - v^2)^{1/2}\left(1 + \frac{\rho}{\rho_s}\frac{a}{h}\frac{1}{n}\right)^{-1/2}, \qquad n \gg 1, \qquad \left[\overline{\Omega}_n^{(1)}\left(\frac{c_p}{c}\right)\right]^2. \tag{9.4}$$

Table 9.1 indicates a large increase in V_n/W_n when the shell is submerged. This is due to the decrease in $\Omega_n^{(2)}$ and results in a large value of

$(V_n/W_n)^{(2)}$, which in turn produces a large modal mass. This is apparent from the modal mass per unit area of the shell vibrating in vacuo:

$$M_n = \frac{\rho_s h}{2n + 1} \left[1 + \frac{V_n^2}{W_n^2} (n^2 + n) \right]. \tag{9.5}$$

Like V_n/W_n, the modal mass differs for the upper and lower branches of modes.

The radiation damping of the normal modes of the submerged spherical shell will be expressed in dimensionless form, in terms of a modal loss factor. It is recalled that the loss factor of a simple oscillator of mass M, undamped natural frequency ω_n, stiffness K, and resistance r is

$$\eta = \frac{r\omega}{K} \tag{9.6a}$$

$$= \frac{r\omega}{\omega_n^2 M}. \tag{9.6b}$$

A frequency-independent resistance requires that η vary linearly with frequency. Since an uncoupled mode is equivalent to a simple oscillator, its loss factor can be similarly defined in terms of the modal mass, resistance, and natural frequency. For the submerged shell, and ignoring structural losses, the radiation loss factor at resonance is

$$\eta_n = \frac{r_n (2n + 1)^{-1}}{\bar{\omega}_n [M_n + m_n (2n + 1)^{-1}]}, \tag{9.7}$$

where r_n is the specific acoustic resistance [equation (6.31)], m_n the accession to inertia per unit area [equation (6.32)], M_n the modal mass per unit area associated with the shell proper [equation (9.5)], and $\bar{\omega}_n \equiv \bar{\Omega}_n c_p/a$ the natural frequency of the submerged shell. Like these parameters, the radiation loss factor differs for the two families of modes. The loss factor can be written more explicitly as

$$\eta_n = \frac{r_n}{(\bar{\Omega}_n c_p/a) \{\rho_s h [1 + (V_n/W_n)^2 (n^2 + n)] + m_n\}}. \tag{9.8}$$

The loss factor is also tabulated in table 9.1. It is seen that for a given n,

the loss factor of either branch can predominate. The reason is that the conditions that favor a small loss factor (specifically, for the upper family of modes, the large values of $\bar{\Omega}_n$ and of V_n/W_n) counteract the conditions that tend to boost η_n [the large r_n and small m_n associated with the relatively large values of $(ka)_n \equiv \bar{\Omega}c_p/c$ of this family]. If structural losses are significant, the resulting loss factor is simply the sum of the acoustic and structural loss factors.

Having determined the characteristics of the normal modes of the submerged shell, we can compute their amplitude when a force is applied to the shell. Finally, the pressure field in the ambient medium is obtained.

9.4 Response and Pressure Field of Point-Excited Submerged Spherical Shells

We shall specialize the equations of motion to an excitation in the form of a concentrated force F applied at $\theta = 0$. Admittedly, flexural stresses in the shell are required to sustain such a load, but in the frequency range considered here, the resulting error in the shell response and in the pressure field in negligible. This driving force combined with the radiation loading in equation (9.1) gives rise to a resultant forcing term, to be used in equation (7.117):

$$f_n = \frac{(2n + 1)F}{4\pi a^2} - z_n \dot{W}_n.$$

(9.9)

The forced velocity response can now be written formally as

$$\dot{w}(\theta) = \sum_{n=0}^{\infty} \frac{f_n}{Z_n} P_n(\cos\theta),$$

(9.10)

where Z_n is the in vacuo modal impedance. For the present purpose, this impedance is obtained from equation (7.121) by dropping the flexural β^2 term:

$$Z_n = -i\frac{h}{a}\rho_s c_p \frac{[\Omega^2 - (\Omega_n^{(1)})^2][\Omega^2 - (\Omega_n^{(2)})^2]}{\Omega^3 - \Omega(n^2 + n - 1 + v)},$$

(9.11)

where $\Omega_n^{(1,2)}$ are the two in vacuo natural frequencies associated with the two roots of equation (7.114). Substituting the expression for f_n, equation

(9.9), in equation (9.10), the modal velocity amplitude becomes

$$\dot{W}_n = \frac{(2n + 1)F}{4\pi a^2 Z_n} - \frac{z_n}{Z_n} \dot{W}_n.$$

This can be solved for $\dot{W}_n$:

$$\dot{W}_n = \frac{(2n + 1)F}{4\pi a^2 (Z_n + z_n)}. \tag{9.12}$$

When this is substituted in equation (9.10), the response of the submerged shell is finally obtained:

$$\dot{w}(\theta) = \frac{F}{4\pi a^2} \sum_{n=0} \frac{2n + 1}{Z_n + z_n} P_n(\cos\theta). \tag{9.13}$$

The sound field is constructed by multiplying each term in the above series by $i\rho c h_n(kR)/h_n'(ka)$ [see equation (6.19)]. Specializing the resulting expression to the shell surface, the surface pressure is found in terms of the specific acoustic impedance:

$$p(a, \theta) = \frac{F}{4\pi a^2} \sum_{n=0} \frac{(2n + 1)z_n}{Z_n + z_n} P_n(\cos\theta).$$

The corresponding far-field pressure is [equation (6.21)]

$$p(R, \theta) = \frac{F\rho c e^{ikR}}{4\pi a^2 kR} \sum_{n=0} \frac{(-i)^n(2n + 1)P_n(\cos\theta)}{(Z_n + z_n)h_n'(ka)}, \qquad kR \gg 1. \tag{9.14}$$

This series converges slowly at high frequencies. An asymptotic form applicable to this frequency range is developed in chapter 12.

At a modal resonance $Z_n + z_n$ reduces to the resistance. When structural damping is negligible, this resistance consists effectively of the specific acoustic resistance. From equation (6.31)

$$\left.\begin{aligned} Z_n + z_n &\approx r_n \\ &= \rho c [ka|h_n'(ka)|]^{-2} \end{aligned}\right\} \qquad \begin{aligned} &\Omega = \bar{\Omega}_n^{(j)} = (ka)\left(\frac{c}{c_p}\right), \\ &\eta_s \ll \eta_n. \end{aligned}$$

The pressure amplitude of the resonant mode, $|p_n|$, is obtained by substituting this expression equation in (9.14):

$$\left.\begin{aligned}|p_n(R, \theta = 0)| &= \frac{F(2n+1)k|h_n'(ka)|}{4\pi R} \\[2ex] &= \frac{F(2n+1)}{4\pi Ra\,(r_n/\rho c)^{1/2}}\end{aligned}\right\} \quad \begin{aligned} &ka = \bar{\Omega}_n^{(j)}\left(\frac{c_p}{c}\right), \\[2ex] &\eta_s \lll \eta_n. \end{aligned} \tag{9.15}$$

Hence, resonant modes controlled by radiation rather than by structural damping generate large far-field pressures if their acoustic resistance is small. Thus, referring to table 9.1, one concludes that the $n = 0$ resonance, which in a submerged thin shell is typically close to critically damped, will generate lower pressures than the $n = 1$ or $n = 2$ resonances, which are characterized by more marked resonance peaks. One should, however, not conclude that the degenerate modes of the lower ($j = 1$) branch will display marked pressure peaks when excited at resonance, merely because their radiation loss factor is negligibly small. This paradoxical situation does not, in practice, arise, because in a real shell such poorly radiating resonances are controlled by structural damping.

To study the pressure radiated by a resonant mode controlled by structural damping, we define a structural resistance r_s, related to the structural loss factor η_s by expressions similar to the equations relating the specific acoustic resistance r_n to the radiation loss factor η_n, equations (9.7) and (9.8). Like the radiation resistance, r_s depends on the mode order, but the corresponding subscript will be suppressed. The impedance ($\mathcal{Z}_n + z_n$) of a resonant mode can now be approximated as r_s since $r_n \ll r_s$. The pressure amplitude radiated by the resonant mode thus becomes

$$\left.\begin{aligned}|p_n(R, \theta = 0)| &= \frac{F\rho c(2n+1)}{4\pi a^2 kR|h_n'(ka)|r_s} \\[2ex] &= \frac{F(2n+1)(r_n/\rho c)^{1/2}}{4\pi aR\,(r_s/\rho c)}\end{aligned}\right\} \quad \begin{aligned} &ka = \bar{\Omega}_n^{(j)}\left(\frac{c_p}{c}\right), \\[2ex] &\eta_s \ggg \eta_n. \end{aligned} \tag{9.16}$$

Hence, for *resonant modes controlled by structural rather than by radiation damping, the pressure is larger for modes with higher radiation resistance.* This situation is therefore precisely the reverse of that encountered for resonances controlled by radiation damping. In table 11.1 these results are summarized and

compared with the pressures radiated by shells excited by an incident plane wave. The respective importance of structural and radiation damping is analyzed more closely in section 9.11.

To conclude the study of spherical shells, we shall illustrate the effect of a fluid contained in a spherical shell.

9.5 Normal Modes of Fluid-Filled Spherical Shells

The analysis will be restricted to the $(n = 0)$ mode, as this example adequately illustrates the effect of a contained fluid as opposed to an ambient fluid. The higher order modes can be analyzed by an identical technique.

The pressure $p(a)$ acting on the inside of a spherical container vibrating uniformly was derived in equation (2.47) as a function of ka, which in the present notation equals $\Omega c_p / c$. Equation (7.118) now takes the form

$$[-\Omega^2 + 2(1 + v)]W_0 = \frac{a^2 p(a)}{\rho_s c_p^2 h}. \tag{9.17}$$

The pressure $p(a)$ was expressed in chapter 2 as an effective spring constant, equation (2.49). This physical interpretation of the fluid reaction as a spring force holds for small ka only, that is, in the low-frequency range relevant to the gas-bubble vibrations studied in section 3.9. This physical interpretation is not germane to the natural frequency range of spherical metal shells filled with water. The mathematical formulation will nevertheless be retained, as it is convenient, it being understood that k_0 can be either an effective spring stiffness or mass, depending on the frequency range. Setting

$$p(a) = -k_0 W_0,$$

$$2(1 + v) = \Omega_0^2$$

in equation (9.17), and dividing by W_0, one obtains the characteristic equation of the fluid-filled sphere:

$$\frac{\rho_s c_p^2 h}{3\rho c^2 a}(\Omega^2 - \Omega_0^2) = \frac{k_0 a}{3\rho c^2}. \tag{9.18}$$

This is a transcendental equation, because k_0 is frequency dependent. The function on the right of this equation, already plotted in figure 2.3, is replotted in figure 9.1. Its intersections with the function on the left of

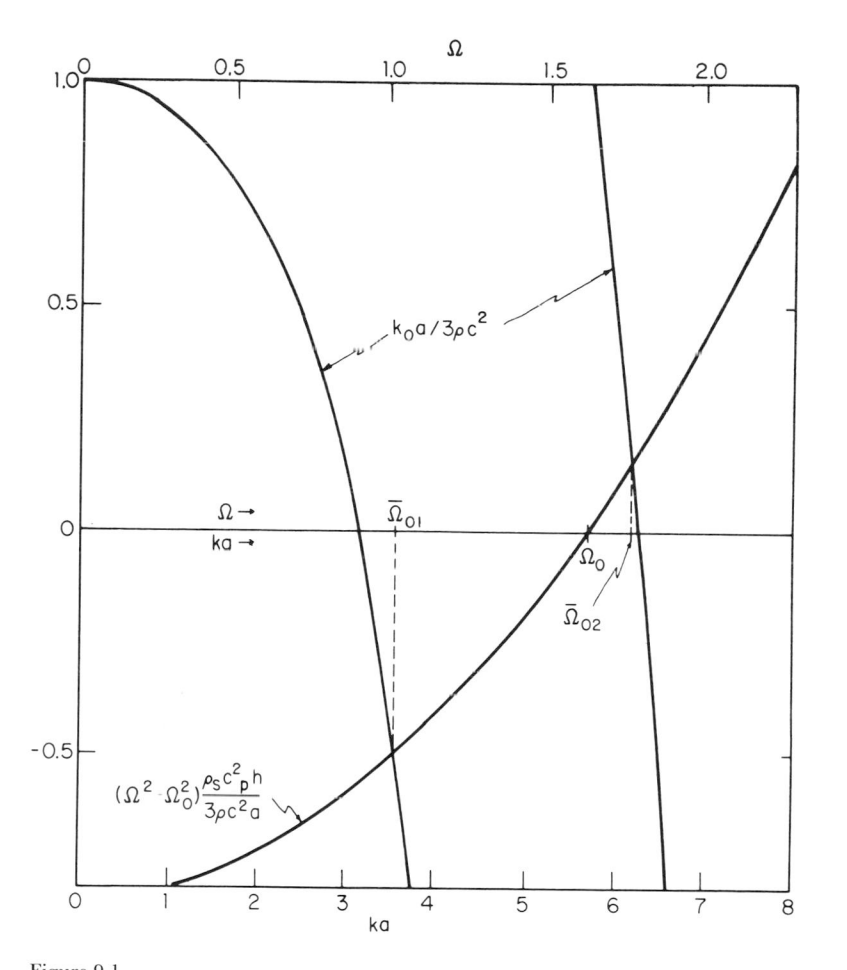

Figure 9.1
Construction of the natural frequencies $\bar{\Omega}_{0j}$ of the spherically symmetric $(n = 0)$ modes of a fluid-filled spherical shell, for the shell and fluid parameters listed in table 9.1 [equation (9.17)].

equation (9.18) yield the natural frequencies of a water-filled fluid shell. Thus, instead of a natural frequency associated with the single $n = 0$ mode obtained when the shell is surrounded by liquid, a family of resonances $\bar{\Omega}_{0j}$ is obtained for the coupled system consisting of the shell and the enclosed fluid. Impedance measurements performed on the outer surface of the shell do not reveal which resonance can be traced to the in vacuo resonance of the shell. It can be seen from the curves in figure 9.1 that, for the parameters selected, a reduction in h/a results in a decrease of the fundamental natural frequency $\bar{\Omega}_{01}$ and an increase in $\bar{\Omega}_{02}$. Each mode is characterized by a different number of concentric, spherical nodal surfaces associated with the zeros of $j_0(kR)$. The pressure field can be envisioned as radial standing waves with a focal point at the center. The fluid-filled spherical shell vibrating in its $(n = 0)$ modes thus constitutes a convenient tool for studying high-amplitude sound waves in the body of the fluid, off the vibrating surface. There is of course no radiation damping, as already pointed out in chapter 3.

We now turn our attention to cylindrical shells surrounded by an acoustic fluid.

9.6 Normal Modes of the Infinite, Submerged Cylindrical Shell

The equations of motion of the shell in vacuo were derived in chapter 7 for an arbitrary forcing function p_a [equation (7.80). In vacuo, these equations admit solutions for the axial, circumferential, and radial components of the standing-wave displacement vector in the form of equation (7.94). As in the case of spherical shells, it will be found that the these normal mode shapes are preserved when the shell is submerged, but that the ratios of tangential to radial displacement and natural frequencies are altered. The pressure associated with the radial component was derived in chapter 6 [equations (6.41) and (6.42)]. The pressure loading on the shell surface will be expressed in terms of the impedance components defined in equation (6.44), (6.45), and (6.46). The loading p_a in equation (7.80) can thus be written as

$$p_a(\phi, z) = -\sum_{m,n} W_{mn}(-\omega^2 m_{mn} - i\omega r_{mn}) \cos k_m z \cos n\phi. \qquad (9.19)$$

The equations of motion of the tangential components remain unaltered from their in vacuo form. The equation governing the radial component is constructed as in the case of the spherical shell, by combining the accession in inertia term from equation (9.19) with the inertial term of the in vacuo

equation of motion. Equation (7.95) thus becomes

$$[-\Omega^2 + k_m^2 a^2 + \tfrac{1}{2}(1 - v)n^2]U_{mn} + [\tfrac{1}{2}(1 + v)nk_m a]V_{mn} + vk_m aW_{mn} = 0,$$

$$[\tfrac{1}{2}(1 + v)nk_m a]U_{mn} + [-\Omega^2 + \tfrac{1}{2}(1 - v)k_m^2 a^2 + n^2]V_{mn} + nW_{mn} = 0,$$

$$vk_m aU_{mn} + nV_{mn} + \left[-\Omega^2\left(1 + \frac{m_{mn}}{\rho_s h}\right) - i\Omega \frac{a}{h}\frac{r_{mn}}{h\rho_s c_p} \right. \tag{9.20}$$

$$\left. + 1 + \beta^2(k_m^2 a^2 + n^2)^2 \right]W_{mn} = 0$$

9.7 Natural Frequencies of Infinite Submerged Cylindrical Shells

The undamped natural frequencies of the submerged shell are the positive roots of the real component of the coefficient determinant. All but one of the coefficients retain their in vacuo value [equation (7.95)]. Only the coefficient in the third row, third column is altered by the accession to inertia term in equation (9.19) and now reads.

$$-\Omega^2\left(1 + \frac{m_{mn}}{\rho_s h}\right) + 1 + \beta^2(k_m^2 a^2 + n^2)^2. \tag{9.21}$$

For n, $k_m \neq 0$, this equation admits three roots, corresponding to three natural frequencies $\bar{\Omega}_{nm}^{(j)}$, with $j = 1$, 2, and 3, each associated with its coefficient ratios $(U_{mn}/W_{mn})^{(j)}$ and $(V_{mn}/W_{mn})^{(j)}$. For a given combination of n and $k_m a$, the lowest of the three natural frequencies is associated with small values of these ratios, except for the transverse, $n = 1$, modes in the small $k_m a$ range, where the radial and circumferential components are approximately equal. The typical $(j = 1)$ mode, being primarily radial, is excited best by the radial forces considered here. Predominantly radial dynamic configurations are also well coupled to the ambient liquid. Since the subsequent development is mostly concerned with this family of modes, the superscript (j) will be omitted when $j = 1$.

The natural frequencies are reduced below their in vacuo values by the inertial radiation loading embodied in the coefficient $[1 + (m_{mn}/\rho_s h)]$, which multiplies the last Ω^2 term in equation (9.20). An approximate relation between the natural frequencies of a predominantly radial mode of the shell vibrating, respectively, in vacuo and in a liquid medium can be constructed. Since only the $\dot{w}^2$ term of kinetic energy is increased,

$$\frac{\Omega_{mn}}{\bar{\Omega}_{mn}} = \left\{ 1 + \frac{m_{mn}}{\rho_s h} \left[1 + \frac{U_{mn}^2}{W_{mn}^2} + \frac{V_{mn}^2}{W_{mn}^2} \right]^{-1} \right\}^{1/2} . \tag{9.22}$$

The accession to inertia lowers the natural frequency, but this decrease is partly neutralized by the inertia forces associated with the two tangential displacement components, which are not enhanced by the accession to inertia.

The in vacuo natural frequencies of the predominantly radial axisymmetric $(n = 0)$ modes are modified by radiation loading in a manner radically different from spherical shell modes or from cylindrical shell beam $(n = 1)$ and lobar $(n > 1)$ modes to be examined further on: In vacuo, the lower-order $m, n = 0$ predominantly radial modes display natural frequencies[designated by Ω_w in equation (7.98c)], which are effectively degenerate, i.e., m independent for thin shells, since $k_m a$ must approach $\beta^{-1/2}$ before a significant rise in natural frequency is observed. However, when the shell is submerged, the inertial reactance diverges at coincidence, $k_m = k$ (see the $n = 0$ curve in figure 6.6), thereby giving rise to two natural frequencies located on either side of the coincidence frequency:

$$(ka)_1 = k_m a - \varepsilon_1, \qquad (ka)_2 = k_m a + \varepsilon_2, \tag{9.23}$$

where ε is small. These two resonance frequencies coincide with the intersection of the effectively m-independent $n = 0$ shell stiffness and of the inertial reactance maximum (figure 9.2). They vary nearly linearly with $k_m a$ and are therefore nondegenerate. For values of c_p/c representative of metal shells, a third natural frequency is observed somewhat below the in vacuo resonance $ka \simeq c_p/c$. Radiation loading therefore has the effect of spreading the $(m, 0)$ resonances over a wide frequency range.

To illustrate the effect of radiation loading on lobar modes, the natural frequencies of the lowest (that is, predominantly radial) $n = 2$ modes are shown in figure 9.3 for three ratios of thickness to radius. These graphs were computed by Warburton[3] from equations of motion that include a v-dependent flexural term omitted in equation (9.20), but whose effect is negligible for the relatively small values of h/a. Also shown are natural frequencies computed from the approximate natural frequency in vacuo, equations (7.99), modified to account for radiation loading in the manner introduced by Smith:[4] Accession to inertia is approximated by the latter of equations (5.63), where the structural wave number is equated to the helical wave number, equation (7.100). As in equation (7.99), the inertial contri-

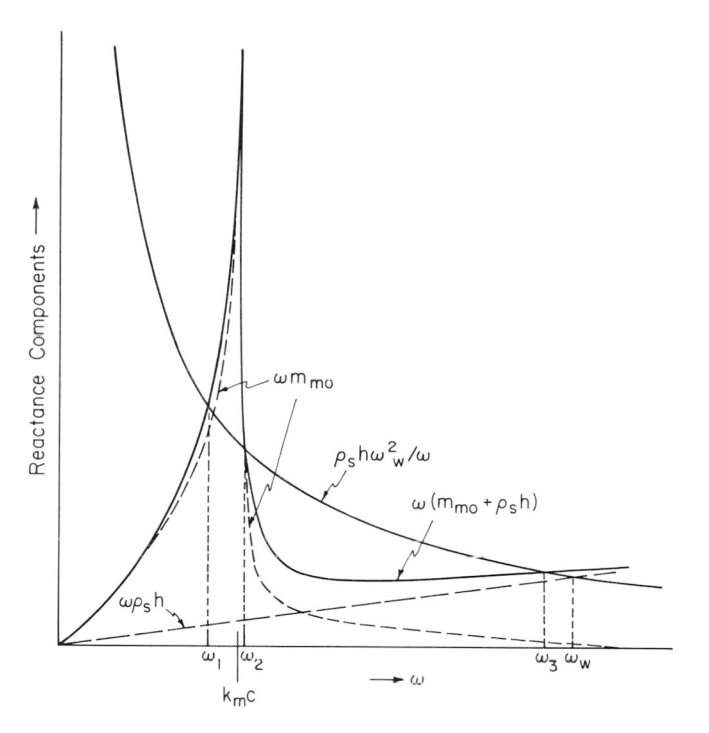

Figure 9.2
Schematic plot of the stiffness and inertial reactance components, of a predominantly radial axisymmetric $(m0)$ mode of a submerged cylindrical shell illustrating the introduction of nondegenerate natural frequencies ω_1 and ω_2 by the coincidence peak of the inertial reactance (figure 6.6). ω_1 and ω_2 are given in dimensionless form in equation (9.23), and ω_w by equation (7.98c). [Reproduced from Junger and Garrelick[2]]

bution of the tangential modes in equation (9.22) is simulated by the factor $(1 + n^{-2})$ valid for $k_m^2 a^2) \ll 1$. Combining equations (7.99) and (9.22), we now have

$$\bar{\Omega}_{mn} = \frac{[(1 - \nu^2)\,(k_m a/k_s a)^4 + \beta^2 k_s^4 a^4]^{1/2}}{[1 + n^{-2} + (\rho/\rho_s k_s h)]^{1/2}}, \qquad n > 0. \tag{9.24}$$

This is plotted in figure 9.3 for $n = 2$.

The results obtained with these equations indicate that the discrepancies between Warburton's rigorous thin-shell theory[3] and equation (9.24) result primarily from the errors inherent in the approximate in vacuo natural

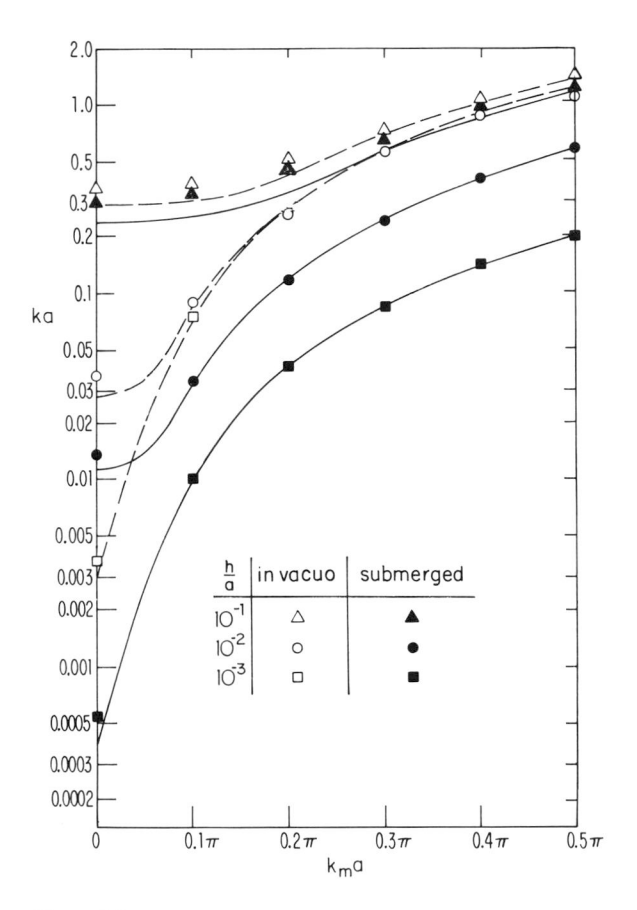

Figure 9.3
Dimensionless natural frequency ($ka \equiv \Omega c_p/c$) for the predominantly radial family of resonances of the two-lobed ($n = 2$) modes of an infinite cylindrical submerged steel shell ($\rho_s/\rho = 7.84$, $c_p/c = 3.71$, $v = 0.3$) as a function of the axial wave numbers. The curves were computed by Warburton[3] from thin shell theory. The dots are approximate results computed for the shell in vacuo from equation (7.99) and for the submerged shell from equation (9.24). Warburton's solid and dashed curves are, respectively, for the shell submerged and in vacuo.

frequencies, equation (7.99), rather than in the approximation to m_{mn} in equation (9.24). These approximate expressions will be used for all modes other than the axisymmetric mode. The insensitivity to the value of h/a displayed by the $(n = 2)$ modes for short axial wavelengths in vacuo is the result of predominantly membrane-type stresses. In contrast, when the shell is submerged, the natural frequencies corresponding to these two shell thicknesses are markedly separated, because the accession to inertia decreases the natural frequencies of the thinner shell more than those of the thicker shell.

Because the radiation resistance, equation (6.46d), is zero for $k_m > k$, a meaningful far-field pressure cannot be constructed for the infinite shell, nor can an acoustic loss factor be computed. For this purpose, we must turn to a finite shell configuration, such as the simply supported finite cylindrical shell.

9.8 Intermodal Fluid Coupling in the Submerged Finite Cylindrical Shell

The simply supported cylindrical shell extended by two semi-infinite, rigid cylindrical baffles (figure 6.1), even though a relatively crude mathematical model, provides a useful analytical approximation to the pressure field radiated by finite cylindrical shells or approximately cylindrical shells of revolution. This model will be used except for the low-frequency range of the free-floating finite shell. For the sake of brevity, we select an excitation producing a shell response symmetric with respect to the ϕ- and z-coordinate axes: a concentrated radial force F applied at $z = 0$, $\phi = 0$. The in vacuo modes of such a shell are in the form of equation (7.94) with k_m as stated in equation (6.57a). In our approximate analysis, we shall assume that the radial configuration of the submerged shell can still be written as a Fourier series:

$$
\begin{aligned}
\dot{w}(z) &= 0, & |z| &> L \\
&= \sum_{m,n} \dot{W}_{mn} \cos n\phi \cos k_m z, & |z| &\le L.
\end{aligned}
\tag{9.25}
$$

The errors inherent in this approximation are associated with the radiation loading near the ends of the shell. It was noted, in connection with equation (6.64), that for $k_m > k$ the far-field is associated with the peripheral regions of the cylindrical radiator, near $z = \pm L$, which thus become energy sinks. Consequently, in addition to standing waves, the dynamic configura-

tion of the shell must embody traveling waves conveying structure-borne energy to these peripheral regions. The coefficients in equation (9.25) must therefore be complex. The fact that the finite shell does not, like our mathematical model, embody two semiinfinite rigid cylinders further alters the radiation loading near the shell ends. In spite of these shortcomings, more exact numerical calculations and natural frequency measurements indicate that a solution based on the series representation in equation (9.25) can predict the shell response reliably enough to be of practical value. In contrast, the associated pressure cannot be approximated by a Fourier series, but requires a transform representation if meaningful results are to be predicted. The reason for this difference in formulation is that the traveling-wave nature of the pressure field is one of its essential characteristics, while it is only a second-order effect in the predominantly standing-wave response of the shell. Consequently, the pressure associated with the configuration in equation (9.25) must be formulated as an inverse Fourier transform, equation (6.49).

The Fourier transform of the dynamic configuration of the cylindrical boundary [designated by the symbol $\tilde{f}_n(\gamma)$ in equation (6.2)] is an integral, which was evaluated in equation (6.57b). The cosine transform of the velocity in equation (9.25) thus finally becomes

$$\tilde{\dot{w}}(\phi; \gamma) = 2 \sum_{n=0} \sum_{m=0} \dot{W}_{mn} \frac{k_m(-1)^m}{k_m^2 - \gamma^2} \cos \gamma L \cos n\phi. \tag{9.26}$$

The corresponding pressure is in the form of an integral, equation (6.49), which will be expressed in terms of a radiation impedance transform constructed by replacing k_m with γ in equation (6.42b):

$$\tilde{z}_n(\gamma) = \frac{i\rho c k H_n[(k^2 - \gamma^2)^{1/2} a]}{(k^2 - \gamma^2)^{1/2} H_n'[(k^2 - \gamma^2)^{1/2} a]}. \tag{9.27}$$

The surface pressure, [equation (6.49) with $r = a$], specialized to the standing-wave configuration in equation (9.25) can now be expressed in terms of $\tilde{z}_n$:

$$p(a, \phi, z) = \frac{2}{\pi} \sum_{n=0} \sum_{m=0} \dot{W}_{mn} k_m(-1)^m \cos n\phi \int_0^\infty \frac{\cos \gamma L \cos \gamma z \, \tilde{z}_n(\gamma) \, d\gamma}{k_m^2 - \gamma^2}. \tag{9.28}$$

The forcing term p_a in the shell equations of motion, equation (7.80), can be stated as the sum of the applied exciting force and of the above radiation loading:

$$p_a = \frac{1}{a} F \delta(z) \delta(\phi) - p(a, \phi, z), \tag{9.29}$$

where δ is the one-dimensional Dirac delta function, whose properties are analogous to those of the three-dimensional delta function defined in equations (4.2).

The surface pressure generated by a mode (m, n) is represented by a continuous spectrum of wave numbers γ [equation (9.28)], encompassing all the axial wave numbers in equation (9.25). This (m, n) mode is thereby coupled to all modes $(m \pm 1, n)$, $(m \pm 2, n)$, ... of the same index number n. Hence, even if a distributed applied force is chosen in the form $P \cos n\phi \cos k_m z$ so as to excite only one of the modes in equation (9.25), the radiation loading embodied in equation (9.28) couples the directly excited mode to all modes of the same order n, whatever their axial wave number. Having recognized the inherent limitations of our approach, we proceed to evolve approximate formulations of the response of finite shells.

We can construct the radiation-coupled equations of motion by substituting equations (9.28) and (9.29) in equation (7.80). The two equations of motion for axial and circumferential displacement components remain homogeneous, and therefore retain the same form as for in vacuo vibrations. The inhomogeneous equation for the radial displacement component becomes

$$\frac{c_p^2 \rho_s h}{a^2} \{ \nu k_m a U_{mn} + n V_{mn} + [-\Omega^2 + 1 + \beta^2 (k_s a)^4] W_{mn} \}$$

$$= \frac{2\varepsilon_n}{\pi L} \int_0^\pi \int_0^L p_a \cos n\phi \cos k_m z \, d\phi \, dz. \tag{9.30}$$

Here k_s is defined in equation (7.100) and $\varepsilon_n = 1$ for $n = 0$ and 2 for $n \geq 1$. When substituting equation (9.29) in the above integral, one must divide the driving force amplitude F by four, the range of integration being confined to one-quarter of the shell surface and the integral being multiplied by four.

Adapting the notation used for the forced planar vibrations of the cylindrical shell [equation (7.87)] to three-dimensional vibrations, the

forcing term on the right side of equation (9.30) is designated as f_{mn}. This modal driving force can now be evaluated by substituting equation (9.29) in equation (9.30). The ϕ integration of the surface pressure uncouples modes of different order n and yields π/ε_n. The notation k_q will be used to designate the axial wave numbers in the integral representation of the pressure, equation (9.28), to distinguish them from the wave number k_m of the shell modes in equation (9.25):

$$
f_{mn} = \frac{\varepsilon_n F}{2\pi La} - \frac{2}{\pi L} \sum_{q=0} \dot{W}_{qn} k_q (-1)^q
$$
$$
\cdot \int_0^L \cos k_m z \int_0^\infty \frac{\cos \gamma L \cos \gamma z \, \tilde{z}_n(\gamma)}{k_q^2 - \gamma^2} \, d\gamma \, dz.
$$
(9.31)

By analogy with the analysis of the infinite cylindrical shell, the q series in equation (9.31) will be expressed in terms of modal specific acoustic impedances. However, because of intermodal radiation coupling, a distinction must be drawn between self-impedance and mutual impedances:

$$
f_{mn} = \frac{\varepsilon_n F}{2\pi La} - \sum_{q=0} \dot{W}_{qn} z_{qmn}.
$$
(9.32)

Here z_{qmn} represents the loading on mode (m, n) by the pressure generated by mode (q, n), or vice versa, because $z_{qmn} = z_{mqn}$. Because of the orthogonality of the $\cos n\phi$ functions, no coupling exists between modes of different circumferential configuration. Physically, this results from the fact that the pressure is truly periodic in ϕ, in contrast to its nonperiodic z dependence. Comparing equations (9.31) and (9.32) and performing the z integration by means of equation (5.46), one obtains the mutual impedance in the form of a single infinite integral:

$$
z_{qmn} = \frac{4}{\pi L} k_q k_m (-1)^{q+m} \int_0^\infty \frac{\cos^2 \gamma L \, \tilde{z}_n(\gamma)}{(k_q^2 - \gamma^2)(k_m^2 - \gamma^2)} \, d\gamma.
$$
(9.33)

The self-impedance $q = m$ can be expected to be larger than the mutual impedances because of the double root displayed by the denominator of the integrand, when $k_m = k_q \to \gamma$, in contrast to the single root, when $k_m \neq k_q \to \gamma$. In fact, as the number of structural wavelengths along the cylinder tends to infinity, the mutual impedances tend to zero. This can be

verified by noting that in equation (9.31)

$$\lim_{k_m L \to \infty} \int_0^L \cos k_m z \cos \gamma z \, dz = \frac{\pi}{2} [\delta (k_m - \gamma) + \delta (k_m + \gamma)].$$

From detailed studies on flat plates,[5,6] it can be surmised that the reactive components of the mutual impedances are negligible even for radiators that are relatively small in terms of wavelengths in contrast to the mutual resistances that are of the same order of magnitude as the corresponding self-resistance. Unfortunately, explicit expressions for the modal amplitudes of the submerged shell cannot be evolved unless one ignores the mutual impedances ($q \neq m$), only the self-impedances ($q = m$) being retained. These can now be identified more concisely by only two subscripts, as z_{mn}. More exact solutions require the numerical solution of integral equation formulations.[7,8]

9.9 Approximations to the Radiation Loading of Finite Cylindrical Shells

Even when a numerical solution is undertaken, approximate but explicit formulations of the radial shell response associated with sound radiation are useful for the purposes of checking asymptotic limits and visualizing the effect of various parameters on the acoustics. For the purpose of generating these approximate results we must ignore intermodal coupling introduced by radiation loading. The radial shell response is governed by the inhomogeneous equation in equations (9.30). The two homogeneous equations that yield U_{mn} and V_{mn} in terms of W_{mn} are not altered by radiation loading and are therefore still given by equations (7.80a) and (7.80b). Rather than inverting the third-order coefficient matrix to obtain the radial components, we use the small-$k_m a$ approximation to account for coupling between tangential and radial motions embodied in the $(1 + n^{-2})$ factor in equation (7.101). For a simply supported shell, point excited at midspan, the radial modal velocity amplitudes become

$$\dot{W}_{m0} = \frac{F}{2\pi L a} \left| -i\omega (\rho_s h + m_{m0}) \left(1 - \frac{\bar{\omega}_{m0}^2}{\omega^2} \right) + \langle r_{m0} \rangle + r_s \right|^{-1},$$

$$\dot{W}_{mn} = \frac{F}{\pi L a} \left| -i\omega [\rho_s h (1 + n^{-2}) + m_{mn}] \left(1 - \frac{\bar{\omega}_{mn}^2}{\omega^2} \right) + \langle r_{mn} \rangle + r_s \right|^{-1}, \quad n > 0,$$

$$(9.34)$$

where $\bar{\omega}_{mn} \equiv \bar{\Omega}_{mn} c_p / a$. This result will be combined with the ratios of far-field pressure to acceleration amplitudes derived in chapter 6. Here, the notation $\langle r_{mn} \rangle$ indicates that the resistance is spatially averaged over the shell surface. Loads not symmetric with respect to the midspan cross section will of course excite antisymmetric modes. Modal generalized forces corresponding to arbitrary load distributions are given by the integral in equation (9.30), supplemented by $\sin k_m z$ and $\sin n\phi$ terms.

The predominantly inertial radiation loading m_{mn} of slow, $(k_s > k)$ modes associated with the localized interaction of the vibrating boundary and the fluid is governed by the Laplace rather than the wave equation. Being insensitive to remote interactions, the added masses of slow modes can be approximated by means of a mathematical model in the form of the infinitely periodic cylinder, equation (6.44), or even, as we already did in formulating the natural frequencies [equation (9.24)], the added mass of the infinitely periodic plate, equation (5.63).

In contrast, the small radiation resistance of these same slow modes, being associated with the shell extremities [see the discussion of equation (6.60)] cannot be evaluated without accounting for the shell's finite length. In fact, the resistance of the infinitely periodic cylinder was seen to vanish when $k_m > k$, equation (6.46). Sandman[8] showed that the difference between the resistance of the finite cylinder and the cylinder bracketed between semi-infinite rigid cylindrical baffles is relatively minor. The latter model is therefore adequate. Rather than solving the integral equation to obtain the surface pressure and hence the self- and mutual resistances, as Sandman[8] did, we can integrate the far field intensity to obtain power, equation (3.27), and hence the space averaged resistance, as already illustrated in equation (6.67). This yields

$$\langle r_{mn} \rangle = \frac{8\rho c L}{\pi^2 (m + \frac{1}{2})^2 a} \int_0^{\pi/2} \frac{\cos^2(kL\cos\theta)\, d\theta}{\sin\theta\{1 - [k^2 L^2 / \pi^2 (m + \frac{1}{2})^2]\cos^2\theta\}^2 |H_n''(ka\sin\theta)|^2}. \tag{9.35}$$

Explicit asymptotic expressions for $k^2 L^2 \ll \pi^2$ are presented in equations (6.70a) and (6.70b) for $k^2 a^2 \ll 1$, and in equation (6.70c) for $k^2 a^2 \gg 1$. In contrast to Sandman's rigorous integral equation solution, equation (9.35) does not distinguish between self- and mutual resistances.

In conclusion, it is noted that the large resistance of supersonic modes $(k_s < k)$ is, like the inertial reactance of slow modes, a local phenomenon that can be approximated by the infinitely periodic cylinder. The physics

can be envisioned by noting that when $k_s < k$, the fluid is locally compressed, the vibration of the boundary reversing its direction too rapidly to allow the fluid to be accelerated to neighboring regions of rarefaction.

Below the coincidence frequency, predominantly flexural shell modes, i.e., modes whose circumferential mode order exceeds $n_{\min}$, equation (7.100a), are inefficient sound radiators at their resonance frequency. For such modes, structural damping embodied in the structural loss factor tends to predominate over radiation damping. The structural resistance in equations (9.34) is given by

$$
\left.
\begin{aligned}
r_s &= \frac{\omega_{m0}^2 \rho_s h \eta_s}{\omega}, \qquad \omega_{m0} = \frac{c_p}{a} \\[2mm]
&= \rho c \, (ka)^{-1} \left(\frac{c_p}{c}\right)^2 \left(\frac{\rho_s}{\rho}\right)\left(\frac{h}{a}\right) \eta_s
\end{aligned}
\right\} \quad n = 0
$$
$$
= \frac{\omega_{mn}^2 \rho_s h \eta_s \left(1 + n^{-2}\right)}{\omega}, \qquad n > 0.
\tag{9.36}
$$

The structural loss factor η_s is typically of order 10^{-3}–10^{-2} for bare metal shells, rising to 10^{-1} for damped shells or shells containing structures.

It is only when the structural losses are so small that the radiation resistance of even poorly radiating modes predominates that we need to evaluate the resistance of such slow modes. The magnitude of the acoustic resistance ratio is illustrated for some representative modes in table 9.2. When realistic structural damping is present, we can ignore radiation damping. This allows us to use the infinitely periodic cylinder to approximate both the radiation resistance for fast $k > k_s$ modes and the added mass for slow $k < k_s$ modes. This is the procedure followed in the next section.

9.10 The Far-Field of Point-Excited Cylindrical Shells

9.10.1 THE SIMPLY SUPPORTED SHELL

The response of the simply supported shell driven at midspan is computed from the equations of motion in equation (9.20), which, it is recalled, are applicable to the infinitely periodic shell as well as to simply supported boundary conditions. The radial generalized force introduced on the right side of the last of these equations, normalized to conform to equation (9.20), is

Table 9.2
Approximate natural frequencies and acoustic resistance ratios of lower-order axially slow $(k_m > k)$ modes of a submerged simply supported cylindrical steel shell $(h/a = 10^{-2}, L/a = 1, \rho_s/\rho = 7.8, c_p/c = 3.6)$

Modal index[a]		Dimensionless natural frequency $ka = \bar{\Omega}_{mn} c_p/c$ [equation (9.24)]	Surface-averaged acoustic resistance ratio $\theta_{mn} = \langle r_{mn} \rangle / \rho c$	
m	n		Exact [equation (9.35)]	Asymptotic [equation (6.70c)[b]]
0	2	0.524	0.82×10^{-5}	—
0	1	0.821	0.049	—
1	1	1.53	0.096	0.046
2	1	1.81	0.032	0.018
2	2	1.96	0.012	—
3	1	2.04	0.014	0.10
1	2	1.51	0.0041	—
3	2	2.31	0.015	0.0135

a. Mode displays $(2m + 1)$ axial half-wavelengths $\{k_m L = k_m a = [(2m + 1)/2)]\pi\}$ and n circumferential wavelengths. It is recalled that L is the half-length of the shell.
b. Even though the large ka condition, $ka \gg n^2 + 1$, inherent in that equation is not satisfied for any of these modes, we have applied the asymptotic expression to modes for which $ka \gg n$ to illustrate that it yields the resistance within a factor of two when the large-ka condition is only marginally met.

$$\frac{\varepsilon_n F a^2}{2\pi L h \rho_s c_p^2}.$$

This forcing term represents the F-dependent portion of equation (9.32). For reasons explained above, the radiation loading terms have been approximated by those of the infinitely periodic cylinder, equations (6.44) and (6.46). The far-field pressure is computed from equation (6.58) or (6.61), i.e., from the mathematical model consisting of the cylindrical radiator bracketed between two semiinfinite cylindrical baffles. This far-field pressure is plotted in figure 9.4 for a steel shell in water. The corresponding pressure for the infinite plate for the same parameter values, equation (8.32) is also shown. For the frequency range considered here, or more precisely for the corresponding ratio $\omega \rho_s h/\rho c$ $[= ka (\rho_s/\rho) (h/a)]$, the plate-radiated pressure (equation (8.32)] does not drop significantly below its low-frequency limit, $p = F/\lambda R$. For the normalization procedure used in

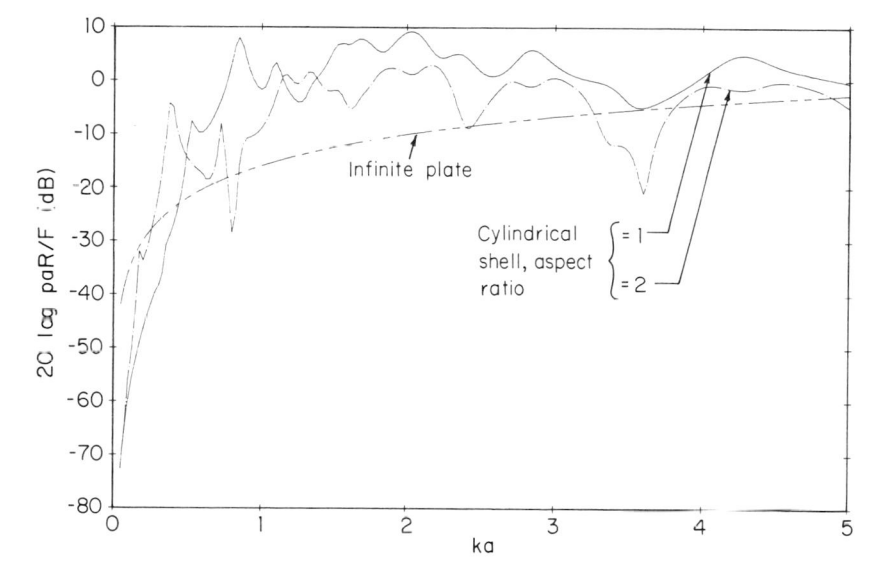

Figure 9.4
Sound pressure at field points on the normal through the drive point radiated by (i) simply supported cylindrical shells excited by a concentrated force at midspan and (ii) by an effectively infinite plate of the same thickness ($\rho_s/\rho = 7.8$, $c_p/c = 3.6$, $h/a = 10^{-2}$, $\eta_s = 0.1$).

figure 9.4, this long wavelength asymptotic pressure is

$$\frac{paR}{F} = \frac{ka}{2\pi} = \frac{a}{\lambda}.$$

It is recalled that this is twice the dipole pressure radiated by a unit force applied directly to the fluid medium, equation (6.34a).

In this book's first edition, the contributions of individual shell modes were examined in detail. Since, for realistic load distributions, these contributions combine in the far-field, it appears more profitable to discuss the physics of sound radiation by the combined modal responses as determined by shell curvature and boundary conditions.

9.10.2 INTERPRETATION OF THE FAR-FIELD RESULTS
The far-field versus frequency curve of the thin simply supported shell displays these characteristics:

1. For $ka > 4$, the shells radiate approximately like the plate, particularly the shell with the larger aspect ratio, i.e., the smaller range of $k_m a$ values.

2. At moderately low and middle frequencies, the shell radiates substantially higher pressures than the plate.

3. The shell with the smaller aspect ratio, i.e., higher values of $k_m a$, radiates more efficiently and drops more slowly to the plate's far-field pressure with increasing frequencies.

4. In the extreme low frequency range, the shell's far-field drops below twice the dipole pressure $F/\lambda R$ radiated by the plate.

We now examine the reasons underlying these four observations:

1. The convergence of the sound radiating properties of the shell and the plate at high frequencies results from the fact that the shell's radius of curvature becomes large in terms of acoustic wavelengths and that the shell's response is, like the plate's, dominated by flexural stresses. This convergence from shell to plate will be analyzed mathematically in chapter 12 by means of short-wavelength asymptotic techniques.

2. It is recalled from the discussion in section 7.16 that nonplanar vibrations of cylindrical shells are dominated by membrane rather than flexural stresses. Compared with modes whose circumferential modal order exceeds $n_{\min}$, equation (7.100a), these membrane modes display a combination of relatively high natural frequencies and small helical wave numbers. The resulting substantial ratios of k/k_s, being larger than k/k_f, lead to more effective sound radiation than the plate's flexural response is capable of.

3. The smaller the shell's aspect ratio, the higher its smallest axial wave number $k_0 a$. Referring once again to equation (7.100a), we see that this increases the number of lobar modes controlled by membrane stresses, i.e., modes that radiate more efficiently than flexural modes.

4. In the extreme low-frequency limit, the response of a simply supported shell or plate is controlled by the stiffness reactance of the structure, which diverges as the frequency tends to zero. Since the radiation resistance decreases simultaneously, the ratio of radiated pressure to exciting force tends more rapidly to zero than the $p = F/\lambda R$ limit of the infinite plate. The latter is an unrealistic model at low frequencies, as anticipated from the inequality in equation (8.98).

9.10.3 LOW-FREQUENCY SOUND RADIATION BY FREE-FREE CYLINDRICAL SHELLS

Having compared the simply supported shell with the infinite plate, let us analyze long-wavelength sound radiation by the neutrally buoyant free-free shell. In this extreme low-frequency range, radiation loading is overwhelmingly reactive and therefore compatible with normal modes—a point to be examined closely in the last section of this chapter. At low frequencies, it is the lowest order n modes that display the highest radiation resistance, viz., the small-$k_m a$, $n = 0$ and $n = 1$ modes, which radiate with, respectively, monopole efficiency and dipole efficiency. The former, axisymmetric modes are stiffness controlled and therefore endowed with a reactance which diverges as $\omega \to 0$. In contrast, the two lowest $n = 1$ modes, viz., rigid-body translation and rotation, are characterized by zero natural frequencies and therefore by inertial reactances that tend to zero as $\omega \to 0$. Consequently, the smaller reactance of the transverse modes compensates for their smaller dipole like radiation efficiency, thereby making them the predominant sound radiators in the low-frequency limit. We shall therefore analyze their far-field.

Since we can assume orthogonality between normal modes in the low-frequency limit where radiation damping is small, the $n = 1$ mode shapes f_{1m} obey the normal-mode relation

$$\int_{-L}^{L} f_{1m}(z) f_{1p}(z) \frac{dM}{dz}\, dz = 0 \qquad \text{for } m \neq p. \tag{9.37}$$

Here dM/dz is the shell mass per unit length, which is constant for a uniform cylindrical shell possibly containing a z-independent ballast distribution to achieve neutral buoyancy. For the rigid-body translational mode, the mode shape f_{10} is similarly z independent. Consequently, selecting $p = 0$ in the above orthogonality relation,

$$\int_{-L}^{L} f_{1m}(z)\, dz = 0, \qquad m > 0. \tag{9.38}$$

The pressure radiated by this transverse mode in the small ka range, equation (6.55), specialized to $n = 1$ displays a maximum pressure at $\theta = \pi/2$, $\phi = 0$:

$$p_{1m}(R, \theta = \pi/2, \phi = 0) = \frac{i\rho a e^{ikR} \ddot{W}_{1m} ka}{2R} \tilde{f}_{1m}(0), \tag{9.39}$$

where

$$\tilde{f}_{1m}(0) = \int_{-L}^{L} f_{1m}(z) \, dz.$$

Noting the similarity between this transform and the orthogonality relation in equation (9.38), we conclude that none of the elastic transverse modes contributes to the pressure maximum, which is therefore exclusively associated with the translation rigid body mode, $f_{10} \equiv 1$. If, however, the cylindrical shell embodies a nonuniform mass distribution, the orthogonality relation in equation (9.37) specialized to $p = 0$, must be used in place of equation (9.38):

$$\int_{-L}^{L} f_{1m}(z) \frac{dM}{dz} \, dz = 0, \qquad m > 0. \tag{9.40}$$

This no longer coincides with the transform $\tilde{f}_{1m}(0)$, thereby enabling a beam mode resonance to contribute to the pressure maximum.

Returning to the case of a uniform shell, the transform in equation (9.39) reduces to $2L$ for f_{10}. The corresponding pressure maximum is

$$p_{10}(R, \theta = \pi/2, \phi = 0) = \frac{-ik\rho V \ddot{W}_{10}}{2\pi R} e^{ikR}, \tag{9.41}$$

where $V = 2\pi a^2 L$ is the volume of the cylindrical shell. Let us now determine its amplitude of translational motion in response to a concentrated force. It is recalled (section 6.15) that a transversely vibrating cylinder of large aspect ratio displays an accession to inertia equal to the mass of fluid it displaces. Consequently, for a neutrally buoyant cylindrical shell,

$$\ddot{W}_{10} = \frac{F}{2\rho V}. \tag{9.42}$$

Combining equations (9.41) and (9.42) gives

$$p_{10}(R, \theta = \pi/2, \phi = 0) = \frac{-ikF}{4\pi R} e^{ikR}. \tag{9.43}$$

This coincides with the pressure radiated by the force F applied directly to

the fluid medium, equation (6.34). It equals half the pressure radiated by the infinite plate, given by the latter of equations (8.32). As anticipated, an infinite structure is not a suitable low frequency model.

9.10.4 LOW-FREQUENCY SOUND FIELD OF FREELY FLOATING NONCYLINDRICAL SHELLS OF REVOLUTION

The large reactance of the axisymmetric, stiffness-controlled modes at low frequencies is not limited to cylindrical shells but holds, e.g., for spheroidal shells. The rigid-body transverse modes of these noncylindrical shells, which similarly display a reactance that vanishes as $\omega \to 0$, therefore again control the low-frequency far-field with their dipolelike sound radiation. The pressure radiated by translational motion is given by equation (6.86), where the θ axis has been rotated 90° from the cylindrical axis. The translation acceleration in response to the force is

$$\ddot{W}_{10} = \frac{F}{\rho V(1 + \gamma)}. \tag{9.44}$$

Substituting this acceleration in equation (6.86), we retrieve the result obtained for the cylinder, equation (9.43). We therefore conclude that no matter what the geometry, the low-frequency sound field of neutrally buoyant bodies of revolution is the force dipole. In Lamb's words,[9] "If the solid were removed, and its place supplied by fluid, the motion at a distance would be very approximately the same as would be produced by a suitable force from without, acting on the substituted matter."

In summary, the *pressure radiated by a freely floating body tends in the low-frequency limit to the pressure radiated by the same exciting force applied directly to the fluid, equation (6.34). This pressure is half as large as the pressure radiated by an effectively infinite plate, equation (8.32), and arbitrarily larger than the pressure radiated by a simply supported shell. The infinite plate model is invalid in the long wavelength limit,* as anticipated from the fact that the criterion in equations (8.98) and (8.99) is necessarily satisfied as $k_f^2 A \to 0$.

9.11 The Effect of Structural Damping on Sound Radiation

While radiation resistance is relevant to the contribution of a given shell mode at all frequencies, structural damping is of particular relevance to the contribution of resonant modes, most particularly to that of modes whose radiation resistance is small. Clearly, if a mode is effectively radiating at resonance, as is the case of the lower order membrane-type modes of the

cylindrical shell, the radiation loss factor, introduced for the spherical shell in equation (9.7), may exceed the structural loss factor. Defining the acoustic resistance ratio $\theta_{mn} = \langle r_{mn} \rangle / \rho c$, the modal radiation loss factor at resonance is

$$
\begin{aligned}
\eta_{mn} &= \frac{\theta_{mn} \rho c}{\omega \rho_s h (1 + \gamma)} \\
&= \frac{\theta_{mn} \rho a}{(ka)_{mn} \rho_s h (1 + \gamma)},
\end{aligned}
\tag{9.45}
$$

where γ accounts for accession to inertia. For a thin $(h/a = 10^{-2})$ cylindrical steel shell in water, and selecting a dimensionless natural frequency $(ka)_{mn} \simeq 1.5$ and $\gamma \simeq 2$ as representative of the lower-order membrane modes in table 9.2, we conclude that even a modest value of $\theta_{mn} = 3 \times 10^{-2}$ yields a radiation loss factor of $O(10^{-1})$, which exceeds all but the largest structural loss factors of a highly damped shell. Consequently, the contribution of resonant modes for which $k_m > k$, and which therefore have been assigned no radiation damping, is overestimated even for the large structural loss factor of 0.1 assumed in figure 9.4. The question which must be answered is this: If even modes radiating with modest efficiency have resonance amplitudes controlled by radiation rather than structural damping, does structural damping have a significant effect on sound radiation? To formulate an answer to this question, we substitute equations (9.43) in equation (6.67) generalized to $n > 0$. We now obtain an explicit expression for the acoustic power. Defining $X_{mn} \equiv -i \operatorname{Im}(\mathcal{Z}_{mn})$, where $\mathcal{Z}_{mn} = F/\dot{W}_{mn}$ is the in vacuo modal impedance of the shell, the acoustic power associated with mode m, n finally becomes

$$
\begin{aligned}
\Pi &= \pi a L \langle r_{mn} \rangle \frac{\dot{W}_{mn}^2}{\varepsilon_n} \\
&= \frac{\varepsilon_n F^2 \langle r_{mn} \rangle}{4 \pi a L [\langle r_{mn} \rangle^2 + r_s^2 + 2 \langle r_{mn} \rangle r_s + (X_{mn} - \omega m_{mn})^2]}.
\end{aligned}
\tag{9.46}
$$

This power is maximum when

$$
\frac{\partial \Pi}{\partial \langle r_{mn} \rangle} = 0.
$$

The solution of this equation is

$$\langle r_{mn}\rangle = [r_s^2 + (X_{mn} - \omega m_{mn})^2]^{1/2}.$$ (9.47)

When $\langle r_{mn}\rangle$ exceeds this quantity, an increase in $\langle r_{mn}\rangle$ reduces acoustic power. The relevance of the above result to the effect of structural damping is revealed by examining the modal resonance:

$$X_{mn} - \omega m_{mn} = 0, \qquad \omega = \bar{\omega}_{mn}.$$

Substituting this relation in equation (9.47), one concludes that the power radiated by a resonant mode is a maximum if at the dimensionless natural frequency, $\langle r_{mn}\rangle = r_s$. The maximum power thus obtained at resonance is

$$\Pi = \frac{\varepsilon_n F^2}{16\pi a L r_s}, \qquad r_s = \langle r_{mn}\rangle, \qquad \omega = \bar{\omega}_{mn}.$$ (9.48)

When structural damping is increased, the condition for maximum power is satisfied by a different resonant mode characterized by a larger acoustic resistance $\langle r_{mn}\rangle$. The corresponding maximum power will, however, have been reduced. It is thus concluded that *in a shell embodying little structural damping, for example, a bare shell with welded joints, inefficiently radiating resonances, such as those of the $n > 2$ modes, will radiate more power than the efficiently radiating resonances of the $n = 0$ and 1 modes.* The equality of structural and radiation resistances as a condition of maximum performance is encountered in other situations. It is, for example, the requirement for maximizing the absorption cross section of Helmholtz resonators where the two resistances embody, respectively, losses in the resonator neck and sound radiation by the virtual piston coinciding with the neck cross section.

For shells vibrating in the atmosphere, η_{mn} is negligible for any practical values of η_s. Consequently, sound radiation by resonant modes is always reduced by an increased η_s. However, overall power is not reduced if η_s is sufficiently large to make supersonic mass-controlled modes the predominant contributors to the sound field. This criterion was formulated for plates in equation (8.99)

We conclude this chapter by exploring whether the preservation of normal modes in a submerged shell is limited to the two geometries studied here: spherical and infinite cylindrical shells.

9.12 Uncoupled Modes in a Submerged Structure

Radiation coupling of the modes of vibration of the finite cylindrical shell and of the finite plate was explained by the fact that each mode gives rise to a surface pressure in the form of a transform over all wave numbers, including the discrete axial wave numbers k_m corresponding to the entire family of the original in vacuo normal modes of vibration. In contrast, the infinite cylindrical and the spherical shell were found to preserve their uncoupled in vacuo normal modes, the nodal lines characterizing the radial displacement of a mode being identical with those of the corresponding surface harmonic.

There are two asymptotic situations for which a structure endowed with in vacuo normal modes retains normal modes in a acoustic fluid, whatever its geometry. We first note that normal modes always exist in undamped structures and, exceptionally, in damped structures. In the latter case, the requirement for the existence of normal modes is that Rayleigh's "dissipation function" be of the same form as, or a linear combination of, the kinetic and potential energies.[10] In what appears to be the only jocular statement in his classical treatise on the *Theory of Sound*, Rayleigh comments that the former situation "occurs frequently, in books at any rate." The present monograph is no exception to Rayleigh's statement, as demonstrated by the use of a structural loss factor η_s, which ensures that the dissipation function is formally similar to the potential energy.

Rayleigh's requirement can be described in physical terms by stating that, when damping is introduced, the distributed damping force associated with the configuration of a normal mode of the original undamped structure must have the same spatial configuration as the displacement characterizing the normal mode. In this case, the damped structure admits the same normal-mode configurations as the undamped structure. For submerged structures, Rayleigh's criterion is clearly met by the spherical and the infinite cylindrical shell, but not by the finite cylinder, whose edge-radiating modes $(k_m > k)$ are associated with radiation damping localized in the regions adjoining the ends of the shell. The required flow of structure-borne energy toward these peripheral regions was shown to be incompatible with a shell response embodying exclusively standing waves.

Rayleigh's criterion is satified without any restriction as to the structure's geometry by the ρc radiation loading corresponding to the asymptotic short-wavelength plane-wave solution. This asymptotic solution is obviously more frequently applicable to metal structures vibrating in an acoustic medium characterized by a small sound velocity, such as the atmo-

sphere, than in water. Of course, even when the plane-wave approximation is not applicable, the small characteristic impedance of air results in weak radiation coupling.

Another situation of practical interest relevant to submerged structures, whereby their normal modes are preserved in the presence of radiation loading, arises when the radiation loading can be approximated by the hydrodynamic or incompressible asymptotic solution discussed in section 4.6. When the acoustic medium is water, this situation is so common that when analyzing hull vibrations the preservation of normal modes is generally taken for granted. A fundamental difference between the hydrodynamic asymptotic condition and the plane-wave asymptotic situation (or the spherical or infinite cylindrical shell), is that, for the hydrodynamic approximation, the normal modes of the submerged shell generally have a different configuration than in vacuo. The reason is that the inertial forces associated with the hydrodynamic near field need not have the same spatial distribution as those of the in vacuo modes of vibration. The "normal modes" of the submerged shell are, however, resistance coupled. Thus, iterative determination of the modes of a submerged, clamped, circular plate illustrates that uncoupling the reactive coefficient matrix does not uncouple the resistive radiation loading.[5] Fortunately in this long-wavelength limit, the latter is negligible compared to the reactive loading.

Normal modes, whose configuration also generally differs from the corresponding in vacuo configuration, can always be found when the acoustic medium is *enclosed* in a shell-like structure. In this situation, the radiation loading is purely reactive (see section 9.5), without the long-wavelength restriction associated with the hydrodynamic approximation.

To conclude this section, we shall explore whether there are nonasymptotic situations whereby structures other than the spherical and infinite cylindrical shell preserve their normal modes when submerged. From Rayleigh's normal-mode condition stated earlier, it is obvious that such structures are restricted to geometries compatible with the discrete-wave number pressure field characteristic of structural standing waves. This, however, is not a sufficient condition. Thus, the surface pressure on a prolate spheroidal radiator is, like that on the sphere or the infinite cylinder, expressible as a wave-harmonic series characterized by a discrete wave number spectrum.[11] These surface harmonics do not, however, match the nontorsional normal-mode configurations of the analytically tractable spheroidal shell geometry, for which the shell is defined by two confocal prolate spheroids of slightly different eccentricity.[12] Each normal mode

therefore gives rise to a surface pressure distribution, which is not orthogonal to the other normal modes, thus producing radiation coupling between modes.[13,14] It is of course possible, in principle, to determine a stiffness and mass distribution that endows the in vacuo shell with normal modes whose configuration matches that of a single surface harmonic. Shell modes will thus remain uncoupled when the shell is submerged. In conclusion, excluding such artificial situations, or mathematical models that are unrealistic because they stipulate a structure of infinite extent (the infinite plate and cylinder), the preservation of normal modes in a submerged structure appears to be confined to the spherical shell and to the long- and short-wavelength asymptotic situations already described.

References

1. M. C. Junger, "Normal Modes of Submerged Plates and Shells," In J. E. Greenspon, ed., *Fluid-Solid Interaction* (Proc. ASME Colloquium, New York, 1967), p. 95.

2. M. C. Junger and J. M. Garrelick, "Multiple Modal Resonances of Thin Cylindrical Shells Vibrating in an Acoustic Medium," *J. Acoust. Soc. Am.* 75:1380–1382 (1984).

3. G. Warburton, "Vibration of a Cylindrical Shell in an Acoustic Medium," *J. Mech. Eng. Sci.* 3:69–79 (1961).

4. P. W. Smith, Jr., "*Underwater Sound Radiation from a Finite Cylinder: Statistical Analysis*, Bolt Beranek and Newman Report 1292, Contract Nonr-44-4476(00), 30 Jan. 1967.

5. M. Lax, "The Effect of Radiation on the Vibration of a Circular Diaphragm," *J. Acoust. Soc. Am.* 16:5–13 (1944).

6. H. G. Davies, "Low Frequency Random Excitation of Water-loaded Rectangular Plates," *J. Sound and Vibration* 15:107–126 (1971).

7. M. L. Baron, "Sound Radiation from Submerged Cylindrical Shells of Finite Length," *Trans. ASME* 87; 393 (1965).

8. B. L. Sandman, "Fluid Loading Influence Coefficients for a Finite Cylindrical Shell," *J. Acoust. Soc. Am.* 60:1256–1264 (1976).

9. H. Lamb, *The Dynamical Theory of Sound*, 2nd ed. (New York: Dover, 1960), pp. 240–241.

10. Lord Rayleigh, *Theory of Sound*, 2nd ed. (New York: Dover, 1945), Vol. I, pp. 130, 131.

11. G. Chertock, "Sound Radiation from Prolate Spheroids," *J. Acoust. Soc. Am.* 33:871–880 (1961).

12. F. DiMaggio and R. Rand, "Axisymmetric Vibrations of Prolate Spherioidal Shells," *J. Acoust. Soc. Am.* 40:179–186 (1966).

13. T. Yen and F. DiMaggio, "Forced Vibrations of Submerged Spheroidal Shells," *J. Acoust. Soc. Am.* 41:618–626 (1967).

14. S. Hayek, "Complex Natural Frequencies of Vibrating Submerged Spheroidal Shells," *J. Solid Structures* 6:333–351 (1970).

10.1 Scattering and Echo Formation

Except for a brief formulation of the Kirchhoff integral equation modified to accommodate scattered fields (section 4.9) we have so far been concerned with an acoustic medium containing a single vibrating boundary: the radiating surface. We now introduce a second, nonvibrating boundary defining a semi-infinite or finite solid. This alters the original sound field radiated by the first boundary. This alteration is the *scattered field* associated with the second boundary. The ratio of scattered to incident pressure is a function of the scatterer configuration, its dimensions in acoustic wavelengths, and its orientation or "aspect" as seen, respectively, from the direction of incidence and from the field point. The scattered field thus conveys information on the scatterer or "target."

Additional information is obtained if the incident sound field is in the form of a transient pulse encompassing a finite number of periods. The signal scattered back to the source takes the form of an echo whose time history is related to the discontinuities presented by the scatterer cross section along the direction of propagation of the incident wave.[1] The time elapsed at the source between the outgoing pulse and the backscattered echo gives information on the range from the source to the scatterer. The direction of incidence of the echo received at the source indicates the location of the target. The evaluation of transient effects requires a Fourier or Laplace transform formulation in the frequency domain. Fortunately, the magnitude of the transiently scattered pressure field can be approximated with the steady-state scattered field when the pulse duration encompasses at least twice the travel time of an incident wave front past the target, thus ensuring that all portions of the scatterer contribute simultaneously to the central portion of the scattered pulse. Furthermore, the pulse must consist of a sufficiently large number of periods to concentrate the spectrum of the incident acoustic energy effectively at a single frequency.

Echo ranging thus provides a technique for detecting, locating, and classifying objects. It is commonly used in the animal world[2] as well as in modern technology. While examples of echo ranging with airborne sound (for example, by bats) and with waterborne sound (for example, by dolphins) are about equally frequent in nature, it is only under exceptional circumstances, specifically when radar techniques are ineffectual because of "clutter," that the acoustician is concerned with echo-ranging techniques in the atmosphere. Echo ranging is principally used in the ocean, in applications as diverse as depth sounding and the localization of schools of fish and of submarines, and in elastic solids, for flaw detection in castings and for geological prospecting.

Because of the larger characteristic impedance of water, submerged echo-ranging targets do not behave as rigid reflectors. The scattered sound field must therefore be corrected for the vibrations of the structure excited by the incident sound waves. As in the case of sound radiation by elastic structures, the scattering action of a submerged elastic structure can be conveniently studied by analyzing separately the elastic response of the structure in vacuo, and the scattering action of a rigid object whose boundary coincides with the water-structure interface. The latter, purely acoustical problem will be analyzed in this chapter; the interaction problem, in the next. This study will be confined to steady-state situations. In practice this presents a minor restriction, because underwater echo ranging is generally performed with pulses encompassing typically 50–1,000 cycles. Under these conditions, the echo level is effectively that of the steady-state backscattered pressure.

Since the reader is by now familiar with the analysis of sound radiation processes, we shall exploit the similarity between the two phenomena and construct the scattered field by a procedure that parallels the construction of the radiated field. Scatterers in the form of infinite plane reflectors, spheres, and cylinders are analyzed rigorously (sections 10.3–10.6). An asymptotic analysis of bodies of revolution whose geometry need not be compatible with separation of variables is developed (section 10.7). The Kirchhoff approximation is developed and illustrated for rectangular baffles large in terms of acoustic wavelengths (sections 10.8 and 10.9). Finally, the Helmholtz reciprocity principle is illustrated for plane and spherical boundaries. The assumption of infinitely rigid boundaries made throughout this chapter will be dropped in the next chapter.

10.2 Formulation of the Scattering Problem

Consider a scatterer whose surface S_0 is defined by the position vector $\mathbf{R}_0$. A distant source generates a train of sound waves that impinge on the scattering boundary. At large range the incident wave front is effectively plane. The decrease in incident pressure amplitude by geometric spreading over the dimensions of the scatterer can be ignored. In the notation of figure 10.1, which illustrates a spherical scatterer, this approximation amounts to setting $R^{-1} \simeq (R + 2a)^{-1}$. In the vicinity of the scatterer, the incident pressure can thus be approximated by a plane wave of infinite extent, equation (2.23). This representation need not be specialized to any particular coordinate system if one defines a vector wave number $\mathbf{k}_i$ whose amplitude is k, and whose direction is the direction of propagation of the wave:

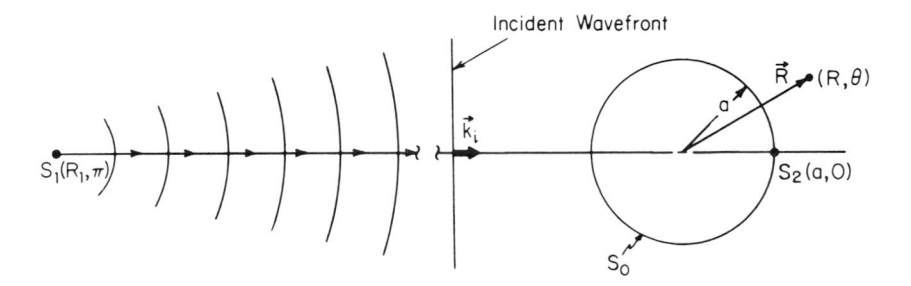

Figure 10.1
Spherical scattering surface S_0 irradiated by a distant point source S_1.

$$p_i(\mathbf{R}, t) = P_i \exp(i\mathbf{k}_i \cdot \mathbf{R} - i\omega t). \tag{10.1}$$

In rectangular coordinates, for a wave traveling in the negative z direction, and incident on the $(z = 0)$ plane at an angle θ to the normal (figure 10.2),

$$\mathbf{R} \equiv x\hat{\xi} + z\hat{\zeta},$$

$$\mathbf{k}_i \equiv k \sin \theta \hat{\xi} - k \cos \theta \hat{\zeta},$$

where $\hat{\xi}$ and $\hat{\zeta}$ are unit vectors. Dropping the time-dependent factor, $\exp(-i\omega t)$, the incident wave is now represented by

$$p_i(x, z) = P_i \exp[ik(x \sin \theta - z \cos \theta)]. \tag{10.2}$$

Similarly, in spherical coordinates, a wave incident from the $(\theta = \pi)$ direction is represented by (figure 10.1)

$$p_i(R, \theta) = P_i \exp(ikR \cos \theta). \tag{10.3}$$

Finally, in cylindrical coordinates, a wave traveling perpendicularly to the z axis and incident from the $(\phi = \pi)$ direction is formulated as

$$p_i(r, \phi) = P_i \exp(ikr \cos \phi). \tag{10.4}$$

The introduction of a rigid boundary $S_0(\mathbf{R}_0)$ produces a disturbance of the pressure field $p_i(\mathbf{R})$ designated as the scattered pressure $p_{s\infty}(\mathbf{R})$.

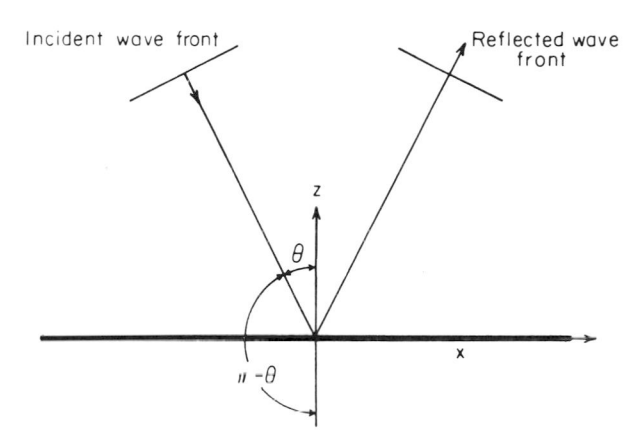

Figure 10.2
Infinite planar reflector.

Symbols related to the field scattered by rigid boundaries will be identified
by the subscripts s∞, the symbol ∞ referring to the infinite impedance of
the scattering surface. This cumbersome notation will be required when
utilizing the results developed here to study elastic scatterers. The resultant
pressure in the presence of a rigid scatterer was stated in equation (4.30).
The boundary being rigid, the resultant particle acceleration must have a
zero component along the normal $\hat{\xi}_0$ to boundary:

$$\ddot{w}_{s\infty}(\mathbf{R}_0) + \ddot{w}_i(\mathbf{R}_0) = 0 \qquad \text{on } S_0(\mathbf{R}_0). \tag{10.5}$$

$\ddot{w}_i$ is the normal fluid particle acceleration that would be observed on the
control surface S_0 in the absence of the scatterer, i.e., when the pressure field
is identical to the incident pressure. This acceleration is given by the component of (2.27) normal to the surface S_0:

$$\frac{\partial p_i}{\partial \xi_0} = -\rho \ddot{w}_i \qquad \text{on } S_0(\mathbf{R}). \tag{10.6}$$

For an incident plane wave, whose pressure is given in equation (10.1), this
becomes

$$\rho \ddot{w}_i(\mathbf{R}_0) = -P_i \frac{\partial}{\partial \xi_0} \exp(i\mathbf{k}_i \cdot \mathbf{R}_0)$$

$$= -i\hat{\xi}_0 \cdot \mathbf{k}_i p_i(\mathbf{R}_0).$$

(10.7)

If the angle between the normal to the scattering surface and the direction of propagation of the incident wave is $(\pi + \theta)$, the dot product in this expression reduces to $-k\cos\theta$ (figure 10.2). Except for planar reflectors, the orientation of the unit vector $\hat{\xi}_0$, and hence the angle of incidence θ, is a function of location $\mathbf{R}_0$.

When one combines equations (10.5) and (10.6) it is apparent that the scattered pressure $p_{s\infty}$ is a solution of the homogeneous Helmholtz equation subject to a boundary condition that is equivalent to the boundary condition specifying a virtual acceleration distribution over a vibrating boundary, equation (4.9b):

$$\rho \ddot{w}_{s\infty}(\mathbf{R}_0) = \frac{\partial p_i(\mathbf{R}_0)}{\partial \xi_0}$$

$$= i\hat{\xi}_0 \cdot \mathbf{k}_i p_i(\mathbf{R}_0), \qquad \text{on } S_0(\mathbf{R}_0).$$

(10.8)

The boundary condition required to construct the scattered field has now been formulated in terms of the incident pressure. We can proceed to construct this field for various boundary configurations.

10.3 The Infinite Plane Reflector   (rigid)

The main analytical task in adapting the radiation solution to the scattering problem consists in expanding $[-\ddot{w}_i(\mathbf{R}_0)]$ in a suitable series. For the infinite plane reflector, this task is trivial, the series being reduced to a single term.. Differentiating the expression for the incident wave in rectangular coordinates, equation (10.2) with respect to z, and setting $z = 0$, one obtains the acceleration distribution:

$$\ddot{w}_{s\infty}(x) = \frac{1}{\rho} \frac{\partial p_i(x,z)}{\partial z}, \qquad z = 0$$

$$= -\left(\frac{ikP_i}{\rho}\right)\cos\theta \exp(ikx\sin\theta).$$

(10.9)

The solution evolved for an infinite train of straight-crested waves, equation

(5.60), can be applied directly by setting

$$k_s = k \sin \theta,$$

$$(k^2 - k_s^2)^{1/2} = k \cos \theta, \qquad\qquad (10.10)$$

$$\ddot{W} = \frac{-ikP_i}{\rho} \cos \theta.$$

When these results are substituted in equation (5.60), the scattered field becomes

$$p_{s\infty}(x, z) = P_i \exp[ik(x \sin \theta + z \cos \theta)]. \qquad\qquad (10.11)$$

The only difference between the expression for the scattered field and equation (10.2) for the incident field is that the sign of the z coordinate is reversed. Hence, the direction in which the reflected wave travels is $(2\pi - \theta)$: the wave is *specularly reflected* (figure 10.2). No energy is scattered back to the direction of incidence, except for normal incidence. In the next chapter, which deals with elastic reflectors and scatterers, the resultant pressure $(p_i + p_{s\infty})$ will be required. To obtain this pressure equations (10.2) and (10.11) are combined by means of Euler's formula:

$$p_i + p_{s\infty} = 2P_i \cos(kz \cos \theta) \exp(ikx \sin \theta). \qquad\qquad (10.12)$$

This result indicates pressure doubling on the reflecting boundary. These results are not valid for grazing incidence $(\theta = \pi/2)$ because equation (5.60) is meaningless when $k = k_s$. For this case, p_i is z independent, thus requiring the virtual acceleration, equation (10.9), and the scattered field to vanish. For the finite-impedance reflectors analyzed in the next chapter, the solution displays a gradual transition as $\theta \to \pi/2$.

We now turn to the more realistic case of finite, convex scatterers, starting with the sphere.

10.4 The Spherical Scatterer   (rigid)
In order to adapt the solution of the radiation problem evolved in the preceding chapter, the normal surface acceleration distribution over the scatterer must be expressed as a series of Legendre functions in the form of equation (6.4). For an incident plane wave, only the $m = 0$ terms of this double series need be retained. For this purpose the exponential, equation

(10.3), is expanded in spherical harmonics by means of a suitable addition theorem:[3]

$$p_i(R, \theta) = P_i \sum_{n=0} (2n + 1) i^n P_n(\cos \theta) j_n(kR). \qquad (10.13)$$

The corresponding acceleration distribution, equation (10.8), is

$$\ddot{w}_{s\infty}(\theta) = -\ddot{w}_i(\theta) = \frac{1}{\rho} \frac{\partial p_i(R, \theta)}{\partial R}, \qquad R = a \qquad (rigid)$$

$$(10.14)$$

$$= \frac{kP_i}{\rho} \sum_{n=0} (2n + 1) i^n P_n(\cos \theta) j'_n(ka).$$

Comparing this expression with the series for the acceleration distribution used in the analysis of the radiation problem, equation (6.4), we can identify ✓ the series coefficients:

$$\ddot{W}_{0n} = P_i\left(\frac{k}{\rho}\right)(2n + 1) i^n j'_n(ka), \qquad ✓$$

$$(10.15)$$

$$\ddot{W}_{mn} = 0, \qquad m > 0.$$

6.19 & 10.14 into 4.22

When substituted in equation (6.19), the scattered pressure becomes

$$p_{s\infty}(R, \theta) = -P_i \sum_{n=0} (2n + 1) i^n P_n(\cos \theta) \frac{j'_n(ka)}{h'_n(ka)} h_n(kR), \qquad (10.16)$$

The scattered far-field equation (6.21), is

$$p_{s\infty}(R, \theta) = \frac{iP_i e^{ikR}}{kR} \sum_{n=0} (2n + 1) P_n(\cos \theta) \frac{j'_n(ka)}{h'_n(ka)}, \qquad kR \gg n^2 + 1. \quad (10.17)$$

The backscattered pressure is obtained by setting $\theta = \pi$, that is, $P_n = (-1)^n$:

$$p_{s\infty}(R, \pi) = \frac{iP_i e^{ikR}}{kR} \sum_{n=0} (2n + 1)(-1)^n \frac{j'_n(ka)}{h'_n(ka)}, \qquad kR \gg n^2 + 1. \quad (10.18)$$

In the low-frequency limit, convergence is rapid, as noted in connection

with the small-argument asymptotic expression for h'_n, equation (6.22). However, even though h'_0/h'_1 is of order ka, this discrepancy is compensated by the fact that $j'_1/j'_0 = (ka)^{-1}$, equation (2.45). Consequently, the first two terms are commensurate, while the $n = 2$ term is only of order $k^5 a^5$:

$$p_{s\infty}(R, \theta) \approx \frac{P_i e^{ikR}}{3R} k^2 a^3 \left(\frac{3}{2} \cos \theta - 1 \right), \qquad k^3 a^3 \ll 1, \qquad kR \gg 1. \qquad (10.19)$$

The backscattered pressure is obtained by setting $\theta = \pi$:

$$p_{s\infty}(R, \pi) \approx -5 P_i e^{ikR} k^2 a^3 / 6R, \qquad k^3 a^3 \ll 1, \qquad kR \gg 1. \qquad (10.20)$$

It is convenient to express this result in terms of the scatterer volume V:

$$\left. \begin{aligned} p_{s\infty}(R, \theta) &\cong -\frac{P_i k^2 V e^{ikR}}{4\pi R} \left(1 - \frac{3}{2} \cos \theta \right) \\ &\cong -\frac{5 P_i k^2 V e^{ikR}}{8\pi R}, \qquad \theta = \pi \end{aligned} \right\} \qquad k^3 V \ll 1. \qquad (10.21)$$

The presence of commensurate monopole and dipole terms proportional to $k^2 V$ is representative of all small scatterers of nonvanishing impedance. They are known as *Rayleigh scatterers* and will be studied for finite-impedance boundary conditions in the next chapter. When the scatterer impedance vanishes, an entirely different frequency-independent scattered pressure obtains (section 11.6).

The results in equations (10.18) and (10.20) are plotted in figure 10.3, together with those of two asymptotic formulations developed in, respectively, sections 10.7 and 10.9, viz., the cylindrical baffle formulation for bodies of revolution at any frequency and the high-frequency *Kirchhoff* approximation. The oscillatory behavior of the backscattered pressure for ka exceeding approximately 1.5 characterizes the *resonance region*. This phenomenon can be interpreted in terms of creeping waves (chapter 12). Having circumnavigated the sphere, these waves combine with the primary contribution from the illuminated region, this addition being either destructive or constructive depending on the difference in acoustic path lengths. These oscillations decrease with increasing frequency as the creeping wave attenuation becomes more marked.

The analysis of the scattering action of an *elastic* spherical shell will be

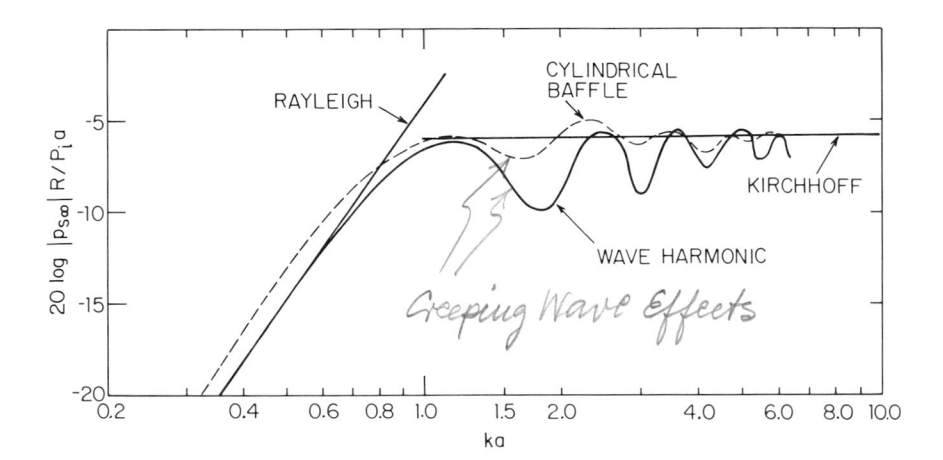

Figure 10.3
Pressure backscattered by a sphere: wave harmonic from equation (10.18), Rayleigh from equation (10.21) cylindrical baffle from equation (10.36), Kirchhoff-geometric acoustics from equation (10.50). [Reproduced from Junger[4]]

seen in the next chapter to require the resultant pressure on the surface of the sphere. Adding equations (10.13) and (10.16), we obtain

$$p(a, \theta) = p_{s\infty} + p_i, \qquad R = a$$

$$= P_i \sum_{n=0} (2n + 1) i^n P_n(\cos \theta) \left[j_n(ka) - \frac{j_n'(ka) h_n(ka)}{h_n'(ha)} \right]. \qquad (10.22)$$

When the two terms in brackets are reduced to the same denominator, one obtains a ratio whose numerator is in the form of the Wronskian in equation (6.14). Substituting this result in the above equation, the <u>surface pressure</u> becomes

$$p(a, \theta) = \frac{iP_i}{(ka)^2} \sum_{n=0} \frac{(2n + 1) i^n P_n(\cos \theta)}{h_n'(ka)} \qquad (10.23)$$

$$\approx P_i[1 + (i_2^3) ka \cos \theta], \qquad k^2 a^2 \ll 1.$$

10.5 The Infinite Cylindrical Scatterer (rigid)

We first consider the case of an infinite cylinder irradiated by a plane wave approaching from the ($\phi = \pi$) direction and incident normally to the cylin-

drical axis. Because the incident pressure in equation (10.4) is z independent, it can be expanded in a Fourier series in ϕ by means of the addition theorem[5]

$$P_i \exp(ikr \cos \phi) = P_i \sum_{n=0} \varepsilon_n i^n J_n(kr) \cos n\phi. \tag{10.24}$$

The corresponding surface acceleration distribution is obtained from equation (10.8):

$$\ddot{w}_{s\omega}(\phi) = \frac{k}{\rho} P_i \sum_{n=0} \varepsilon_n i^n J_n'(ka) \cos n\phi. \tag{10.25}$$

Like the incident pressure, the acceleration is z independent. Consequently, in adapting the radiation solution only the ($k_m = 0$) terms need be retained in the Fourier series, equation (6.3):

$$\ddot{W}_{0n} = \frac{kP_i}{\rho} \varepsilon_n i^n J_n'(ka),$$

$$\ddot{W}_{mn} = 0, \qquad m > 0. \tag{10.26}$$

The scattered pressure is obtained upon substituting these coefficients in equation (6.41): *or use 4,23 & 10,25 in 4,22*

$$p_{s\infty}(r, \phi) = -P_i \sum_{n=0} \varepsilon_n i^n \frac{J_n'(ka) H_n(kr)}{H_n'(ka)} \cos n\phi. \tag{10.27}$$

The derivation of the asymptotic expressions for the low- and high-frequency limits parallels the analysis of the spherical scatterer and will not be repeated here.

Once again, when analyzing the scattering action of elastic cylinders, we require the resultant pressure on the cylindrical surface. Adding equations (10.24) and (10.27), we obtain

$$p(a, \phi) = p_i + p_{s\infty}, \qquad r = a$$

$$= P_i \sum_{n=0} \varepsilon_n i^n \left[J_n(ka) - \frac{J_n'(ka) H_n(ka)}{H_n'(ka)} \right] \cos n\phi.$$

When the terms in brackets are reduced to the same denominator, they can

be condensed by means of the Wronskian relation in equation (6.45d):

$$p(a, \phi) = \frac{2P_i}{\pi k a} \sum_{n=0} \frac{\varepsilon_n i^{n+1}}{H'_n(ka)} \cos n\phi$$

$$\cong P_i(1 + 2ika \cos \phi), \qquad k^2 a^2 \ll 1. \tag{10.28}$$

10.6 The Cylindrical Scatterer of Finite Length

The results derived in the preceding section provide a useful approximation to the surface pressure on a finite cylinder, which will be required in the next chapter. Unfortunately, the far-field cannot be constructed from equation (10.27), because the range is never large compared with the infinite length of the cylinder. For sound waves incident normally on the cylindrical axis, the scattered field generated by a finite cylinder of length $2L$ can be approximated by confining the prescribed acceleration distribution to the length of the cylinder:

$$\ddot{w}_{s\infty}(z, \phi) = \sum_{n=0} \ddot{W}_{0n} \cos n\phi, \qquad |z| < L$$

$$= 0, \qquad\qquad\qquad |z| > L,$$

where the coefficients $\ddot{W}_{0n}$ are given in equation (10.26). The scattered far-field can now be constructed by means of the Fourier transform technique developed in chapter 6 for finite cylindrical radiators. For this purpose, we require the z transform of the above acceleration distribution evaluated at the point of stationary phase:

$$\tilde{\ddot{w}}_{s\infty}(\bar{\gamma}; \phi) = \sum_n \ddot{W}_{0n} \tilde{f}_m(\bar{\gamma}) \cos n\phi, \qquad \bar{\gamma} = k \cos \theta.$$

When the transform $\tilde{f}_m$, which was evaluated in equation (6.56), is substituted in equation (6.53), the far-field "rigid-body" scattered pressure becomes

$$p_{s\infty}(R, \theta, \phi) = \frac{i2Le^{ikR} P_i j_0(kL \cos \theta)}{\pi R \sin \theta} \sum_{n=0} \frac{\varepsilon_n J'_n(ka) \cos n\phi}{H'_n(ka \sin \theta)}. \tag{10.29}$$

The backscattered pressure is obtained by setting $\theta = \pi/2$, $\phi = \pi$, and hence $\sin \theta = 1$, $j_0 = 1$, and $\cos n\phi = (-1)^n$:

$$p_{s\infty}(R, \theta = \pi/2, \phi = \pi) = \frac{i2Le^{ikR}P_i}{\pi R} \sum_{n=0} \frac{\varepsilon_n \mathcal{J}_n'(ka)(-1)^n}{H_n'(ka)}. \tag{10.30}$$

The long-wavelength limit is obtained by substituting the small-argument expression for H_n', equation (6.54), and of $\mathcal{J}_n'$:

$$\mathcal{J}_0'(x) = -\mathcal{J}_1(x) = -x/2, \qquad \mathcal{J}_1'(x) = 1/2, \qquad x^2 \ll 1. \tag{10.31}$$

As in the case of the sphere, the lead terms are of order $n = 0$ and 1. To emphasize the similarity with the sphere, the products La^2 will be expressed in terms of the scatterer volume as $V/2\pi$. The scattered pressure in the Rayleigh region now becomes

$$p_{s\infty}(R, \theta, \phi) \approx \frac{-P_i k^2 V j_0(kL\cos\theta)e^{ikR}}{4\pi R}(1 - 2\sin\theta\cos\phi), \tag{10.32}$$

$$k^2 a^2 \ll 1, \qquad kR \gg 1.$$

If $k^2 L^2 \cos^2\theta \ll 1$, the directivity factor j_0 also becomes unity. Specializing this result to the backscattered direction gives

$$p_{s\infty}(R, \theta = \pi/2, \phi = \pi) = -\frac{3P_i k^2 V e^{ikR}}{4\pi R}, \qquad k^2 a^2 \ll 1, \qquad kR \gg 1. \tag{10.33}$$

While the analysis of the infinite cylindrical scatterer can be readily extended to obliquely incident plane waves, the approximate finite-cylinder analysis becomes rapidly invalid unless it is modified to account for contributions from the end caps and their edges. Results for normal incidence are plotted in figure 10.4, together with the Kirchhoff pressure to be derived later in this chapter. First, however, we shall use this formulation to approximate the pressure scattered by noncylindrical slender bodies of revolution.

10.7 Asymptotic Formulation of the Scattered Field of Slender Bodies of Revolution

It is useful, at this stage, to recapitulate the factors controlling the scattering action in various frequency ranges. In the Rayleigh region, where the scatterer is small in terms of λ, it is its volume rather than its shape that controls the scattered pressure. In the resonance range, where the scatterer

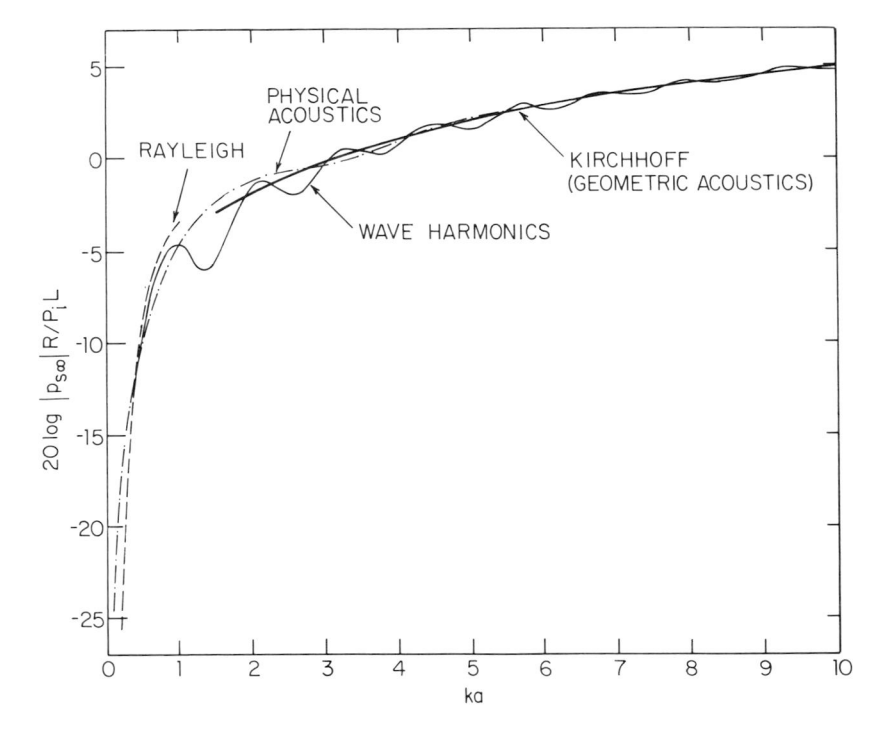

Figure 10.4
Pressure backscattered by a cylinder of length $2L$ insonified with a normally incident plane wave [wave-harmonic solution from equation (10.30), Rayleigh from equation (10.33), Kirchhoff-physical acoustics from equation (10.44), and Kirchhoff-geometric acoustics from equation (10.45)].

diameter measures more than one λ, creeping waves (chapter 12) diffracted around the scatterer through the shadow zone combine with the pressure backscattered by the illuminated surface with varying phases depending on the acoustic length of the diffracted path, thus giving rise to alternatively constructive and destructive interference as one scans over frequency. Eventually, at sufficiently high frequencies, the creeping waves are attenuated to the point where they make a negligible contribution to the backscattered pressure. Consequently, in this short-wavelength or Kirchhoff limit (sections 10.8–10.10), only the illuminated region of the scatterer need be considered. Asymptotically, each infinitesimal area element can be considered to scatter sound as though it were located in an infinite planar baffle.

In the far-field the scatterer shape makes itself felt in the difference in path length and hence in phase between contributions from various area elements.

One anticipates that, for beam-aspect incidence, the cylindrical mathematical model developed in section 6.15 is applicable to all three frequency ranges: Starting with the high-frequency region, it is noted that Kirchhoff's plane-baffle approximation is relaxed, only one of the two radii of curvature being assumed infinite (see figure 6.8). This extends the validity of the cylindrical model into the resonance range, since the cylindrical wave harmonics adjusted to the local diameter properly predict the creeping wave contribution from various scatterer regions. In the Rayleigh region, the baffling action of adjoining length elements is negligible, thus making their precise diameter unimportant. Here the small-argument asymptotic form of cylindrical wave harmonics predicts differential scattered pressure contributions determined by the local scatterer cross section. When integrated over scatterer length, these contributions properly predict a scattered pressure proportional to $k^2 V$.

To extend equation (10.32) to bodies of variable cross section, the z integral associated with the transform $\tilde{f}_m(\bar{\gamma})$ in the preceding section is modified to account for the z dependence of the radius. The scatterer has effectively become an amplitude-shaded line array:

$$p_{s\infty}(R, \theta, \phi) = \frac{2iP_i e^{ikR}}{\pi R \sin \theta} \sum_{n=0} \varepsilon_n \cos n\phi \int_0^L \frac{J'_n(ka)}{H'_n(ka \sin \theta)} \cos(kz \cos \theta) \, dz. \quad (10.34)$$

In the small-argument or Rayleigh limit, it is convenient to replace the cross section $\pi a^2(z)$ by the derivative of the displaced volume, dV/dz:

$$p_{s\infty}(R, \theta, \phi) = -\frac{P_i k^2 e^{ikR}}{2\pi R}(1 - 2\sin\theta\cos\phi) \int_0^L \frac{dV}{dz} \cos(kz \cos \theta) \, dz,$$
$$k^2 a^2 \ll 1. \quad (10.35)$$

The backscattered pressure computed from equation (10.35) is

$$p_{s\infty}(R, \theta = \pi/2, \phi = \pi) = \frac{2iP_i e^{ikR}}{\pi R} \sum_{n=0} \varepsilon_n(-1)^n \int_0^L \frac{J'_n(ka)}{H'_n(ka)} \, dz. \quad (10.36)$$

In the Rayleigh limit specialized to the backscattered direction, the integral

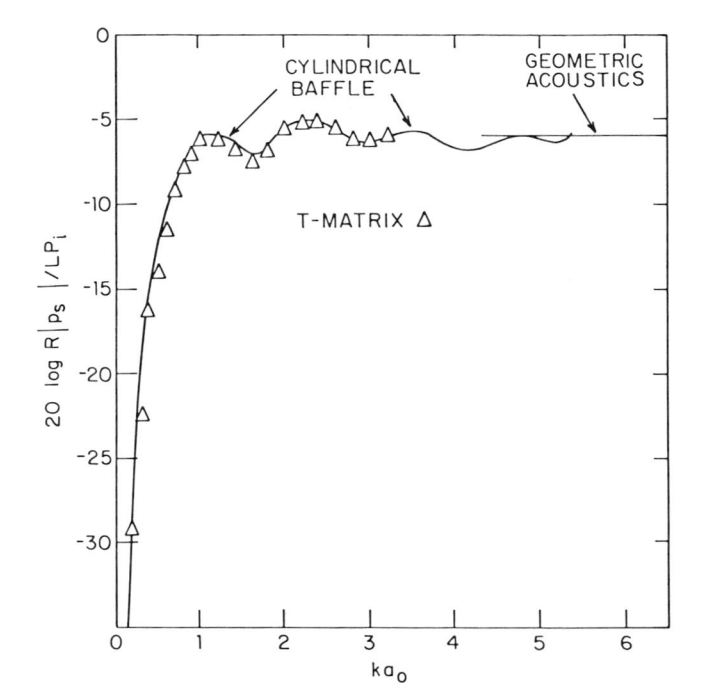

Figure 10.5
Pressure backscattered by a 2:1 prolate spheroid irradiated from "beam aspect," computed
from equation (10.36) and the radius given by equation (10.38), which becomes independent
of eccentricity when pressure and frequency are normalized as indicated. The T-matrix results
were generated by Varadan et al.[6] [Reproduced from Junger[4]]

in equation (10.35) reduces to $V/2$:

$$p_{s\infty}(R, \theta = \pi/2, \phi = \pi) = -\frac{3P_i k^2 V e^{ikR}}{4\pi R}, \qquad k^2 a^2 \ll 1. \tag{10.37}$$

A gross upper limit of the error implicit in the mathematical model is
provided by the sphere, which clearly is not slender. Comparing (10.37)
with the result found earlier for the sphere, (10.21) the error is seen to be
20% in the Rayleigh limit. A physical interpretation of this error will be
attempted in section 11.9. Setting

$$a(z) = (a_0^2 - z^2)^{1/2}$$

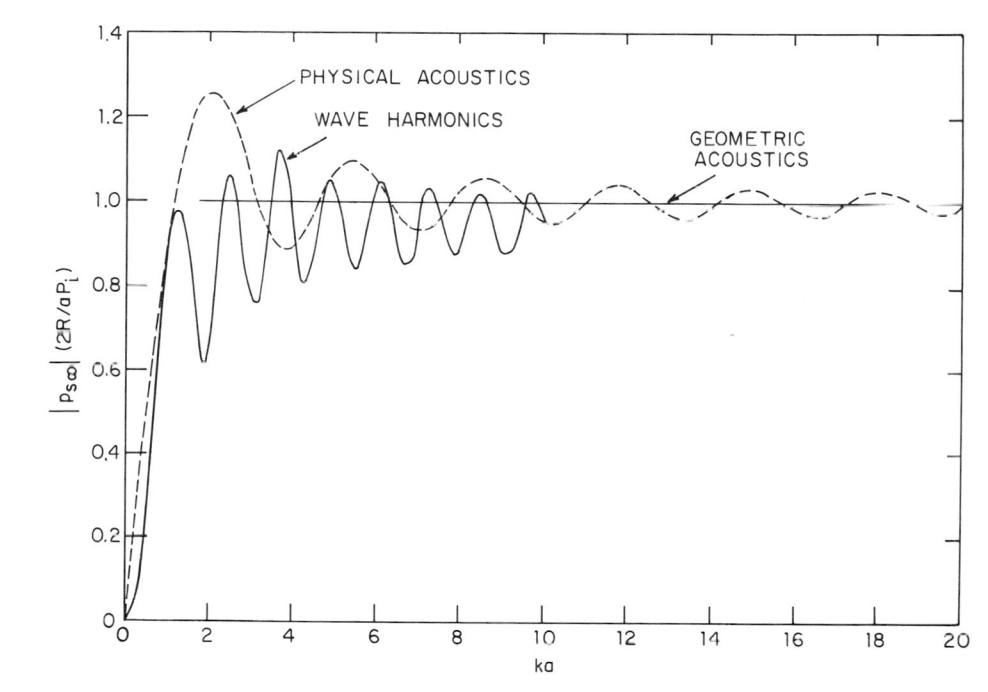

Figure 10.6
Pressure backscattered by a sphere computed from the wave-harmonic series [equation (10.18)], from the Kirchhoff-physical acoustics approximation [equation (10.49), reproduced from Neubauer,[7] and from the Kirchhoff-geometric acoustics formulation [equation (10.50)].

in equation (10.36), it is verified that the cylindrical-baffle model goes to the proper high-frequency limit and provides a reasonable approximation to the resonance region (figure 10.3). Here it matches the period of the fluctuations fairly well, which the physical acoustics approximation does not, as illustrated in figure 10.6.

Finally, the cylindrical baffle model is applied to the prolate spheroid. Setting

$$a = a_0 \left(1 - \frac{z^2}{L^2} \right)^{1/2} \tag{10.38}$$

in equation (10.36), one obtains the plot in figure 10.5, which compares well with the T-matrix computer solution[6], even though the scatterer is only

moderately slender. The Kirchhoff result,

$$|p_{s\infty}(R, \theta = \pi/2, \phi = \pi)| = \frac{P_i L}{2R},$$

is also illustrated. It is this high-frequency asymptotic technique that will now be introduced.

10.8 Nature of the Kirchhoff Approximation; Surface Pressure

The slow convergence of the wave-harmonic series, in the short-wavelength limit, whether associated with a scatterer or a radiator, has provided the incentive for generating asymptotic solutions. If the scatterer (or radiator), in addition to being large in terms of wavelengths, is of a geometry not compatible with the separation of the wave equation, the numerical solution of the Helmholtz integral equation for scatterers [equation (4.31)] becomes similarly unwieldy. The advantage of the asymptotic formulation described here is that it is not restricted to scatterers that admit a wave-harmonic solution.

The plane-wave approximation for large radiators described in section 4.6 is equally applicable to scattering problems, where it is associated with Kirchhoff's name, by extension of his asymptotic solution of the diffraction of waves through an aperture in a baffle. In the next section, the Kirchhoff approximation is applied to the backscattered far-field of arbitrary scatterer geometries. Detailed results are worked out for the sphere, the cylinder, and the finite rectangular plane reflector. The Kirchhoff approximation can be accepted intuitively, because one expects that individual surface elements tend to interact with the incident wave as though they were located in a plane baffle when the radii of curvature of the boundary, measured in terms of acoustic wavelengths, tend to infinity. The pressure scattered by an area element can therefore be approximated as though this virtual source were located in an infinite plane baffle, equation (3.7):

See Eq. 4.29

$$dp_{s\infty}(\mathbf{R}) = \frac{\rho \ddot{w}_{s\infty}(\mathbf{R}_0)\, dS(\mathbf{R}_0) \exp(ik|\mathbf{R} - \mathbf{R}_0|)}{2\pi R}, \qquad R \gg R_0. \qquad (10.39)$$

To obtain an explicit expression for the virtual acceleration, we substitute the incident pressure,

$$p_i(\mathbf{R}) = P_i \exp(ik|\mathbf{R} - \mathbf{R}_1|),$$

in equation (10.8):

$$\ddot{w}_{s\infty}(\mathbf{R}_0) = \frac{iP_i}{\rho}\hat{\xi}(\mathbf{R}_0)\cdot\mathbf{k}_i\exp(ik|\mathbf{R}_0-\mathbf{R}_1|). \tag{10.40}$$

When this is substituted in equation (10.39), and integrated over the illuminated zone, i.e., the region whence the source generating the incident sound waves is visible, one obtains the scattered field. The integral representation of the backscattered pressure, at $R = R_1$, thus becomes

$$p_{s\infty}(\mathbf{R}) = -\frac{iP_i}{2\pi R}\int_{s_0}\hat{\xi}(\mathbf{R}_0)\cdot\mathbf{k}_i\exp(2ik|\mathbf{R}_0-\mathbf{R}|)\,dS(\mathbf{R}_0), \qquad R \gg R_0. \tag{10.41}$$

10.9 Kirchhoff Scattering from Cylinders and Spheres; Fresnel Zones

The evaluation of the integral has been variously approached. The *physical acoustics* procedure consists in integrating over the illuminated scatterer area.[7] The resulting scattered pressure versus frequency curve displays oscillations associated with successively constructive and destructive interference of contributions from the limits of integration. This phenomenon results from the use of an unrealistic mathematical model and has therefore no physical meaning, in contrast to similar oscillations associated with creeping waves that have circumnavigated the scatterer. In fact, better agreement with the rigorous formulation is obtained by taking the high-frequency asymptotic limit of the integral, as predicated by *geometric acoustics*. This latter result can also be obtained directly by stationary-phase integration. In some texts only the former, physical acoustics procedure is associated with Kirchhoff's name. Other authors, whose nomenclature we follow here, apply the term to all techniques based on the assumption that surface elements scatter as though located in an infinite plane baffle. Both the physical and geometric acoustics results will now be constructed for two scatterers examined earlier.

For the cylinder, the terms in equation (10.41) are

$$dS = 2aL\,d\phi,$$

$$\hat{\xi}(\mathbf{R}_0)\cdot\mathbf{k}_i = k\cos\phi, \tag{10.42}$$

$$2ik|\mathbf{R}_0-\mathbf{R}| = 2ik[a(1-\cos\phi) + R].$$

It is convenient to orient the ϕ axis to have the plane wave impinge on the cylinder at $\phi = 0$. For the sake of brevity, only the absolute value of the integral will be considered. The backscattered pressure amplitude is

$$\frac{|Rp_{s\infty}|}{P_i} = \frac{kaL}{\pi} \left| \int_{-\pi/2}^{\pi/2} \cos\phi \exp\left(-2ika \cos\phi\right) d\phi \right|. \tag{10.43}$$

The real component of the integral yields the Weber function E_1, which reduces to the Neumann function at high frequencies:[8]

$$\int_{-\pi/2}^{\pi/2} \cos\phi \cos\left(2ka \cos\phi\right) d\phi = \pi E_1\left(2ka\right)$$

$$= \pi Y_1\left(2ka\right) + O\left(2ka\right)^{-2}.$$

The imaginary component is readily evaluated in terms of a Bessel function:[9]

$$-i \int_{-\pi/2}^{\pi/2} \cos\phi \sin\left(2ka \cos\phi\right) d\phi = i\pi J_1\left(2ka\right).$$

The scattered pressure thus finally becomes

$$R|p_{s\infty}|/P_i = kaL\left[E_1^2\left(2ka\right) + J_1^2\left(2ka\right)\right]^{1/2}$$

$$\cong 2kaL/\pi, \qquad k^2a^2 \ll 1 \tag{10.44}$$

$$\cong kaL|H_1\left(2ka\right)|, \qquad k^2a^2 \gg 1.$$

The above physical acoustics solution varies linearly with ka in the long-wavelength limit, while the correct Rayleigh limit indicates a k^2a^2 dependence.

The geometric acoustics solution can be obtained as the high-frequency limit of equation (10.44), the Hankel function being specialized to its high-frequency value, equation (5.19b). Alternatively, the geometric acoustics solution can be constructed directly, by evaluating equation (10.43) by the method of stationary phase applicable to integrals of the form of equation (5.18). The stationary-phase value of ϕ being zero, one sets $\cos\phi$ equal to unity. Equation (5.21) now yields

$$R|p_{s\infty}|/P_i = (ka/\pi)^{1/2} L. \tag{10.45}$$

It can be verified by comparison with the wave-harmonic result in figure 10.4 that this result is good within 1 dB for $ka > 3$. The error inherent in the more complex physical acoustics curve is generally larger.

It is interesting to interpret this result in terms of *Fresnel zones*, defined as regions small enough to generate scattered pressures not displaying phase cancellation. A Fresnel zone therefore subtends an arc resulting in a change of π radians in the exponential of equation (10.43). Successive Fresnel zones contribute scattered pressures of opposite sign, which therefore tend to cancel each other. One expects the backscattered pressure to be associated primarily with the Fresnel zone centered on the line of tangency $\phi = 0$ of the incident plane wave on the cylinder or the point of tangency $\theta = 0$ on the sphere. If the area of this first Fresnel zone is S_F, one anticipates from equation (10.43) or, more generally, from equation (10.41) that a geometric acoustics backscattered pressure takes the form

$$|p_{s\infty}| = \frac{kP_i\alpha S_F}{2\pi R},\tag{10.46}$$

where α is a numerical factor that depends on the scatterer geometry and frequency.

For the cylinder, the first Fresnel zone subtends an arc $2\phi_1$ such that, be definition,

$$\pi = 2ka(1 - \cos\phi_1)$$
$$\cong ka\phi_1^2, \qquad \phi_1^2 \ll 1.$$

Consequently, the arc subtended by the first Fresnel zone is

$$2\phi_1 = 2(\pi/ka)^{1/2}, \qquad k^2a^2 \gg 1.\tag{10.47}$$

The area of this zone is

$$S_F = 4aL\phi_1 = 4aL(\pi/ka)^{1/2}.\tag{10.48}$$

Substituting this area in equation (10.46), and comparing the result with equation (10.45), it is seen that for the cylinder $\alpha = 1/2$.

To conclude this section, consider the sphere. The latter two relations in equation (10.42) apply, provided ϕ is replaced by θ. The area element is

$dS = 2\pi a^2 \sin\theta \, d\theta.$

Equation (10.43) now becomes

$$\frac{R|p_{s\infty}|}{P_i} = ka^2 \left| \int_0^1 e^{i2kax} x \, dx \right|, \tag{10.49}$$

where $x \equiv \cos\theta$, $dx = -\sin\theta \, d\theta$. The physical acoustics solution requires the precise evaluation of this integral. The result is plotted in figure 10.6 and need not be repeated here. The geometric acoustics result can be obtained from equation (10.49) by integration by parts and dropping higher-order terms:[10]

$$I \cong (2ka)^{-1}, \qquad k^2 a^2 \gg 1.$$

Substituting in equation (10.49), the geometric acoustics result becomes

$$\frac{R|p_{s\infty}|}{P_i} = \frac{a}{2}. \tag{10.50}$$

The scattered pressure in equation (10.50) will now be formulated in terms of the area of the first Fresnel zone. This region coincides with a spherical cap defined by a vertex angle θ_1 that equals the cylindrical angle ϕ_1, equation (10.47). Consequently,

$$\begin{aligned} S_F &= \pi(a \sin\theta_1)^2 \\ &\cong \pi a^2 \theta_1^2 = \pi^2 a/k, \qquad \theta_1^2 \ll 1, \qquad k^2 a^2 \gg 1. \end{aligned} \tag{10.51}$$

When this is substituted in equation (10.46) and compared with equation (10.50), one concludes that for the sphere $\alpha = \pi^{-1}$.

Comparing the scattered pressures of the sphere and the cylinder, it is noted that the former is independent of the wave number, while the latter varies as $k^{1/2}$. The reason can be found in the wave number dependence of S_F. For the sphere, S_F varies as k^{-1}, thus canceling the k factor in the numerator of equation (10.46). For the cylinder, the slower $k^{-1/2}$ dependence of S_F only partly cancels the k factor. For a planar scatterer normally irradiated with a plane wave, the first Fresnel zone effectively coincides with the entire scatterer and therefore displays a k-independent area S_F. One

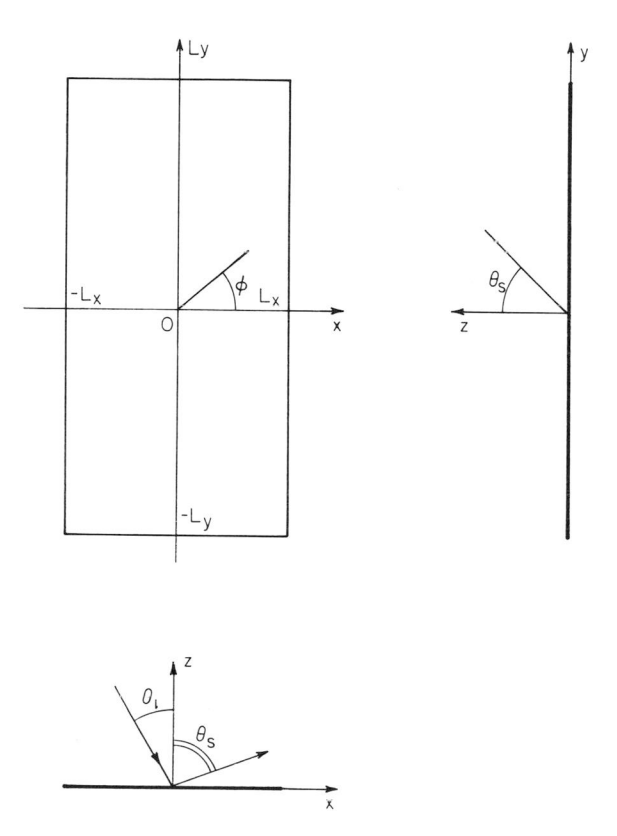

Figure 10.7
Rectangular reflector.

therefore anticipates a scattered pressure increasing linearly with k. This will now be verified.

10.10 Reflection from a Rectangular Baffle

The Kirchhoff approximation will now be used to approximate the scattered pressure field generated by a finite rectangular baffle located in the (x, y) plane in the region $-L_x < x < L_x$, $-L_y < y < L_y$. The incident plane wave is given by equation (10.2), but because of the finite dimensions of the baffle, a nonspecular distribution-in-angle will be obtained for the scattered field. Consequently the angle of incidence must be identified with a subscript, as θ_i, to distinguish it from the angle θ_s of the field point radius vector (figure 10.7). This distinction is not necessary for the infinite plane reflector,

because the only scattered pressure occurs in the direction of specular reflection. The solution will be generalized to account for an incident wave arbitrarily oriented with respect to the baffle. Thus generalized, and formulated in spherical coordinates, the virtual acceleration over the scattering surface becomes

$$\ddot{w}_{s\infty}(x_0, y_0) = \frac{-ik}{\rho} P_i \cos \theta_i \exp[ik \sin \theta_i (x_0 \cos \phi_i + y_0 \sin \phi_i)]. \tag{10.52}$$

Because of the simple scatterer geometry, the integral in equation (10.42) can be evaluated analytically. By virtue of Kirchhoff's mathematical model, the field scattered by the baffle in figure 10.7 is constructed like the radiated field of a rectangular source located in a rigid baffle, and displaying an acceleration given by the above virtual acceleration. The far-field of such a radiator is given in equation (5.31) in terms of the double Fourier transform defined in equation (5.29). When equation (10.52) is substituted in that definition, the acceleration transform becomes

$$\tilde{w}_{s\infty}(\gamma_x, \gamma_y) = \frac{-i4k}{\rho} P_i L_x L_y \cos \theta_i$$
$$\cdot j_0[kL_x(\sin \theta_i \cos \phi_i + \gamma_x)] j_0[kL_y(\sin \theta_i \sin \phi_i + \gamma_y)]. \tag{10.53}$$

Substituting this result in equation (5.31), we obtain the scattered far-field:

$$p_{s\infty}(R, \theta, \phi) = \frac{-i2k}{\pi R} e^{ikR} P_i L_x L_y \cos \theta_i$$
$$\cdot j_0[kL_x(\sin \theta_i \cos \phi_i + \sin \theta \cos \phi)] j_0[kL_y(\sin \theta_i \sin \phi_i$$
$$+ \sin \theta \sin \phi)]. \tag{10.54}$$

The backscattered pressure is given by $\theta = \theta_i$, $\phi = \phi_i$. The maximum occurs in the specular direction, $\theta = \theta_i$, $\phi = \phi_i + \pi$, for which the argument of the spherical Bessel functions is zero, making these functions unity. For normal incidence, $\theta_i = 0$, the backscattered and specular directions coincide, thus making $\cos \theta_i$ as well as the j_0 equal to unity and generating the largest value of scattered pressure.

The directivity patterns obtained in the $(\phi = 0)$ plane for $\phi_i = 0$ are illustrated in figure 10.8 for normal incidence $(\theta_i = 0)$ and in figure 10.9 for oblique incidence $(\theta_i = 45°)$. Inspection of equation (10.54) indicates that

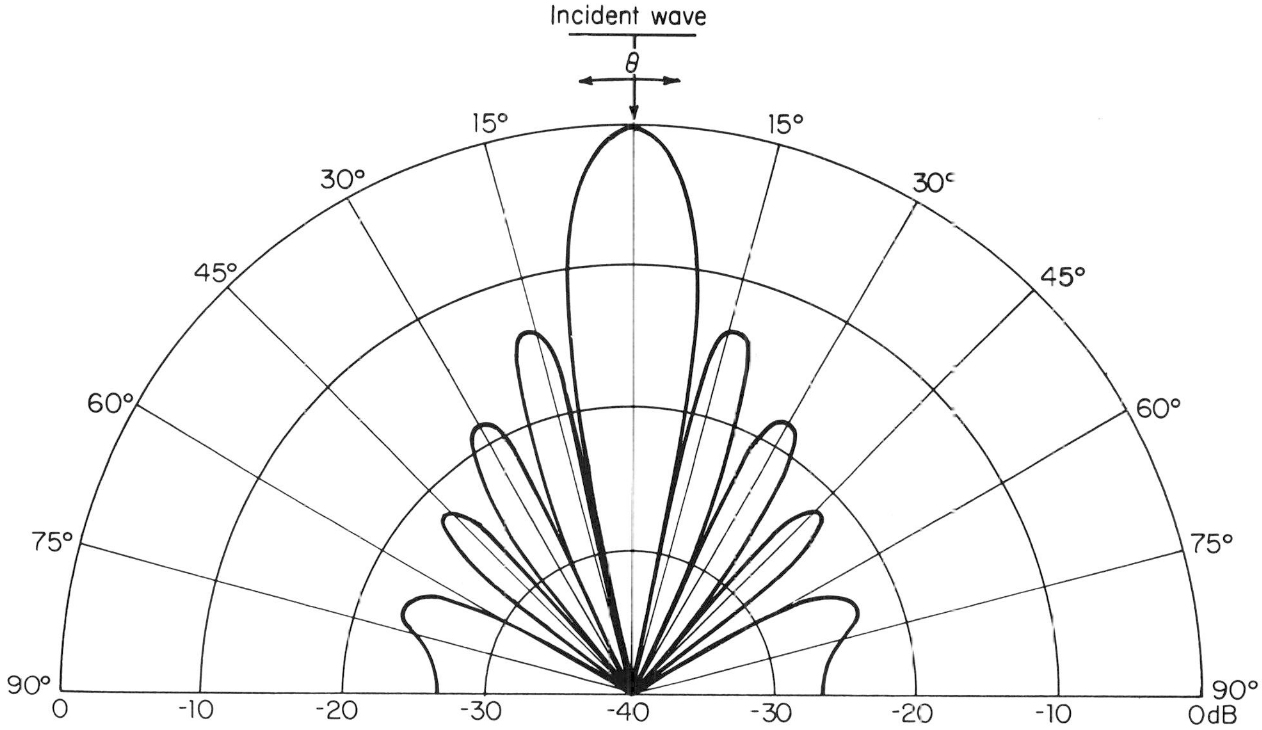

Figure 10.8
Distribution in angle of the sound pressure scattered in the (x, z) plane by the rectangular reflector depicted in figure 10.7, for a normally incident wave [equation (10.54) with $\phi_i = \theta_i = 0$], with $kL_x = 4.8\pi$ normalized to the backscattered pressure.

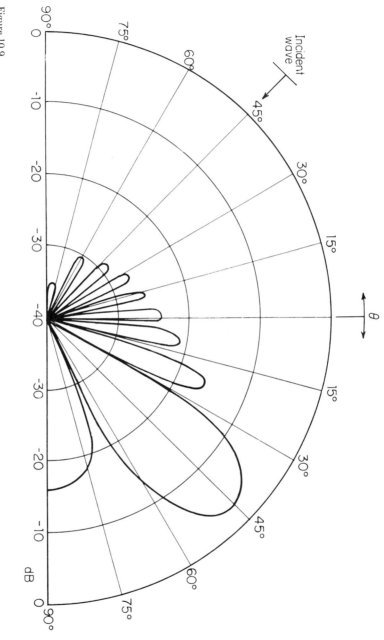

Figure 10.9
Same as figure 10.8 with $\theta_i = 45°$ [equation (10.54)], still normalized to the backscattered pressure for a normally incident wave.

these highly directive multilobe patterns are a consequence of the large value of kL_x selected. A small baffle dimension results in a nondirective pattern. Thus for $kL_x < \pi$, the directivity pattern does not display a null in the (x, z) plane. Because the pressures in figures 10.8 and 10.9 have been normalized to the backscattered pressure produced by a normally incident wave, the L_y dimension need not be specified. As the baffle dimensions are increased, the pattern becomes increasingly directive until, in the limit of the infinite plane, the scattered pressure is concentrated entirely in the direction of specular reflection. In other respects, equation (10.54) does not converge to those obtained for the infinite baffle as one lets $L_x L_y \to \infty$. The reason is that they represent a scattered *far-field*, a condition that is never satisfied if one approaches the infinite reflector, by letting $L_x L_y \to \infty$, because the range R, however large, cannot satisfy the inequality $R \gg L_x, L_y$. Significantly, the scattered field of the infinite plane reflector, equation (10.11), does not satisfy the first of the far-field criteria formulated in section 3.7.

We conclude this chapter by illustrating a reciprocity principle that has applications in the area of acoustic measurements.

10.11 The Helmholtz Reciprocity Principle

The free-space Green's function, equation (4.8), which depends only on the absolute value of the vector $|\mathbf{R} - \mathbf{R}_0|$, is symmetrical in $\mathbf{R}$ and $\mathbf{R}_0$. Consequently, in free space, the locations of a source and of a field point can be interchanged, without altering the pressure measured at the field point. We shall now extend this reciprocity relation to include the presence of scatterers. Rather than providing a general proof, the reciprocity relation will be illustrated for two geometries familiar to the reader: the rectangular baffle and the sphere. Consider a point source S_1, located at $(x_1 = 0, y_1 = 0, z_1)$, a distance z_1 from an infinite rigid baffle coinciding with the $(z = 0)$ plane (figure 10.10). This source generates the incident wave that, at a field point $(x_2, y_2, 0)$ in the plane of the plane wave, takes the value

$$p_i(x_2, y_2, 0) = \frac{\rho \ddot{Q}}{4\pi R} \exp(ikR), \qquad R = (x_2^2 + y_2^2 + z_1^2)^{1/2}.$$

Proceeding as in section 3.4, we simulate the rigid boundary by an image source located at $(0, 0, -z_1)$, which generates in the $z = 0$ plane a pressure equal to p_i. The resultant pressure at a field point in this plane is therefore

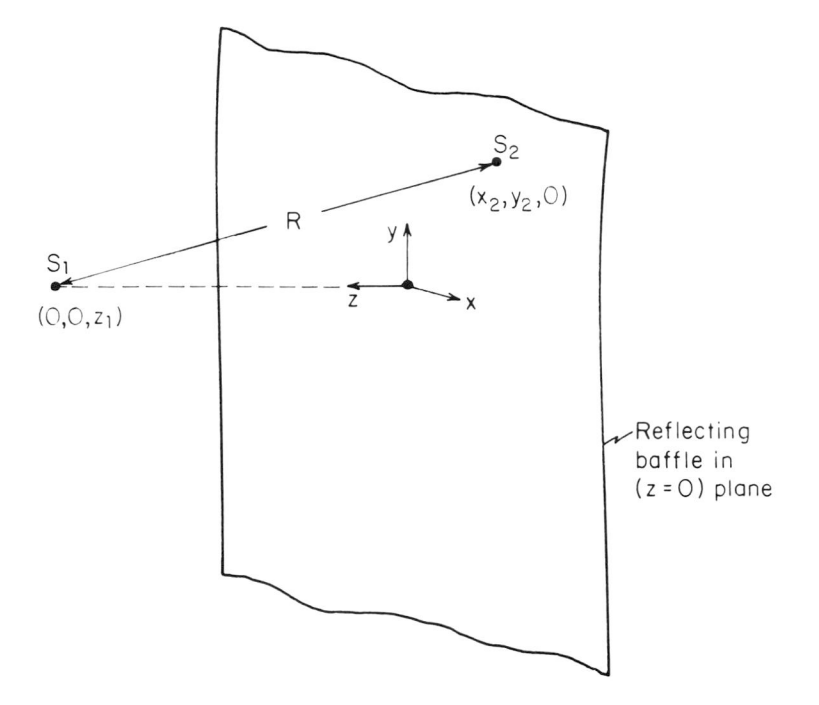

Figure 10.10
Point source near an infinite plane baffle. Illustration of the Helmholtz reciprocity principle for a plane baffle.

$$p(x_2, y_2, 0) = 2p_i(x_2, y_2, 0).$$

The baffle is seen to produce pressure doubling, as in the case of the incident plane wave studied in section 10.3.

Now consider a source S_2 located on the baffle at point $(x_2, y_2, 0)$, whose volume acceleration is identical to that of source S_1. As shown in section 3.4, it radiates twice the pressure of a source that is located in free space. Consequently the pressure radiated by S_2 to the location of S_1 is given by $2p_i(x_2, y_2, 0)$. Thus the pressure at S_1 generated by S_2 equals the pressure generated by S_2 at S_1. In other words, a rigid baffle doubles the signal sensed at a point on the baffle, as it doubles the signal generated by a source located in the baffle. This conclusion is, of course, limited to baffles large in terms of wavelengths. When the baffle is small, its effect is negligible, as noted in connection with equation (6.27b).

The foregoing geometry is so simple as to make this an almost trivial example of the reciprocity principle. Let us therefore consider a more interesting situation and analyze the effect of a spherical baffle. A point source S_2 is located on a spherical scatterer at $\theta = 0$, figure 10.1. The far field radiated by this source is given in equation (6.27a). It is specialized to the direction $\theta = \pi$, where the incident wave originates by setting $P_n(\cos\theta = (-1)^n$. These factors can be merged with the factors ($i)^n$ in equation (6.27a), to yield $(i)^n$:

$$p(R_1, \theta = \pi) = \frac{ie^{ikR}\rho\ddot{Q}}{4\pi a^2 k^2 R_1} \sum_{n=0}^{\infty} \frac{(2n+1)i^n}{h_n'(ka)}, \qquad kR_1 \gg n^2 + 1. \tag{10.55}$$

We next consider the resultant surface pressure produced by a plane wave incident from $\theta = \pi$, equation (10.23), specialized to the location of the point source S_2 at $\theta = 0$, where $P_n = 1$:

$$p(R = a, \theta = 0) = \frac{iP_i}{(ka)^2} \sum_{n=0}^{\infty} \frac{(2n+1)i^n}{h_n'(ka)}. \tag{10.56}$$

The terms under the summation sign are the same as in equation (10.55). To relate the factors multiplying the two series, we assume that the incident pressure P_i in the latter equation is produced by a point source S_1 located in free space, at (R_1, π). The volume acceleration of this source is the same as that of the point source located on the spherical baffle at $\theta = 0$. The pressure it radiates is given by equation (3.1), with $|\mathbf{R} - \mathbf{R}_0| = (R_1 + a)$ in the phase angle, and $|\mathbf{R} - \mathbf{R}_0| \approx R_1$ in the denominator, since $R_1 \gg a$. This pressure, evaluated at $R_0 = a, \theta = 0$, is the incident pressure at the source S_2:

$$p_i(a, 0) = \frac{\rho\ddot{Q}e^{ik(R_1+a)}}{4\pi R_1}.$$

When this is set equal to equation (10.3) evaluated at $(R = a, \theta = 0)$, we can solve for the amplitude P_i of the incident wave:

$$P_i = \frac{\rho\ddot{Q}}{4\pi R_1}e^{ikR_1}. \tag{10.57}$$

Substituting this quantity into equation (10.56), we obtain precisely the

same expression as equation (10.55). Hence, a source located on a spherical baffle at $(a, 0)$ produces a pressure at a field point located in free space at (R_1, π) that equals, in both amplitude and phase, the pressure generated on the spherical baffle at the field point $(a, 0)$ by a source located in free space at (R_1, π). Generalized, this reciprocity relation can be restated as follows: *Whether scatterers are present or not, the locations of the source and field point can be interchanged without altering the pressure at the field point.* Helmholtz's reciprocity principle was extended by Rayleigh to vibrating elastic structures. It was formulated by contemporary researchers for an acoustic fluid containing elastic scatterers and baffles, a situation that constitutes the subject of the next chapter.

References

1. A. Freeman, "A Mechanism of Acoustic Echo Formation," *Acustica* 12:10–21 (1962).

2. R. G. Busnel and J. F. Fish, eds., *Animal Sonar Systems* (New York: Plenum Press, 1980).

3. M. Abramowitz and I. A. Stegun, *Handbook of Mathematical Functions* (Washington, D.C.: NBS, Applied Math. Series 55, June 1964) p. 440, formula (10.1.47).

4. M. C. Junger, "Scattering by Slender Bodies of Revolution," *J. Acoust. Soc. Am.* 72:1954–1956 (1982).

5. Abramowitz and Stegun, *Handbook*, p. 361, formulas (9.1.44) and (9.1.45).

6. V. K. Varadan, V. V. Varadan, L. R. Dragonette, and L. Flax, "Computation of Rigid Body Scattering by Prolate Spheroids Using the T-matrix Approach," *J. Acoust. Soc. Am.* 71:22–25 (1982). The T-matrix computer results were kindly made available by Dr. Dragonette.

7. W. G. Neubauer, "A Summation Formula for Use in Determining the Reflections from Irregular Bodies," *J. Acoust. Soc. Am.* 35:279–285 (1963).

8. Abramowitz and Stegun, *Handbook*, p. 498, formula (12.3.3), with $v = 1$, $\theta = \phi + \pi/2$, and formulas (12.3.9) and (12.1.31).

9. Ibid., p. 360, formula (9.1.22) with $n = 1$ and $\theta = \phi + \pi/2$.

10. B. O. Peirce, *A Short Table of Integrals*, 3rd ed., (Boston: Ginn, 1929), formula (402).

Elastic Scatterers and Waveguides

11.1 The Effect of Scatterer Elasticity

In the preceeding chapter, we analyzed the interaction of an incident plane wave with plane and convex boundaries of infinite impedance. The presence of such boundaries was seen to distort the sound field, the difference between the resultant pressure and the incident pressure being defined as the *scattered pressure*. We shall now study this same interaction for the more realistic situation where the boundary is not infinitely rigid but an elastic structure, such as a plate or shell, that responds to the surface pressure. This scatterer response results in radiated pressure field. The excitation of structural vibrations by incident sound waves is a familiar phenomenon. The shattering of glass by the human voice is reported by Chladni[1] in an early acoustics monograph extensively illustrated with figures, now known by the author's name, of experimentally determined mode shapes of plates. Chladni emphasizes the great antiquity of the observation that sound can excite large-amplitude vibrations of shell-like structures by pointing out that Talmudic law stipulates compensation for glass vessels fractured by crowing roosters and other farm animals. In our times, sonic fatigue of aircraft structures has been of greater concern.

Most of this chapter parallels the preceding one in dealing with sound scattering by planar, spherical, and cylindrical structures and more generally by bodies of revolution displaying elastic response to the incident field. The finite elastic plate and the spherical shell are used to illustrate acoustic-structural reciprocity relations. The emphasis here is on the modification of the scattered field by the scatterer's elastic response. The initial sections analyze the scattering by planar plates and sound transmission through plates of finite and effectively infinite extent. Because of its simplicity the sphere has been selected for the purpose of illustrating in detail the features of elastic scattering (sections 11.5–11.8). Rayleigh scattering and scattering by cylindrical shells are covered next. Invoking the reciprocity principle, the scattering solutions are used to construct the far-field of a point source located on elastic planar, spherical, and cylindrical baffles (section 11.11). The concluding sections deal with the interaction of elastic cylindrical shells and an interior sound field, i.e., with sound propagation in fluid-filled pipes, a subject dealt with in elementary fashion in section 2.10. Elastic waveguides and elastic scatterers have in common that they both involve the passive response of a shell-like structure to an incident sound field.

While a rigid scatterer distorts the sound field by interfering with the propagation of the incident wave, the dynamic response of an elastic

scatterer excited by the incident wave further modifies the resultant sound field. The pressure scattered by an elastic boundary will be identified with the symbol p_{se} to distinguish it from the pressure $p_{s\infty}$ scattered by the infinite-impedance boundary coinciding with the surface $S_0(\mathbf{R}_0)$ of the elastic scatter. The resultant pressure p is

$$p \equiv p_i + p_{se}.$$

It is convenient to express the scattered pressure p_{se} as the sum of the solutions $p_{s\infty}$ constructed in the preceding chapter, and of an unknown pressure component p_r:

$$p_{se} \equiv p_{s\infty} + p_r. \tag{11.1}$$

To interpret the meaning of p_r, we note that the boundary condition, equation (4.9b), governing the resultant pressure

$$\frac{\partial p}{\partial \xi_0} = -\rho \ddot{w} \qquad \text{on } S_0(\mathbf{R}_0)$$

is satisfied if

$$\left.\begin{aligned} \frac{\partial p_r}{\partial \xi_0} &= -\rho \ddot{w} \\[2mm] \frac{\partial p_i}{\partial \xi_0} &= -\frac{\partial p_{s\infty}}{\partial \xi_0} \end{aligned}\right\} \quad \text{on } S_0(\mathbf{R}_0).$$

The latter relation is satisfied automatically by virtue of the definition of $p_{s\infty}$. The former equation indicates that p_r is in the nature of a radiated pressure associated with the acceleration $\ddot{w}$ of the elastic structure responding to the surface pressures generated by the incident wave field.

When the structure is an effectively infinite elastic plate, structural resonances do not, of course, occur. Instead, coincidence effects described in chapter 8 account for the discrepancy between the pressures reflected by, respectively, the rigid, plane boundary and the planar elastic baffle or partition. For the sake of uniformity of nomenclature the reflected pressure will be identified with the symbol p_{se} used for the scattered pressure.

11.2 Sound Reflection by an Infinite Elastic Plate

The plate is an effectively infinite planar reflector (figure 10.2). In this section we shall deal with a plate exposed to the acoustic medium on only one side. This is a proper approximation for a plate, separating a liquid and a gas.

The situation where the plate separates two fluid-filled half-spaces will be analyzed in the next section. In both cases, the expressions for p_i, $p_{s\infty}$, and their sum are given in equations (10.2), (10.11), and (10.12), respectively. The exciting pressure acting on the plate is obtained by specializing the latter expression to $z = 0$:

$$p_i(x, z = 0) + p_{s\infty}(x, z = 0) = 2P_i \exp(ikx \sin \theta), \tag{11.2}$$

where, as usual, the time-dependent factor $\exp(-i\omega t)$ has been suppressed. The dynamic response of the plate must, like the excitation, be y independent. It consists of structure-borne waves propagating in the x direction with the same wave number as the surface pressure in equation (11.2), the wave fronts being parallel to the y axis:

$$\ddot{w}(x) = \ddot{W} \exp(ikx \sin \theta). \tag{11.3}$$

The phase velocity of the forced structural waves is

$$c_s = \frac{\omega}{k \sin \theta} = \frac{c}{\sin \theta}. \tag{11.4}$$

This velocity is never less than the sound velocity. Referring to section 5.11, we conclude that, for $c_s \geq c$, the radiation impedance is always resistive. To solve the equations of motion of the plate, equation (7.59), we require the resultant surface load acting on the plate. One component, equation (11.2), is in the nature of a forcing term, being independent of the plate response. The other component is the radiated pressure that is proportional to $\ddot{W}$. The relation between p_r and $\ddot{W}$ is obtained from equation (5.60), with

$$k_s = k \sin \theta,$$

$$(k^2 - k_s^2)^{1/2} = k \cos \theta.$$

Hence,

$$p_r(x, z) = \frac{i\rho \ddot{W}}{k \cos \theta} \exp[ik(x \sin \theta + z \cos \theta)].$$ (11.5)

When this expression is specialized to $z = 0$ and added to equation (11.2), one obtains the resultant pressure acting on the plate.

The equation of motion of the plate, equation (7.59), will be rewritten by factoring the bending stiffness D out of the acceleration term, and making use of the flexural wave number defined in equation (7.62). Noting that, for the present, y-independent configuration, we can reduce the surface differential operator to

$$\nabla_\sigma^4 = \frac{\partial^4}{\partial x^4} = k^4 \sin^4 \theta,$$

we find that the equation of motion becomes

$$\frac{D}{\omega^2}(k^4 \sin^4 \theta - k_f^4) \ddot{W} \exp(ikx \sin \theta) = [2P_i + \frac{i\rho}{k \cos \theta} \ddot{W}] \exp(ikx \sin \theta).$$ (11.6)

Structural damping can, as usual, be simulated by a loss factor. This equation can be solved for $\ddot{W}$. It is convenient to divide each term by the exponential and by the ratio $Dk_f^4/\omega^2 \cos \theta$ and to express the resulting ratio $(k/k_f)^4$ in terms of the coincidence frequency, as $(\omega/\omega_c)^2$, using equations (8.2) and (8.3). Setting $Dk_f^4/\omega^2 = \rho_s h$, one can solve for the complex acceleration amplitude:

$$\ddot{W} = \frac{-2P_i \cos \theta}{\rho_s h\{(i\rho/\rho_s kh) + [1 - (\omega/\omega_c)^2 \sin^4 \theta] \cos \theta\}}.$$ (11.7)

The radiated pressure can now be constructed by substituting this result in equation (11.5):

$$p_r(x, z) = \frac{2P_i \exp[ik(x \sin \theta + z \cos \theta)]}{ikh(\rho_s/\rho)[1 - (\omega/\omega_c)^2 \sin^4 \theta] \cos \theta - 1}.$$ (11.8)

When this is combined with $p_{s\infty}$, as given by equation (10.11), one obtains an explicit expression for the reflected pressure:

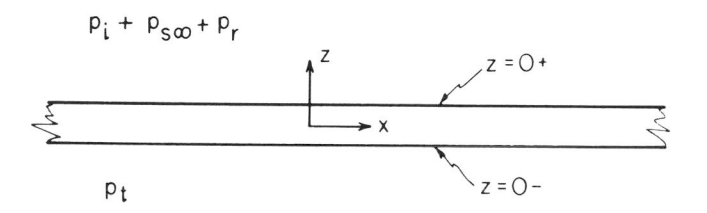

Figure 11.1
Components of the pressure fields in two semiinfinite fluid-filled spaces separated by an elastic plate.

$$p_{se}(x, z) = P_i \exp[ik(x\sin\theta + z\cos\theta)]$$
$$\cdot \frac{ikh(\rho_s/\rho)[1 - (\omega/\omega_c)^2\sin^4\theta]\cos\theta + 1}{ikh(\rho_s/\rho)[1 - (\omega/\omega_c)^2\sin^4\theta]\cos\theta - 1}. \tag{11.9}$$

It can be verified that for the infinite-impedance boundary (obtained by letting $\rho_s \to \infty$), this expression tends to $p_{s\infty}$, equation (10.11). For grazing incidence $(\theta \to \pi/2)$ and for waves incident in the critical or coincidence direction $[\theta = \theta_c \equiv \sin^{-1}(\omega_c/\omega)^{1/2}]$, the plate acts as a "pressure-release" reflector, for which $p_{se} = -p_i$. The same condition is approached for light plates or at very low frequencies, for which $k^2h^2(\rho_s/\rho)^2 \ll 1$: For *any* direction of incidence, $|p_{se}| = P_i$, in other words, only the phase angle is a function of θ. It will be shown that both phase *and* amplitude depend on θ when the plate acts as a boundary between two identical semiinfinite media.

11.3 Sound Transmission through an Infinite Elastic Plate

An additional pressure component p_t, associated with the wave transmitted into the $z < 0$ half-space, must be added to the surface loading term in the equation of motion of the plate, equation (11.6) (figure 11.1). This transmitted pressure is associated with the elastic response of the plate. It is, in fact, the pressure radiated by the plate into the half space $z < 0$. Since the two plate surfaces partake in identical motions, and assuming the same acoustic medium on both sides of the plate, $|p_t| = |p_r|$. The phase is, however, different. Thus, on the plate surface this transmitted pressure $p_t(x, 0-)$ takes the value $-p_r(x, 0+)$, since a plate acceleration $\ddot{w}(x)$ producing a positive pressure on one plate surface generates a negative pressure on the other. The z dependence of p_t must display the fact that this wave propagates in the negative z direction, like the incident wave. Hence

$$p_t(x, z) = |p_t| \exp[ik(x\sin\theta - z\cos\theta)]$$
$$= -p_r(x, z)\exp(-i2kz\cos\theta).$$
(11.10)

The resultant plate loading term is now of the form

$$-p_i - p_{s\infty} - p_r + p_t = -2p_i - 2p_r, \qquad z = 0.$$
(11.11)

This differs from the loading of the plate exposed to fluid on one side only by the factor 2 that multiplies p_r. Noting that the ρ term in equations (11.6) and (11.7) arises from the radiated pressure p_r, these equations can be easily modified to construct, respectively, the equation of motion and the response of the plate exposed to fluid on both sides, by the simple device of multiplying the ρ term in these equations by 2. Substituting the value of $\ddot{W}$ thus obtained in equation (11.5), one constructs an expression for radiated pressure which differs from equation (11.8) only by this same factor 2:

$$p_r(x, z) = \frac{2P_i \exp[ik(x\sin\theta + z\cos\theta)]}{ikh(\rho_s/\rho)[1 - (\omega/\omega_c)^2\sin^4\theta]\cos\theta - 2}.$$
(11.12)

The transmitted pressure is now constructed by substituting this result in equation (11.10):

$$p_t(x, z) = -\frac{2P_i \exp[ik(x\sin\theta - z\cos\theta)]}{ikh(\rho_s/\rho)[1 - (\omega/\omega_c)^2\sin^4\theta]\cos\theta - 2}.$$
(11.13)

The scattered or reflected pressure p_{se} is obtained by combining $p_{s\infty}$, given in equation (10.11), with the radiated pressure in equation (11.12):

$$p_{se}(x, z) = \frac{iP_i kh(\rho_s/\rho)[1 - (\omega/\omega_c)^2\sin^4\theta]\cos\theta}{ikh(\rho_s/\rho)[1 - (\omega/\omega_c)^2\sin^4\theta]\cos\theta - 2}$$
$$\cdot \exp[ik(x\sin\theta + z\cos\theta)].$$
(11.14)

Our mathematical model stipulates an incident plane wave. This assumption is not valid at grazing incidence, since the fluid particle velocity cannot be parallel to boundary displaying a finite impedance. Consequently, the above equations do not hold as $\theta \to \pi/2$.

The transmitted pressure, equation (11.13), becomes identical to the incident pressure for a wave incident along the critical direction

$[\theta = \theta_c \equiv \sin^{-1}(\omega_c/\omega)^{1/2}]$, and when $kh(\rho_s/\rho) \ll 1$, viz., at low frequencies and for light plates whose inertial reactance per unit area, $\omega h \rho_s$, is negligible compared with the characteristic impendance ρc. These features are illustrated in figure 11.2, where transmission loss, defined as $-20\log|p_t|/P_i$, is plotted versus angle incidence for a steel plate in water. Only the thicker of the two plates shows a coincidence effect at the 10 kHz frequency selected for the calculations. However, the spatially selective deterioration of the reflecting qualities of the thick plate at the coincidence angle is compensated by the low transmission loss of the lighter plate at all angles of incidence. The conditions resulting in a small transmission loss are precisely those that give rise to a "pressure-release" type reflection when the plate is exposed to a vacuum on one of its surfaces [equation (11.9)]. The plate response is given by (11.7), where the ρ term is multiplied by 2 to account for the presence of fluid on both plate surfaces. When excited at coincidence, or when radiation loading predominates, the response is expressed more compactly in terms of the plate velocity:

$$\dot{W} = -\frac{P_i \cos \theta}{\rho c}, \qquad \theta = \theta_c \quad \text{or} \quad \rho^2 \gg \rho_s^2 k^2 h^2. \tag{11.15}$$

This amplitude is doubled when only one plate surface is exposed to the acoustic fluid.

11.4 Sound Transmission through Finite Plates; Reciprocity

For plates whose dimensions as measured in flexural wavelengths, and whose structural damping is sufficiently small to ensure the existence of standing rather than traveling waves, structural resonances play a role analogous to that of coincidence effects in extended plates. This equivalence will become apparent from the expression derived here for the ratio of transmitted to incident pressure. This result will be used to extend the reciprocity concept to elastic boundaries.

The transmitted pressure is expressed in terms of the normal modes of the plate excited by the incident sound fields. For a rectangular plate, the acceleration response is formally given by

$$\ddot{w}(x, y) = \sum_{m,n} \ddot{W}_{mn} f_{mn}(x, y). \tag{11.16}$$

For simply supported plates, the mode shapes are given in (7.73). Here,

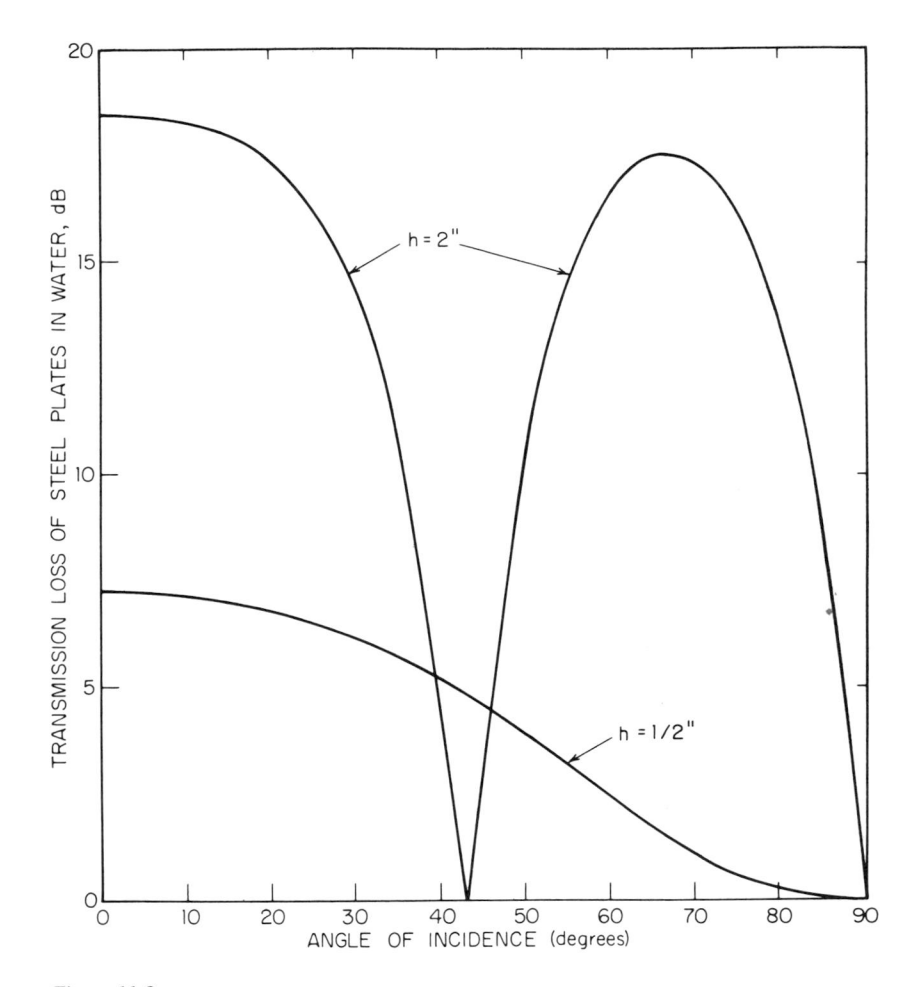

Figure 11.2
Computed transmission loss $-20\log(|p_t|/P_i)$, equation (11.13), plotted as a function of angle of incidence for 2-inch and $\frac{1}{2}$-inch steel plates, for a frequency of 10 kHz.

however, there is no need to specify the boundary conditions. The modal acceleration amplitudes are solutions of the equation of motion, (7.59). For simply supported plates, where all modal masses equal one-quarter the static mass, these amplitudes are given in (7.77), generalized from a point drive to a distributed load. For arbitrary boundary conditions, the modal masses are a function of the mode order. If $f_{mn}(x. y)$ is a mode configuration, its modal mass is

$$M_{mn} = \rho_s h \int_{L_x}^{L_x} \int_{L_y}^{L_y} f_{mn}^2(x, y) \, dy \, dx.$$

The modal acceleration amplitudes now become

$$\ddot{W}_{mn} = \frac{F_{mn}}{M_{mn}[1 - (\omega_{mn}/\omega)^2]}, \tag{11.17}$$

where the generalized forces exerted by the resultant of the various pressure fields are

$$F_{mn} = \int_{-L_x}^{L_x} \int_{-L_y}^{L_y} p(x, y, z = 0) f_{mn}(x, y) \, dx \, dy. \tag{11.18}$$

Structural losses are accounted for by a loss factor η_s such that $\omega_{mn}^2 = |\omega_{mn}|^2(1 - i\eta_s)$. For structural plates located in light acoustic fluids, typically air, the ratio of the characteristic impedance of the ambient medium to that of the plate is so small that the plate vibrations excited by the incident sound field produce a negligible alteration of the incident and rigid-body scattered pressures. This finite-plate analysis may be simplified by specializing it to this situation. The resultant pressure acting on the plate can therefore be approximated by equation (11.2). This expression must be modified to account for the fact that the orientation of the incident wave with respect to the plate boundaries is arbitrary, the incident wave fronts making an angle ϕ_i with planes normal to the x axis:

$$p(x, y, z = 0) = 2P_i \exp[ik \sin \theta_i (x \cos \phi_i + y \sin \phi_i)]. \tag{11.19}$$

To facilitate comparison with the results obtained in section 11.3, a single incident plane wave will be assumed. Since a diffuse field can be formulated

as a superposition of randomly incident plane waves,[2] plate vibrations excited by such a field can be constructed by integrating, over all angles of incidence, the plate response computed here for a single plane wave incident from a specified direction. The transmission loss is based on the power rather than pressure ratio. The transmitted power is obtained by integrating the transmitted pressure over all aspects, as shown in (3.27). The incident power is obtained by multiplying incident diffuse field intensity[2] by the wall area. These power calculations not being necessary for the purpose of illustrating the intended effects, only the pressure ratio will be computed.

When equations (11.18) and (11.19) are combined, one obtains the generalized force:

$$F_{mn} = 2P_i \int_{-L_x}^{L_x} \int_{-L_y}^{L_y} f_{mn}(x, y) \exp\left(ikx \sin\theta_i \cos\phi_i + iky \sin\theta_i \sin\phi_i\right) dx \, dy.$$

$$(11.20)$$

The double integral can be interpreted as the Fourier transform of the plate response, equation (5.29), evaluated at the points of stationary phase corresponding to field points $(\pi - \theta_i, \phi_i)$:

$$F_{mn} = 2P_i \tilde{f}_{mn}(\bar{\gamma}_{xi}, \bar{\gamma}_{yi}).$$

$$(11.21)$$

The modal amplitudes are obtained by substituting this result in equation (11.17). When these amplitudes are combined with (11.16), the resultant plate response is obtained:

$$\ddot{w}(x, y) = 2P_i \sum_{m,n} \frac{\tilde{f}_{mn}(\bar{\gamma}_{xi}, \bar{\gamma}_{yi}) f_{mn}(x, y)}{M_{mn}[1 - (\omega_{mn}/\omega)^2]}.$$

$$(11.22)$$

The transmitted pressure field will be computed for a geometry consistent with the assumption of a single incident plane wave; the elastic panel is analyzed as a window in a rigid baffle between two semiinfinite fluid-filled spaces. This simple, seemingly unrealistic model predicts the principal transmission loss-controlling phenomena generated by the more realistic but laborious model whereby the plate's normal modes are coupled to those of the two reverberant rooms it separates.[3] For the situation assumed here, the pressure transmitted to the far-field is given by (5.31). In the notation of this chapter, this expression takes the form

$$p_t(R, \theta_t, \phi_t) = \frac{\rho e^{ikR}}{2\pi R} \sum \ddot{W}_{mn} \tilde{f}_{mn}(\bar{\gamma}_{xt}, \bar{\gamma}_{yt}), \tag{11.23}$$

where the stationary-phase wave numbers are given in (5.30). Substituting the transform of (11.22), one obtains the ratio of transmitted pressure to incident pressure:

$$\frac{|p_t(R, \theta_t, \phi_t)|}{P_i(\theta_i, \phi_i)} = \frac{\rho}{\pi R} \sum \frac{|\tilde{f}_{mn}(\bar{\gamma}_{xi}, \bar{\gamma}_{yi}) \tilde{f}_{mn}(\bar{\gamma}_{xt}, \bar{\gamma}_{yt})|}{M_{mn}|1 - (\omega_{mn}/\omega)^2|}. \tag{11.24}$$

A basic difference between this result and (11.13) is that in the low-frequency limit, the infinite plate, being mass controlled, is transparent to sound, while the finite panel is stiffness controlled, and therefore provides a high insertion loss. Only the fundmental mode need be retained in the low-frequency expression of the pressure ratio:

$$\frac{|p_t(R, \theta_t, \phi_t)|}{P_i(\theta_i, \phi_i)} \simeq \frac{\rho \omega^2}{\pi R} \frac{|\tilde{f}_{11}(\bar{\gamma}_{xt}, \bar{\gamma}_{yt}) \tilde{f}_{11}(\bar{\gamma}_{xi}, \bar{\gamma}_{yi})|}{\omega_{11}^2 M_{11}}, \qquad \omega^2 \ll \omega_{11}^2. \tag{11.24a}$$

At resonance the transmitted pressure is controlled by the resonant mode, the modal admittances being, electrically speaking, in parallel. Because radiation loading has been ignored, the denominator associated with a resonant mode tends to $-iM_{mn}\eta_s$ as $\omega \to \omega_{mn}$. Had just radiation loading been accounted for, with $\eta_s - 0$, the pressure ratio would tend to unity as it does for the infinite plate when $\omega \to \omega_c$, equation (11.13). For the assumed light acoustic fluid, structural damping will keep the ratio well below unity.

The symmetry of (11.24) and (11.24a) indicates that *the sound source and receiver can be interchanged without altering the pressure ratio*, as already concluded for rigid boundaries in section 10.11. The extension of Helmholtz's reciprocity principle to a situation involving elastic scatterers is due to Rayleigh (1873). Furthermore, comparing (11.21) and (11.23), it is apparent that *a mode that radiates sound efficiently in a given direction is also efficiently excited by a plane wave incident from that same aspect*. This reciprocity principle will be developed further in connection with spherical shells (section 11.7) and sources located on elastic baffles (section 11.11).

11.5 The Spherical Shell as an Acoustic Scatterer

The system illustrated in figure 10.1 is modified only in that the sphere now is a thin elastic, evacuated spherical shell free to translate as a rigid body or

to deflect elastically, instead of a rigid motionless sphere. The shell is thin enough and the frequency low enough to ignore flexural stresses as compared with membrane stresses. The dynamic response of the shell to a surface excitation,

$$p(a, \theta) = \sum_{n=0} p_n P_n(\cos \theta), \tag{11.25}$$

can be written in the form of equation (7.111), the modal velocity amplitudes $\dot{W}_n$ being expressed, as in equation (9.13), by combining the in vacuo modal impedances of the shell Z_n with the radiation impedances z_n:

$$\dot{w}(\theta) = -\sum_{n=0} \frac{p_n P_n(\cos \theta)}{Z_n + z_n}. \tag{11.26}$$

The radiation impedances are given in equations (6.30)–(6.32). The in vacuo shell impedance derived in chapter 7 was expressed in terms of dimensionless driving frequencies $\omega a / c_p \equiv \Omega$, and of the corresponding natural frequencies [equation (7.121)]. The surface pressure in equation (11.25) is the sum $(p_i + p_{s\infty})$ of the incident and "rigid-body scattered" pressures derived in equation (10.23). The radiated pressure, being a function of $\dot{w}(\theta)$, is not treated as an excitation, but is combined with the homogeneous portion of the equations of motion, that is, absorbed in the coefficient matrix, where it is accounted for by the z_n term. Had we proceeded as in the case of the elastic plate and included $p_r(a, \theta)$ in the surface excitation, we would have had to retrace the steps in section 9.4, which led to equation (9.13). Comparing equations (11.26) and (10.23), one sees that

$$p_n = \frac{i^{n+1}(2n + 1) P_i}{(ka)^2 h_n'(ka)}. \tag{11.27}$$

When equations (11.26) and (11.27) are combined, the shell response to the incident and "rigid-body" scattered pressure is obtained:

$$\dot{w}(\theta) = -\frac{P_i}{k^2 a^2} \sum_{n=0} \frac{i^{n+1}(2n + 1) P_n(\cos \theta)}{(Z_n + z_n) h_n'(ka)}. \tag{11.28}$$

The corresponding far-field radiated pressure is, from equation (6.21),

$$p_r(R, \theta) = -\frac{i\rho c e^{ikR} P_i}{kR} \sum_{n=0} \frac{(2n+1) P_n(\cos \theta)}{(\mathcal{Z}_n + z_n) [kah'_n(ka)]^2}, \qquad kR \gg n^2 + 1. \quad (11.29)$$

Finally, according to equation (11.1), the scattered pressure is obtained by combining this result with the "rigid-body" scattered pressure, equation (10.17):

$$p_{se}(R, \theta) = \frac{ie^{ikR} P_i}{kR} \sum_{n=0} \frac{(2n+1) P_n(\cos \theta)}{h'_n(ka)}$$

$$\cdot \left[j'_n(ka) - \frac{\rho c}{(ka)^2 h'_n(ka) (\mathcal{Z}_n + z_n)} \right], \qquad kR \gg n^2 + 1. \quad (11.30)$$

One effect of a modal resonance of the submerged shell is an abrupt change in sign of $\mathrm{Im}(\mathcal{Z}_n + z_n)$ as the incident pressure sweeps through a frequency band encompassing a natural frequency $\bar{\Omega}_n^{(j)}$. This effect is felt in the far-field only if the resonant term predominates, i.e., if structural damping is small enough to make the ρc term of the resonant mode, in equation (11.30), comparable to the series of "rigid-body" scattered terms in equation (10.17) and hence to the series of j'_n terms in equation (11.30). Since the factor multiplying $(\mathcal{Z}_n + z_n)^{-1}$ in equation (11.29) changes slowly with frequency, the curve of p_r versus frequency changes sign abruptly as one sweeps through resonance. In contrast, the "rigid-body" scattered pressure $p_{s\infty}$ changes gradually with frequency. The phase angle between the two pressure components p_r and $p_{s\infty}$ just below resonance depends on the acoustic transfer function relating the structural response $\ddot{w}(\mathbf{R}_0)$ to the radiated pressure $p_r(\mathbf{R})$ at a particular field point $\mathbf{R}$. If this phase angle lies near $0°$, a constructive combination of $p_{s\infty}$ with p_r becomes a destructive interference as the frequency of the incident wave is made to sweep upward through the natural frequency of the structural mode. If the phase angle starts out as $180°$, this sequence is reversed. Whatever the phase angle, a rapid fluctuation of the scattered pressure versus frequency curve is observed provided p_r is comparable to or larger than $p_{s\infty}$. These resonance-induced fluctuations distinguish elastic from rigid scatterers. It must be recalled that a structure appearing as rigid in a gaseous medium may display elastic behavior when submerged. It is again emphasized that the above only applies to resonant modes for which radiation damping predominates over structural damping. For these circumstances, the expression for the contribution of the resonant mode becomes relatively simple. The impedance

$(\mathcal{Z}_n + z_n)$ is reduced to the radiation resistance r_n, equation (6.31), and the contribution of the resonant mode can be easily constructed. The pressure radiated by the resonant mode takes on the simple form

$$p_{rn}(R, \theta) = \frac{-iP_i(2n + 1)P_n(\cos\theta)\exp(ikR - 2i\alpha_n)}{kR},$$

$$\quad (11.31)$$

$$ka = \bar{\Omega}_n^{(j)}\frac{c_p}{c}, \qquad \eta_s \ll \eta_n,$$

where α_n is phase angle of $h_n'(ka)$:

$$\alpha_n \equiv \tan^{-1}\frac{y_n'(ka)}{j_n'(ka)}.$$

This result is noteworthy in that the amplitude of this pressure is independent of ka. The radiated pressure normalized to unit range and incident pressure $R|p_{rn}|/P_i$ is proportional to the acoustic wavelength only. Readers familiar with the theory of noise control techniques recognize the similarity with the pressure scattered by an ideal Helmholtz resonator.[4] The effect of a modal resonance on the scattered pressure is illustrated in figure 11.3 at the resonant frequency of the $(n = 2, j = 1)$ mode of the submerged shell. Only the radiated pressure component of the resonant mode need be taken into account, except near the zeros of $P_2(\cos\theta)$, where this component vanishes.

We shall see in the next section that small pressure-release spheres, specifically spheres measuring less than one wavelength in diameter, produce a more marked distortion of the sound field than small rigid spheres. This is not surprising if we consider that a small compliant cavity effectively "short-circuits" the stiffness (or effective bulk modulus) of the adjacent liquid volume, while a small rigid inclusion only slightly raises the stiffness. The small $(2a \approx \lambda/3)$ resonant sphere whose scattering pattern is depicted in figure 11.3 distorts the sound field even more markedly than a "pressure-release" sphere. The reason is that the latter simulates a shell of vanishing in vacuo impedance, whose motion is inhibited by radiation loading when submerged. In contrast, for the resonant shell, the major, viz., reactive, component of this loading is cancelled, leaving only the small radiation resistance to limit the resonant shell response. The correspondingly large response of the resonant sphere distorts the incident sound field more markedly than a "pressure-release" sphere.

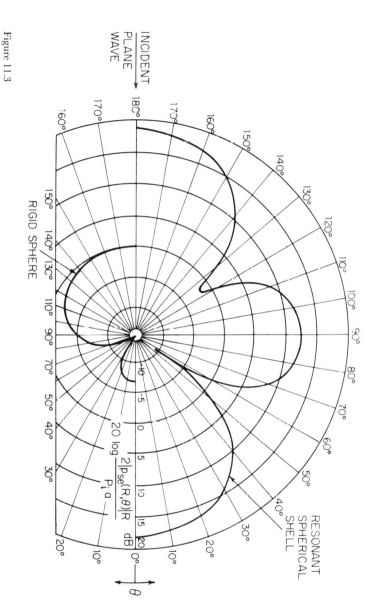

INCIDENT
PLANE
WAVE

RIGID SPHERE

RESONANT
SPHERICAL
SHELL

$$20 \log \frac{2|p_{se}(R,\theta)|R}{p_i\,a}\ \ dB$$

θ

Figure 11.3
Distribution in angle of normalized pressure scattered by a submerged spherical steel shell at the resonant frequency of the lower $n = 2$ mode [$h/a = 10^{-2}$, $\eta_s = 0$; $ka = (c_p/c)\,(\Omega = \bar\Omega_2^{(1)}) = 1.123$, and $\eta_2^{(1)}$ from table 9.1], computed from equation (11.30); same for a rigid sphere of the same diameter computed from equation (10.17).

Table 11.1
Pressure radiated by a resonant mode[a] of a spherical shell

	Excitation	
	Concentrated force F	Incident plane wave of pressure amplitude P_i
Amplitude of modal forcing term	$\dfrac{(2n+1)F}{4\pi a^2}$ [equation (9.9)]	$\dfrac{(2n+1)P_i(r_n/\rho c)^{1/2}}{ka}$ [equation (11.37)]
Pressure amplitude radiated by resonant modes controlled by		
Structural damping $(r_s \gg r_n)^{b}$	$\dfrac{(2n+1)F(r_n/\rho c)^{1/2}}{4\pi a R(r_s/\rho c)}$ [equation (9.16)]	$\dfrac{(2n+1)P_i r_n}{kRr_s}$ [equation (11.32)]
Radiation damping $(r_s \ll r_n)$	$\dfrac{(2n+1)F}{4\pi a R(r_n/\rho c)^{1/2}}$ [equation (9.15)]	$\dfrac{(2n+1)P_i}{kR}$ [equation (11.31)]

a. Natural frequencies for representative shell are tabulated in table 9.1.
b. Radiation loss factors tabulated in table 9.1 are related to radiation resistance in equation (9.8).

This effect of scatterer resonances is exaggerated by our unrealistic assumption of a lossless shell. When the resonant mode is controlled by structural rather than radiation damping, the impedances $(Z_n + z_n)$ in equation (11.29) can be approximated by a structural resistance r_s. The pressure radiated by the resonant mode now becomes

$$
\left.
\begin{aligned}
p_{rn}(R,\theta) &= -\frac{iP_i(2n+1)P_n(\cos\theta)\rho c e^{ikR}}{kR[kah_n'(ka)]^2 r_s} \\
&= -\frac{iP_i(2n+1)P_n(\cos\theta)r_n\exp(ikR - 2i\alpha_n)}{kRr_s}
\end{aligned}
\right\}
\quad
\begin{aligned}
&r_s \gg r_n, \\
&ka = \bar{\Omega}_n^{(j)}\frac{c_p}{c},
\end{aligned}
$$

$$(11.32)$$

where the second form of the equation is in terms of the modal resistance r_n, equation (6.31). When r_n is small compared with r_s, the resonant mode has a negligible effect on the scattered field.

These results are summarized in table 11.1. A comparison with the

pressure radiated by a resonant point-excited shell indicates that poorly radiating modes $(r_n \ll r_s)$ radiate less when excited by an incident plane wave than when excited by a concentrated force. The implications of this fact with regard to the relation between shell response and far-field pressures are analyzed in section 11.8.

The marked frequency dependence of the backscattered pressure that results from the presence of structural resonances is illustrated in figure 11.4. This theoretical curve is based on three-dimensional elasticity theory rather than on the thin-shell theory used here. This frequency dependence on structural resonances, which is predicted by both theories, is confirmed by experiments.[5] Figure 11.4 also shows the curves for rigid spheres and for the "pressure-release" spheres analyzed below. A comprehensive discussion of resonant scatterers can be found in a recent study by Gaunaurd and Überall.[6]

11.6 The Scattering Action of the "Pressure-Release" Sphere

A gas-filled, spherical bubble in a liquid medium excited near resonance can be approximated as a pressure-release sphere. This mathematical model will be used to illustrate the differences between soft and rigid scatterers. The pressure-release boundary condition requires that the incident wave, expressed in spherical harmonics, equation (10.13), equal the negative of the scattered surface pressure formulated in terms of the modal specific impedances z_n, equation (6.29), and of the as yet unknown coefficients describing the velocity distribution associated with the scattered field on the scatterer surface:

$$P_i \sum_{n=0}^{\infty} (2n + 1) i^n P_n (\cos\theta) j_n (ka) = -\sum_{n=0}^{\infty} \dot{W}_{0n} z_n P_n (\cos\theta). \qquad (11.33)$$

Corresponding terms of this series can be solved for the velocity coefficients. Expressing the specific acoustic impedance explicitly, we see that these coefficients are

$$\dot{W}_{0n} = -\frac{P_i}{\rho c} (2n + 1) i^{n-1} j_n (ka) \frac{h_n'(ka)}{h_n(ka)}.$$

The scattered far-field can now be constructed by substituting these coefficients in the series developed for spherical radiators, equation (6.21):

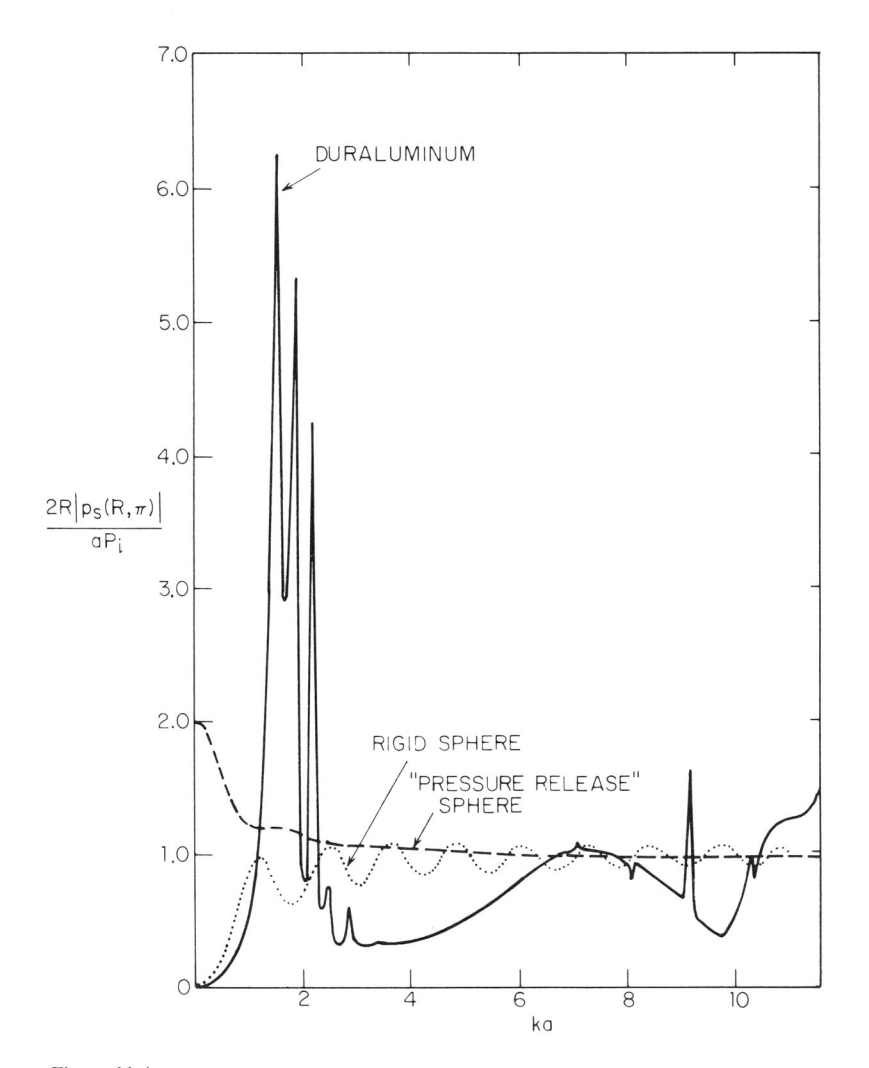

Figure 11.4
Pressure backscattered by a rigid sphere, a "pressure-release" sphere, and an evacuated duraluminum shell in water ($h/a = 1/20$), plotted versus dimensionless frequency. [Reproduced from Hickling and Diercks and Hickling[5]]

$$p_{s0}(R, \theta) = \frac{iP_i e^{ikR}}{kR} \sum_{n=0} (2n + 1) \frac{j_n(ka)}{h_n(ka)} P_n(\cos\theta). \tag{11.34}$$

It is of interest to compare this result with the scattered field of the rigid sphere. In the small-ka limit, $j_n(x)$ varies as x^n, and $h_n(x)$ as $x^{-(n+1)}$. The terms in the above series thus converge as $(ka)^{2n+1}$. In the long-wavelength limit, we need therefore retain only the first term. Using equation (2.45) for j_0, equation (6.11b) for h_0, and $P_0 = 1$, one obtains the scattered far-field:

$$P_{s0}(R) \approx -\frac{a}{R} P_i e^{ikR}; \qquad k^2 a^2 \ll 1. \tag{11.35}$$

This result differs in two important ways from the long-wavelength solution for the rigid scatterer, equation (10.19): (1) the small-ka scattered pressure is frequency independent and converges to a finite value, specifically twice the geometric acoustic result, equation (10.50), in contrast to the solution for the rigid scatterer, which goes to zero with vanishing frequency as $k^2 a^2$; (2) the $\cos\theta$ term, being of order $k^2 a^2$, is negligible compared with the θ-independent term in the solution for the pressure-release sphere. In contrast, for the rigid sphere, equation (10.19), both terms vary as $k^2 a^2$ and must therefore be retained. In the high-frequency limit, the ratios $|j_n/h_n|$ in equation (11.34) vary as $|\cos ka|$ and are therefore of the same order of magnitude as the radios $|j'_n/h'_n|$ in equation (10.17), which vary as $|\sin ka|$. Hence, in the short-wavelength limit, the pressures scattered by rigid and pressure-release spheres are comparable.

We shall now use the results obtained for the spherical shell to illustrate the reciprocity relation between the incident pressure generated near or on the scatterer by a point source located in the far-field and the pressure radiated back to this point source by a resonant structural mode of the elastic scatterer.

11.7 Structure-Acoustic Medium Reciprocity Relation Illustrated for the Spherical Shell

Consider the pressure p_{rn} radiated by an arbitrary mode of velocity amplitude $\dot{W}_n$. From equation (6.21), the far-field modal pressure is

$$p_{rn}(R, \theta) = \frac{\rho c P_n(\cos\theta)(-i)^n e^{ikR}}{kRh'_n(ka)} \dot{W}_n.$$

It is convenient, at this stage, to express this pressure in terms of the specific acoustic resistance, equation (6.31), and to specialize this result to the back-scattering direction, $\theta = \pi$, by setting $P_n(\cos \theta) = (-1)^n$:

$$p_{rn}(R, \theta = \pi) = \frac{(\rho c r_n)^{1/2} \dot{W}_n i^n \exp(ikR - i\alpha_n)}{R/a}. \tag{11.36}$$

Thus, when the dimensionless frequency is substantially smaller than n, r_n, figure 6.3, and consequently the transfer impedance $p_{rn}/\dot{W}_n$ are small.

Now consider the pressure component that excites this mode, equation (11.27). As already noted this pressure component can also be expressed in terms of the specific acoustic resistance:

$$p_n = (2n + 1) P_i \left(\frac{r_n}{\rho c}\right)^{1/2} \frac{i^{n+1} e^{-i\alpha_n}}{ka}. \tag{11.37}$$

The same dependence on r_n is displayed by this modal forcing pressure, as by the radiated pressure, equation (11.36). Hence, an efficiently radiating mode is also efficiently excited by an incident plane wave, in contrast to the situation of a shell excited by a concentrated force, where the applied modal excitation is independent of frequency [see equation (9.9)]. Consequently, for a shell characterized by realistic structural damping, vibration measurements indicating a marked resonance in a shell excited by an incident wave imply a significant resonant effect in the far-field scattered pressure. In contrast, for a force-excited shell, a resonant shell response does not necessarily manifest itself in the far-field.

The incident pressure amplitude P_i was expressed in terms of the volume acceleration $\ddot{Q}$ of a distant point source in equation (10.57). When this relation is substituted in equation (11.37), one obtains an expression for the modal exciting pressure:

$$p_n = \frac{\rho \ddot{Q} i^{n+1} (2n + 1) (r_n/\rho c)^{1/2} \exp(ikR - i\alpha_n)}{4\pi Rka}.$$

This result can now be combined with (11.36) to form the ratio of the modal exciting and radiated pressures:

$$\frac{|p_n|}{|p_{rn}(R, \pi)|} = \frac{(2n + 1) \ddot{Q}}{4\pi a^2 \ddot{W}_n}. \tag{11.38}$$

The modal exciting force per unit area of scatterer and the pressure radiated by the corresponding mode of the scatterer back to the point source generating the incident plane waves are thus simply related in terms of their respective volume accelerations, independently of frequency. The reciprocity principle can be formulated in terms not restricted to a specific geometry.[7] The principle will be developed further in section 11.11.

11.8 Rayleigh Scattering by Compressible, Movable Spheres

A small sphere generates a scattered pressure proportional to $k^2 V/R$ that consists of a monopole term and a dipole term, equation (10.21), and of terms proportional to higher powers of k, which can be ignored. The same was found to hold for other small scatterers, equation (10.37). These results were derived by Rayleigh for the purpose of proving that the scattering action of small bodies constitutes effectively a high-pass filter. Rayleigh thus explained the blue color of the sky associated with the short-wavelength end of the visible spectrum (1871). He extended his results to compressible, movable spheres. Their effective bulk modulus B_s modifies the monopole term, their density ρ_s the dipole term. The scattered pressure can be expressed in the form of (11.1). Identifying the monopole and dipole terms radiated by the scatterer response by, respectively, the subscripts m and d, the scattered pressure becomes

$$p_{se} = p_{s\infty} + p_{rm} + p_{rd}. \tag{11.39}$$

The radiated components are associated with the scatterer response to the monopole and dipole terms in the applied pressure, equation (10.23). The monopole-radiated pressure is the field of a point source, equation (3.2), where, for our present purpose, the volume acceleration is expressed in terms of the scatterer's periodic volume change δV associated with the incident sound wave:

$$\ddot{Q} = -k^2 c^2 \, \delta V. \tag{11.40}$$

The volume change is computed from the monopole component of the applied pressure, equation (10.23), compared with which the scattered pressure is negligible in the low-frequency range. From the definition of the bulk modulus, equation (2.3), we obtain a low-frequency asymptotic expression corresponding to the quasi-static compression of the scatterer:

$$\delta V = -V P_i/B_s, \qquad \omega^2 \ll \omega_0^2, \tag{11.41}$$

where ω_0 is the breathing mode resonance frequency of the scatterer, e.g., for the air bubble, equation (3.34), for which $B_s = \gamma P_s$. An effective bulk modulus can be readily defined for relatively complex scatterers. Thus, for the spherical shell,[8]

$$B_s = -\frac{P_i}{3 W_{00}/a} = \frac{2Eh}{3(1-v)a}.$$

The monopole field can now be explicitly formulated by combining (3.2), (11.40), and (11.41):

$$p_{rm} = \frac{P_i k^2 V \rho c^2 e^{ikR}}{4\pi R B_s}, \qquad \omega^2 \ll \omega_0^2. \tag{11.42}$$

This quasi-static result is not applicable as $B_s \to 0$, i.e., as $f_0 \to 0$, since the assumption of stiffness-controlled pulsations becomes untenable under these conditions. A solution not thus restricted requires the use of a potentially negative effective dynamic bulk modulus, as illustrated in (3.44).

The radiated dipole term in (11.39) is given in (6.24). For the sake of consistency, this result will be formulated in terms of the scatterer volume:

$$p_{rd} = -\frac{i3\rho k V e^{ikR} \ddot{W}_{01} \cos\theta}{8\pi R}. \tag{11.43}$$

The rigid-body acceleration is computed from the scatterer mass and the entrained mass of water. The latter was found, in (6.33), to equal half the displaced mass of water. Consequently, the translational acceleration is

$$\ddot{W}_{01} = \frac{F}{V(\rho_s + 0.5\rho)}. \tag{11.44}$$

The force associated with the dipole component of the applied pressure, equation (10.23), is

$$F = -3i\pi k a^3 P_i \int_0^\pi \cos^2\theta \sin\theta \, d\theta$$

$$= -i1.5kV P_i.$$

When this is substituted in (11.44), one obtains the scatterer acceleration:

$$\ddot{W}_{01} = -\frac{3iP_ik}{2\rho_s + \rho}.$$

The dipole pressure, equation (11.43), can now be stated explicitly:

$$p_{rd} = -\frac{9P_i\rho k^2 Ve^{ikR}\cos\theta}{8\pi R(2\rho_s + \rho)}. \tag{11.45}$$

The resultant scattered pressure can finally be formulated by combining the two radiated pressure components, equations (11.42) and (11.45), with the corresponding $p_{s\infty}$ components, equation (10.21):

$$p_{se}(R,\theta) = \frac{P_ik^2 Ve^{ikR}}{4\pi R}$$

$$\cdot\left[\left(\frac{\rho c^2}{B_s} - 1\right) + \frac{3(\rho_s - \rho)}{2\rho_s + \rho}\cos\theta\right], \qquad \omega^2 \ll c^2/a^2, \omega_0^2. \tag{11.46}$$

The backscattered pressure is obtained by setting $\cos\theta = -1$. The monopole term vanishes when the scatterer matches the compressibility of the acoustic medium, as does the dipole term for neutrally buoyant scatterers. As B_s and $\rho_s \to \infty$, one retrieves the rigid-sphere result, equation (10.21), but (11.42) being restricted to the stiffness-controlled frequency range of the compressible sphere, the above result cannot be used in the asymptotic, vanishing-B_s limit corresponding to a "pressure-release" sphere, equation (11.35).

This formulation will now be extended to other geometries.

11.9 Extension of the Rayleigh Scattering Formulation to Slender Bodies of Revolution

The analysis parallels the development in the preceding section. Consider first a cylinder, still insonified from $\theta = \pi/2$, $\phi = \pi$. The rigid-body scattered pressure, equation (10.32), has the same k^2V dependence as the sphere, equation (10.21), but the numerical coefficients of the dipole terms differ and there is a directivity factor, $j_0(kL\cos\theta)$, which tends to unity when the length as well as the diameter of the cylinder become small in terms of wavelengths. Now consider the pressure radiated by the cylinder's response to the applied pressure. As usual in the Rayleigh limit, the lead terms of

the applied pressure in the latter of equations (10.28) are the monopole and dipole components. The former is identical to that of the sphere, equation (10.23). The volume acceleration, equation (11.40), and the constitutive relation, equation (11.41), are similarly not confined to the sphere, but are generally applicable to small scatterers. The only difference arises from the directivity of the cylinder, which, even though ka is small, is not restricted as to its length. This introduces a line source directivity factor in the form of (3.18), where $\gamma = 0$ for the assumed normal direction of incidence. Multiplying this directivity factor by (11.42), the monopole term takes the form

$$p_{rm} = \frac{P_i k^2 V e^{ikR} \rho c^2 j_0 (kL \cos \theta)}{4\pi R B_s}. \tag{11.47}$$

Both the radial and the axial responses, W_0 and U_0, must be accounted for in computing the bulk modulus of a cylindrical shell:[8]

$$B_s = -\frac{P_i}{(2W_0/a) + (U_0/2L)}$$

$$= \frac{Eh}{2(1.2 - v)a}.$$

The dipole terms for the cylinder and the sphere differ extensively. For beam aspect incidence, the Fourier transform of the acceleration required in the expression for the $(n = 1)$ term of the radiated pressure, equation (6.55), is given by (6.56). Combining the two equations, the dipole pressure becomes

$$p_{rd} = -\frac{i\ddot{W}_1 \rho k V e^{ikR} j_0 (kL \sin \theta) \sin \theta \cos \phi}{2\pi R}, \qquad k^2 a^2 \ll 1. \tag{11.48}$$

The rigid-body acceleration requires first the evaluation of the force exerted by the sound field. On the cylinder, this force is computed from the dipole component of the surface pressure in the latter of equations (10.28):

$$F = -4iP_i kLa^2 \int_0^{2\pi} \cos^2 \phi \, d\phi \tag{11.49}$$

$$= -2iP_i kV.$$

The translational acceleration is formulated as in (11.44). Integrating the entrained mass per unit area, equation (6.44b) with $n = 1$, the resultant entrained mass,

$$M_1 = -2La^2\rho \int_0^{2\pi} \cos^2\phi\, d\phi$$

$$= \rho V,$$

(11.50)

is found to be twice the value per unit volume calculated for the sphere, equation (6.33). Consequently,

$$\ddot{W}_1 = \frac{F}{V(\rho_s + \rho)}.$$

Combining this result with (11.49), the acceleration becomes

$$\ddot{W}_1 = \frac{-2iP_ik}{\rho_s + \rho}.$$

(11.51)

When this is substituted in (11.48), one obtains an explicit expression for the dipole pressure:

$$p_{rd} = -\frac{\rho P_i k^2 V e^{ikR} j_0(kL\cos\theta)\sin\theta\cos\phi}{\pi(\rho_s + \rho)R}.$$

(11.52)

The scattered field is finally constructed by substituting the two radiated pressure components, equations (11.47) and (11.52), as well as the rigid-body scattered pressure, equation (10.32) in equation (11.39):

$$p_{se}(R,\theta,\phi) = \frac{P_i k^2 V e^{ikR}}{4\pi R}\left[\left(\frac{\rho c^2}{B_s} - 1\right) + \frac{2(\rho_s - \rho)}{\rho_s + \rho}\sin\theta\cos\phi\right] j_0(kL\cos\theta).$$

(11.53)

The backscattered pressure is obtained by setting $\sin\theta$ and j_0 equal to unity, and $\cos\phi = -1$.

The asymptotic approach to rigid *noncylindrical* bodies of revolution developed in section 10.7 is readily extended to account for compressibility and translational response by introducing the monopole and dipole terms

derived in section 6.15. The monopole, i.e., spatially uniform, normal acceleration of the scatterer surface induced by the incident pressure can be expressed in terms of the change in volume per unit length:

$$\ddot{W}_0 = -\frac{\omega^{2\cdot}}{2\pi a}\frac{d(\delta V)}{dz}.$$

The time-dependent change in scatterer volume is formulated in terms of the scatterer's bulk modulus defined in (11.41) and the incident sound pressure:

$$\frac{d(\delta V)}{dz} = -\frac{P_i}{B_s}\frac{dV}{dz}$$

$$= -\frac{P_i \pi a^2}{B_s},$$

where the harmonic time dependence has been suppressed. Combining these equations, the acceleration becomes

$$\ddot{W}_0 = \frac{P_i k^2 c^2 a}{2 B_s}.$$

When the above acceleration is substituted in the $n = 0$ term of (6.75), one constructs the monopole pressure field, effectively the field of a broadside line array. In contrast to the spatially uniform monopole contribution of the cylindrical scatterer, the tapered body is equivalent to an amplitude-shaded array. Expressing the a^2 factor resulting when $\ddot{W}_0$ is substituted in equation (6.75) as $(dV/dz)/\pi$, the monopole pressure becomes

$$p_{\text{rm}} = \frac{P_i k^2 \rho c^2 e^{ikR}}{2\pi R B_s}\int_0^L \frac{dV}{dz}\cos(kz\cos\theta)\,dz. \tag{11.54}$$

We now turn to the dipole term. The expression for the transverse force acting on the scatterer is

$$F = -4ikP_i\int_0^{2\pi}\cos^2\phi\,d\phi\int_0^L a^2\,dz. \tag{11.55}$$

This coincides with the corresponding result for the cylinder, equation (11.49), as does the translational acceleration, equation (11.51). The dipole pressure can now be constructed from the $n = 1$ term in equation (6.75):

$$p_{\mathrm{rd}} = -\frac{2P_{\mathrm{i}}k^2 e^{ikR}}{\pi R} \frac{\rho}{\rho_{\mathrm{s}} + \rho} \sin \theta \cos \phi \int_0^L \frac{dV}{dz} \cos(kz \cos \theta)\, dz. \tag{11.56}$$

Combining (11.54) and (11.56) with the rigid-body scattered pressures, equation (10.35), one finally obtains the Rayleigh pressure generalized to slender bodies of revolution:

$$p_{\mathrm{se}}(R, \theta, \phi) = \frac{P_{\mathrm{i}}k^2 e^{ikR}}{2\pi R} \left[\left(\frac{\rho c^2}{B_{\mathrm{s}}} - 1 \right) + \frac{2(\rho_{\mathrm{s}} - \rho)}{\rho_{\mathrm{s}} + \rho} \sin \theta \cos \phi \right]$$
$$\cdot \int_0^L \frac{dV}{dz} \cos(kz \cos \theta)\, dz. \tag{11.57}$$

For backscattering ($\theta = \pi/2, \phi = \pi$), the integral reduces to $V/2$ and coincides with the corresponding result for the cylinder. The monopole terms of the backscattered pressures are identical to the result obtained for the sphere, equation (11.42). The discrepancy in the dipole terms is associated with the difference in accession to inertia, which equals the displaced mass of fluid for the cylinder, equation (11.50), and one-half this mass for the sphere, equation (6.33). The ratio of entrained to displaced mass of fluid, denoted by γ in (6.86), was tabulated by Lamb[9] for bodies of revolution of varying eccentricity. His results can be approximated in terms of the aspect ratio A, which increases from unity for the sphere to infinity for the cylinder:

$$\gamma \simeq 1 - 0.5A^{-1};$$

the smooth variation of γ suggests a heuristic interpolation for bodies of revolution of moderate eccentricity falling between the sphere and the slender body:

$$p_{\mathrm{se}}(R, \theta, \phi) = \frac{P_{\mathrm{i}}k^2 e^{ikR}}{2\pi R} \left[\left(\frac{\rho c^2}{B_{\mathrm{s}}} - 1 \right) + \frac{(\gamma + 1)(\rho_{\mathrm{s}} - \rho)}{\gamma \rho + \rho_{\mathrm{s}}} \sin \theta \cos \phi \right]$$
$$\cdot \int_0^L \frac{dV}{dz} \cos(kz \cos \theta)\, dz. \tag{11.58}$$

The pressure scattered by the slender body of revolution, viz., the cylinder, equation (11.53), is retrieved by setting $\gamma = 1$, $dV/dz = \pi a^2$, that of the sphere by setting $\gamma = 0.5$, $dV/dz = \pi a^2 \sin^2 \theta$. The result in (11.46) is formally retrieved by rotating the axis by 90°, the θ axis of the sphere coinciding with the direction $\theta = \pi/2$, $\phi = 0$ in cylindrical coordinates. Also, since $k^2 L^2 = k^2 a^2 \ll 1$, $j_0 \simeq 1$. Setting ρ/ρ_s and $\rho c^2/B_s = 0$, one obtains the corresponding results for rigid scatterers. In the backscattering direction, this becomes

$$p_{s\infty}(R, \theta = \pi/2, \phi = \pi) = -\frac{P_i k^2 V e^{ikR}}{4\pi R}(2 + \gamma). \tag{11.59}$$

This reconciles the discrepancy between (10.21) and (10.37).

To conclude the subject of scattering by slender bodies of revolution, we consider cylindrical shells not restricted to the long-wavelength region.

11.10 The Cylindrical Shell as a Scatterer

We shall begin by analyzing an effectively infinite cylindrical shell irradiated by a plane wave incident from the $\phi = \pi$ direction, normally to the axis of symmetry of the shell. This configuration admits a rigorous analysis, in contrast to the finite cylindrical scatterer. It is useful in that it constitutes an approximation to the scattered near-field of a uniform, finite cylindrical shell at ranges smaller than the length of the cylinder. This near-field provides the forcing function of the finite, elastic cylindrical scatterer.

For the infinite cylindrical shell, the analytical procedure parallels the one developed for the spherical shell and will therefore be covered rapidly. Since the shell is uniform and the plane wave normally incident, the system is independent of the axial coordinate. To obtain the radiated pressure component that according to equation (11.1) must be added to $p_{s\infty}$ to construct the resultant scattered field p_{se}, we must first evaluate the response $\dot{w}(\phi)$ of the shell excited by the pressure $(p_i + p_{s\infty})$. This shell response can be written formally as

$$\dot{w}(\phi) = -\sum_{n=0}^{\infty} \frac{p_n \cos n\phi}{Z_n + z_n},$$

where the pressure coefficients are the Fourier coefficients in equation (10.28).

The modal structural impedance Z_n is given in equation (7.92) and the corresponding specific acoustic impedance z_n is obtained by specializing

equations (6.42b) to the z-independent situation, by setting $k_m \equiv 0$. When these impedance equations are combined with the Fourier coefficients of applied pressure, equation (10.28), one obtains the Fourier expansion of the shell velocity. The velocity coefficient multiplied by $-i\omega$ yields the acceleration coefficient $\ddot{W}_{on}$ required to compute the pressure radiated by the elastic scatterer response. From equation (6.41) specialized to $k_m = 0$,

$$p_r(r, \phi) = \frac{2P_i\rho c}{\pi k a} \sum_{n=0}^{\infty} \frac{i^n \varepsilon_n H_n(kr) \cos n\phi}{(Z_n + z_n)[H'_n(ka)]^2}. \tag{11.60}$$

For the sake of brevity, the impedances have not been explcitly stated. To obtain the scattered field, this pressure is combined with the "rigid-body" scattered pressure, equation (10.27):

$$p_{se}(r, \phi) = -P_i \sum_{n=0}^{\infty} \frac{i^n \varepsilon_n H_n(kr)}{H'_n(ka)}$$

$$\cdot \left[J'_n(ka) - \frac{2\rho c}{(Z_n + z_n)\pi ka H'_n(ka)} \right] \cos n\phi. \tag{11.61}$$

Near the structural resonance of a shell mode, $\mathrm{Im}(Z_n + z_n) \approx 0$. This resonance is well coupled to the fluid medium if ka is of the same order as or larger than n, thus ensuring that H'_n is not so large as to make the contribution of the nth term negligible compared with the lower-order terms. For such a resonance, radiation damping typically predominates over structural damping. The sum of the two impedances can therefore be approximated by the radiation resistance, equation (6.46a), with $k_m = 0$. The term in brackets in equation (11.61) reduces to $iY'_n(ka)$.

   Under these circumstances, the effectively coupled resonant term is comparable to the sum of the nonresonant terms. The effect of the abrupt phase change of the resonant terms as frequency sweeps through resonances was described in the previous section and illustrated for the sphere. Figure 11.5 presents experimental and theoretical results illustrating the effect of resonances on the pressure field scattered by a cylindrical shell located vertically in a horizontal fluid layer,[10] thus approximately simulating the two-dimensional situation, analyzed earlier, of an infinite cylindrical shell exposed to a wave incident normally to the cylindrical axis. Both graphs correspond to the same dimensionless frequency ka; but because the materials and wall thicknesses are different, they are associated with different

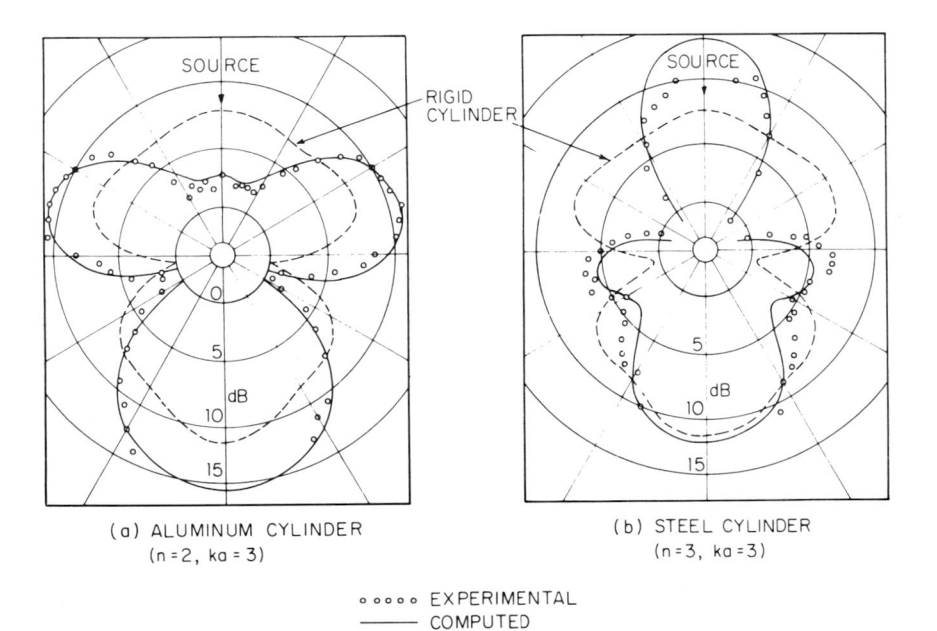

(a) ALUMINUM CYLINDER
(n = 2, ka = 3)

(b) STEEL CYLINDER
(n = 3, ka = 3)

○ ○ ○ ○ ○ EXPERIMENTAL
———— COMPUTED

Figure 11.5
Computed and experimental distributions in angle of pressure scattered by thick-walled elastic cylindrical shells in water, at $ka = 3$: (a) aluminum shell excited at the natural frequency of $(n = 2)$ predominantly radial $(j = 1)$ mode, $a = 0.6$ cm (outer radius), $h = 0.41$ cm, 136 kHz; (b) steel shell excited at the natural frequency of $(n = 3)$ predominantly radial $(j = 1)$ mode, frequency and outer radius as for figure 11.5a, $h = 0.101$ cm. [Reproduced from Hund[10]]

resonant modes. For the $n = 2$ resonance, the respective phases of p_r and $p_{s\infty}$ are such as to reduce the backscattered pressure. In the other case, the backscattered pressure is enhanced by the shell response. The theoretical curves are computed from three-dimensional elasticity theory, which is of course indispensable for the thick-walled shells to which the two graphs in figures 11.5 apply. Qualitatively similar results are obtained for thin shells using the thin-shell theory developed earlier.

The scattered far-field of the rigid cylinder of finite length was constructed in equation (10.30). The pressure p_r radiated by the elastic response of a finite cylindrical shell is computed by means of the approximate technique developed for the radiated field of such shells in sections 9.9 and 9.10. The only difference is the forcing term, which, instead of being a concentrated force F, is now given by the z-independent distributed pressure

$(p_i + p_{s\infty})$. On the shell surface, this pressure can be approximated with the pressure on the infinite cylinder, equation (10.28). This approximation is poor near the ends of the shell; but if we assume a simply supported shell, incapable of motion near the shell ends, the excitation in these two regions is relatively unimportant. In other shell regions, this approximation is valid when the shell is long enough in terms of wavelengths to validate the radiation loading approximations used for the finite shell in chapter 9. When this series expression for $p_i + p_{s\infty}$ is substituted in the surface integral in equation (9.30), the forcing term that takes the place of equation (9.32) is found:

$$f_{mn} = -\frac{4P_i \varepsilon_n i^{n+1} (-1)^m}{\pi^2 ka H_n'(ka) (m + \frac{1}{2})} - \sum_{q=0} \dot{W}_{qn} z_{qmn}. \tag{11.62}$$

The modal amplitudes of the submerged shell are given by equation (9.34), the F term being replaced by the P_i term in equation (11.62). The radiated pressure p_r is then computed from the far-field expressions developed in section 6.12. The procedure need not be repeated here. Finally, the expression for p_r thus obtained is added to the "rigid-body" scattered pressure in equation (10.30) to construct the scattered field p_{se}.

The $H_n'(ka)$ term in the denominator of equation (11.62) can be displayed in the form used for the spherical shell [see equation (11.37) and table 11.1] as $r_n^{1/2}$. Specifically for small dimensionless natural frequencies ($ka \ll n$), this forcing term varies as $(ka)^n$. Thus, it is verified again that the contribution of poorly radiating resonances is relatively less important for elastic scatterers than for point-excited shells, whose forcing term does not depend on their sound-radiating properties. Consequently, the small natural frequencies of the lower, predominantly flexural, branch of resonances of these submerged cylindrical shells reduce their relative importance and enhance that of the predominantly circumferential, membrane-type, resonances, whose dimensionless natural frequencies are of order n, i.e., relatively large. For submerged cylindrical shells acting as scatterers, in contrast both to the point-excited cylindrical shells analyzed in chapter 9 and to the spherical scatterers discussed earlier in this chapter, the scattering action of flexural resonances is typically overshadowed by membrane-type resonances. This statement does not apply to the relatively thick-walled cylindrical shells relevant to figure 11.5, nor to ideally lossless shells, nor to shells vibrating in a low-sound velocity medium (e.g., the atmosphere), for which a given natural frequency corresponds to a larger ka than in water.

11.11 Sound Sources Located on an Elastic Baffle

The effect of baffle elasticity on the far-field of a sound source will be used to illustrate the use of Rayleigh's reciprocity principle. We first verify that principle by computing the far-field of a source located on a planar elastic baffle directly and reciprocally.

11.11.1 SOUND SOURCE LOCATED ON A PLANAR ELASTIC BAFFLE

Sound radiation by a circular piston located on an infinite rigid baffle is one of the classical problems of acoustics. The resulting far-field pressure, identified here by the subscript ∞ to indicate a baffle of infinite impedance, was derived in equation (5.10). In practice, a piston vibrating as a plane (that is, as an effectively rigid surface) can be achieved by means of a circular array of magnetostrictive or piezoelectric elements. Uniform motion is ensured if electrical excitation of individual elements has been adjusted as to amplitude and phase to compensate for surface pressure fluctuations over the array surface, which result in differences in radiation loading between array elements. The physical baffle, excluding the case of a plane of symmetry simulating a rigid baffle, is necessarily of finite impedance and is capable of propagating flexural waves. The baffle therefore responds as an elastic plate to the sound field generated by the array.

The circular array can be idealized as a massless, distributed spring of negligible thickness resting on an infinite elastic baffle. With this mathematical model, the displacement distribution can be stated in cylindrical coordinates as the superposition of the prescribed motion W of the piston and of the forced response w_b of the baffle:

$$w(r) = w_b(r) + WH(a - r). \tag{11.63}$$

Here, W is the amplitude of motion of the piston, a is the piston radius, and H the Heaviside unit function defined by equation (5.2).

The as yet unknown response of the elastic baffle is $w_b(r)$. Using the techniques developed in section 5.3, we require the Hankel transform, equation (5.3a), of the displacement distribution stated in equation (11.63):

$$\tilde{w}(\gamma) = \tilde{w}_b(\gamma) + Wa\frac{J_1(\gamma a)}{\gamma}. \tag{11.64}$$

The transform of the corresponding surface pressure is obtained from equation (5.15) specialized to $z = 0$, that is, to the plane of the piston:

$$\tilde{p}(\gamma,0) = -\frac{i\rho c^2 k^2}{(k^2 - \gamma^2)^{1/2}}\left[\tilde{w}_b(\gamma) + Wa\frac{\mathcal{J}_1(\gamma a)}{\gamma}\right]. \tag{11.65}$$

The Hankel transform of the equation of motion of the plate can now be constructed by substituting equation (11.65) in equation (8.14) and setting the concentrated force $F = 0$. When $\tilde{w}(\gamma)$ is separated into its known and unknown components, the equation of motion becomes

$$(\gamma^4 - k_f^4)\tilde{w}_b(\gamma) = \frac{k_f^4(\rho/\rho_s)}{(\gamma^2 - k^2)^{1/2}h}\left[\tilde{w}_b(\gamma) + \frac{Wa\mathcal{J}_1(\gamma a)}{\gamma}\right].$$

This can be solved for the transform $\tilde{w}_b$ of the baffle response:

$$\tilde{w}_b(\gamma) = -Wa\mathcal{J}_1(\gamma a)\left\{\gamma + \left(\frac{\rho_s}{\rho}\right)h(\gamma^2 - k^2)^{1/2}\gamma\left[1 - \left(\frac{\gamma}{k_f}\right)^4\right]\right\}^{-1}. \tag{11.66}$$

The stationary-phase approximation to the far-field pressure, equation (5.22), requires the evaluation of $\tilde{w}_b$ at $\overline{\gamma} = k\sin\theta$. Using the ratio $(\omega/\omega_c)^{1/2}$ again to express the ratio k/k_f, the stationary-phase value of $\tilde{w}_b$ becomes

$$\tilde{w}_b(k\sin\theta) = -\frac{Wa\mathcal{J}_1(ka\sin\theta)}{k\sin\theta}\left\{1 - ikh\cos\theta\left(\frac{\rho_s}{\rho}\right)\left[1 - \left(\frac{\omega}{\omega_c}\right)^2\sin^4\theta\right]\right\}^{-1}. \tag{11.67}$$

The pressure field is now obtained by substituting this result in equation (5.22):

$$p(R,\theta) = -\rho c^2 kaW\frac{e^{ikR}}{R}\left\{1 - \left[1 - ikh\cos\theta\left(\frac{\rho_s}{\rho}\right)\left(1 - \frac{\omega^2}{\omega_c^2}\sin^4\theta\right)\right]^{-1}\right\}$$
$$\cdot\left\{\frac{\mathcal{J}_1(ka\sin\theta)}{\sin\theta}\right\}. \tag{11.68}$$

The ratio of this pressure to the pressure p_∞ radiated in the absence of any baffle response, equation (5.10a), is

$$\frac{p(R,\theta)}{p(R,\theta)_\infty} = \frac{-ikh(\rho_s/\rho)\cos\theta[1 - (\omega/\omega_c)^2\sin^4\theta]}{1 - ikh(\rho_s/\rho)\cos\theta[1 - (\omega/\omega_c)^2\sin^4\theta]}. \tag{11.69}$$

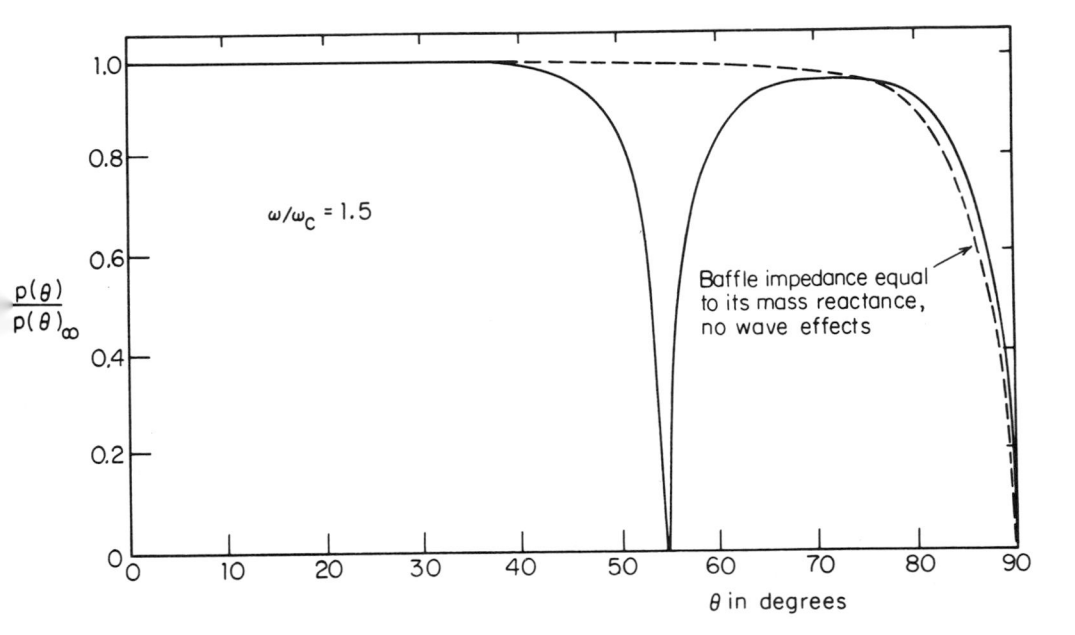

Figure 11.6
Far-field of circular piston set in an elastic, effectively infinite baffle, (11.69). [Reproduced from Feit[12]]

The beam-forming ability deteriorates near grazing as $\theta \to \pi/2$. This particular conclusion can also be reached by means of the simpler mathematical model of a baffle characterized by a finite impedance but incapable of flexural wave propagation.[11] If, as in the present analysis, elastic waves are taken into account, it is found that the pattern also deteriorates in the vicinity of the conical surface defined by the coincidence angle, equation (8.33) (see figure 11.6).

Structural damping limits the pressure drop on the coincidence cone. The structural loss factor is conveniently introduced in terms of the complex coincidence frequency $\omega_c(1 + \frac{1}{2}i\eta_s)$ already used in the derivation of equation (8.37). The pressure at the coincidence angle now becomes

$$\frac{p(R, \theta_c)}{p(R, \theta_c)_\infty} = \frac{\eta_s[1 - (\omega_c/\omega)]^{1/2}}{1 + kh(\rho_s/\rho)[1 - (\omega_c/\omega)]^{1/2}\eta_s}. \tag{11.70}$$

In the absence of structural damping, the pressure radiated along the coincidence cone is, of course, zero.

For the purpose of illustrating the reciprocity principle, it is convenient to reduce the piston to a point source of volume acceleration amplitude $\ddot{Q} = \pi a^2 \omega^2 W$. The pressure radiated by this point source is concisely expressed in terms of the free-field pressure $p_s(R)$ radiated in the absence of the baffle, equation (3.2). Specializing the result derived above for the baffled source, equation (11.68), to $k^2 a^2 \ll 1$ by using the small-argument approximation $\mathcal{J}_1(x) \simeq x/2$, the far-field of the point source located on an elastic baffle now becomes

$$p(R,\theta) = 2p_s(R)\left\{1 + \left[ikh\cos\theta\left(\frac{\rho_s}{\rho}\right)\left(1 - \frac{\omega^2}{\omega_c^2}\sin^4\theta\right) - 1\right]^{-1}\right\}. \qquad (11.71)$$

The factor multiplying p_s embodies the effect of the elastic baffle on the far-field.

Turning to the reciprocal problem, we now consider the resultant pressure on an elastic plate insonified by a distant sound source. This pressure constitutes the right side of equation (11.6):

$$p(x,0) = 2P_i\exp(ikx\sin\theta) + p_r(x,0), \qquad (11.72)$$

where p_r is given in equation (11.8). Setting P_i equal to the pressure $|p_s(R)|$ generated in the plane of the baffle by a point source of strength $\ddot{Q}$ located at the field point R, θ equation (11.72) becomes identical to equation (11.71).

The reciprocity principle, formulated for arbitrary baffle geometries,[7] states that *the pressure generated in the fluid medium at a field point* **R** *by a source of strength* $\ddot{Q}$ *located at* **R₀** *on an elastic structure* [equation (11.71)] *equals the pressure generated on that structure, at point* **R₀**, *by this same source located in the fluid medium at field point* **R** [equation (11.72)], *with* $P_i = |p_s|$ *and* p_r *given by* equation (11.8). This is illustrated in figure 11.7.

11.11.2 SOURCES ON ELASTIC SPHERICAL AND CYLINDRICAL BAFFLES

It is generally less laborious to solve the reciprocal problem, viz., to evaluate the resultant pressure on an elastic scatterer insonified by a distant sound source, than to solve the direct problem when the sound source is located on an elastic baffle. We shall therefore use the results obtained earlier for elastic scatterers in the form of spherical and cylindrical shells to formulate, by virtue of the reciprocity principle, the far-field of sources located on elastic spherical and cylindrical baffles.

Figure 11.7
By virtue of Rayleigh's reciprocity principle, the pressure remains invariant when source and field points are interchanged.

To start, let us compute in this manner the pressure radiated by a source located on a spherical shell at $(R = a, \theta = \pi)$ to a field point at $(R, \theta = \pi)$. By virtue of the reciprocity principle, this pressure equals the pressure at (a, θ) when the shell is insonified by this source located in the far-field at $(R, \theta = \pi)$. The latter pressure is the sum of the incident and rigid-body scattered pressures $p_i + p_{s\infty}$, equation (10.23), with $P_n(\pi) = (-1)^n$, and of the radiation loading p_r. That pressure component is obtained by multiplying each term in the velocity series $\dot{w}$, equation (11.28), by the corresponding modal impedances z_n, equation (6.29):

$$p_r(R = a, \theta = \pi) = -\frac{iP_i}{k^2 a^2} \sum_n \frac{(-i)^n (2n + 1)}{h_n'(ka)} \frac{z_n}{Z_n + z_n}. \qquad (11.73)$$

Comparing this result with the surface pressure $p_i + p_{s\infty}$ on a rigid sphere, equation (10.23) evaluated at $\theta = \pi$, we note that the two quantities differ only by the factor $-z_n(Z_n + z_n)^{-1}$. Adding these two pressure components, one obtains the resultant pressure on the elastic spherical shell:

$$p(a, \pi) = P_i + p_{s\infty}(a, \pi) + p_r(a, \pi)$$

$$= -\frac{iP_i}{k^2 a^2} \sum_n \frac{(-i)^n (2n + 1)}{h_n'(ka)} \frac{Z_n}{Z_n + z_n}. \qquad (11.74)$$

P_i is the pressure $|p_s(R)|$ radiated to range R by the point source in the absence of any structure. Since, by virtue of the reciprocity principle, the above pressure equals the pressure radiated to the field point (R, π) by the point source located at (a, π), the factor multiplying P_i in equation (11.74) represents the modification of the pressure field $p_s(R)$ by the elastic baffle. In the low-frequency limit, only the monopole $n = 0$ term need be retained. Since in this small-ka limit z_0 tends to zero linearly with increasing wavelength, while the shell impedance Z_0, equation (7.121), diverges as ω^{-1}, the impedance ratio tends to unity as $\omega \to 0$. Substituting equation (6.20) for h'_0 specialized to $k^2 a^2 \ll 1$, one verifies that the pressure ratio in equation (11.74) tends to unity, as anticipated when the elastic scatterer or baffle is small in terms of acoustic wavelengths.

A similar procedure applies to the elastic cylindrical baffle. Adding the combined incident and rigid-body scattered pressures, equation (10.28), and the radiated pressure, equation (11.60), specialized to $r = a$, $\phi = \pi$, one obtains the resultant pressure on the insonified shell:

$$p(a, \pi) = \frac{2iP_i}{\pi ka} \sum_n \frac{\varepsilon_n (-i)^n Z_n}{(Z_n + z_n) H'_n(ka)}. \tag{11.75}$$

If structural damping makes a negligible contribution to Z_n, the nth term in both equations (11.74) and (11.75) goes to zero at the corresponding in vacuo natural frequencies. At these resonances, a single one for $n = 0$ and two for $n > 0$, the shell cannot support the pressure associated with the corresponding wave harmonic, and the nth term drops out of the wave-harmonic series.

To conclude this chapter, we consider the interaction of an interior acoustic fluid and of a cylindrical shell.

11.12 Sound Propagation in Fluid-Filled Elastic Waveguides

11.12.1 DISPERSON RELATIONS FOR CYLINDRICAL WAVEGUIDES

In these last three sections, wave propagation in a pipe, studied by elementary methods in section 2.9, will be analyzed with the techniques and insight developed in the preceding chapters. When we first considered sound propagation in pipes, our formulation was deliberately restricted to nearly plane waves, a situation incompatible with small wall impedances associated with substantial radial fluid particle motion. The quasi-plane wave assumption also excludes radial standing waves anticipated when the pipe diameter is commensurate with wavelength. To eliminate these restrictions, the solution

of the Helmholtz equation is formulated here in terms of cylinder functions. In fact, the earlier results will be found to be asymptotic forms of the wave-harmonic solution developed here.

In contrast to the solutions in chapter 6, the sound field formulated here encompasses the cylindrical axis. Bessel functions of the second kind are therefore excluded because they diverge for vanishing argument. Consequently, one anticipates pressure fields in the form of modal series whose radial configuration is given by Bessel functions of the first kind, and whose axial and circumferential configuration is in the form of traveling or standing waves. Selecting axial traveling waves and circumferential standing waves, the modal series becomes

$$p(r, z, \phi) = \sum_{n, m} P_{nm} \mathcal{J}_n(\alpha_{nm} r/a) \exp(\pm i\gamma_{nm} z) \cos n\phi. \tag{11.76}$$

The radial wave numbers are the roots of the transcendental equation constructed by substituting individual terms of the n series in the boundary condition:

$$x_w/\rho c k a = \mathcal{J}_n(\alpha)/\alpha \mathcal{J}_n'(\alpha) \equiv F_n(\alpha). \tag{11.77}$$

The function $F_n(\alpha)$ is plotted in figure 11.8 for axisymmetric modes, $n = 0$. The terms in (11.76) satisfy the Helmholtz equation provided axial wave numbers γ_{nm} are related to the eigennumbers α_{nm} as

$$\gamma_{nm} = [k^2 - (\alpha_{nm}/a)^2]^{1/2}. \tag{11.78}$$

This relation, which has already been encountered in (2.62), is consistent with the radial wave numbers of the cylinder functions in (6.39), where γ_{nm} now takes the place of k_m. When $k < \alpha_{nm}/a$, γ_{nm} is imaginary. It therefore represents an exponentially decaying mode in (11.76). Such a mode is associated with a propagating pressure in the frequency range above its *cutoff frequency*:

$$f_{nm} = c\alpha_{nm}/2\pi a, \qquad \gamma_{nm} = 0. \tag{11.79}$$

The phase velocity of the propagating wave is

$$\begin{aligned} c_{nm} &= \omega/\gamma_{nm} \\ &= c[1 - (\alpha_{nm}/ka)^2]^{-1/2}. \end{aligned} \tag{11.80}$$

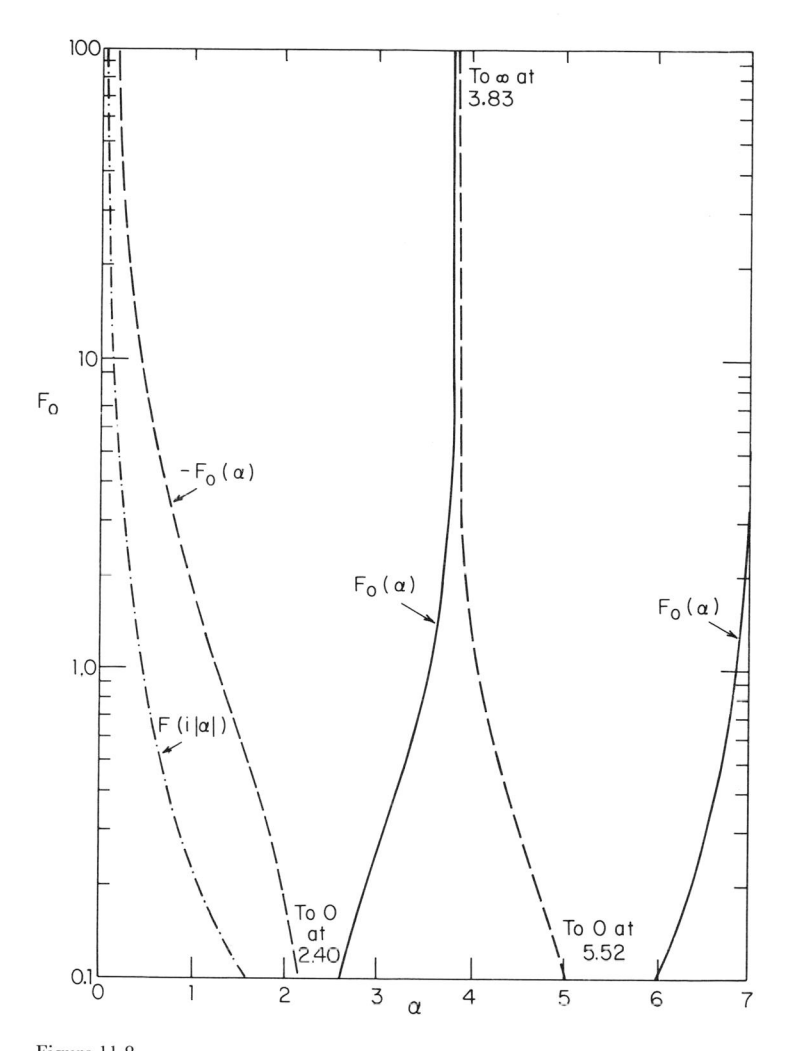

Figure 11.8
Plot of the normalized reactance of a cylindrical fluid column, equation (11.77).

From infinity at cutoff, the phase velocity drops asymptotically to the sound velocity at high frequencies. For an axisymmetric pressure field in a rigid pipe, the eigenvalues of (11.79) are the roots of $\mathcal{J}_0'$, the lowest of which is $\alpha_{00} = 0$. This root is associated with a vanishing cutoff frequency, a wave number $\gamma_{00} = k$, and a nondispersive phase velocity equal to the sound velocity. Noting that $\mathcal{J}_0(0) = 1$, one retrieves the plane wave $P_{00} \exp(\pm ikz)$ as the fundamental term in (11.76).

The effect of a finite wall reactance will now be explored for axisymmetric waves. The characteristic equation, (11.77), becomes

$$x_w/\rho cka = F_0(\alpha). \tag{11.81}$$

In the small-argument limit,

$$F_0(\alpha) \simeq -2/\alpha^2, \qquad \alpha^2 \ll 1. \tag{11.82}$$

The asymptotic small-α form of F_0 corresponds to large, massive wall reactances. Combining (2.65), (11.81) and (11.82), one solves for

$$\alpha = (2\rho a/\rho_s h)^{1/2}. \tag{11.83}$$

When this is substituted in (11.78), one retrieves the elementary solution, (2.67).

Now consider a large stiffness-controlled, i.e., positive, reactance, equation (2.63). Since $F_0(\alpha) < 0$ for small α, none of the terms in (11.76) is capable of generating the Korteweg-Lamb result, (2.58a). This series therefore does not form a complete set when $x_w < 0$. More generally, not limiting ourselves to axisymmetric modes, we note that terms associated with imaginary eigenvalues $\alpha_{n0} = i|\alpha_{n0}|$ must be added[13] when $n - F_n^{-1} < 0$. The inclusion of imaginary radial eigenvalues gives rise to modes whose radial configuration is described by modified Bessel functions:[14,15]

$$I_n(|\alpha|) \equiv i^{-n}\mathcal{J}_n(i|\alpha|). \tag{11.84}$$

The corresponding axial wave number, equation (11.78),

$$\gamma_{n0} = [k^2 + (|\alpha_{n0}|/a)^2]^{1/2}, \tag{11.85}$$

is always real, thereby describing modes that propagate even at the lowest

frequencies. The phase velocity,

$$c_{n0} = c[1 + (|\alpha_{n0}|/ka)^2]^{-1/2}, \tag{11.86}$$

is *smaller* than the sound velocity to which it tends asymptotically at high frequencies. The Bessel function ratio F_n in the characteristic equation is now positive at all frequencies, and therefore matched everywhere to a stiffnesslike wall reactance (figure 11.8). For axisymmetric modes,

$$F_0(i|\alpha|) = I_0(|\alpha|)/|\alpha|I_1(|\alpha|)$$

$$\simeq 2|\alpha|^{-2}, \qquad |\alpha|^2 \ll 1 \tag{11.87}$$

$$\simeq |\alpha|^{-1}, \qquad |\alpha| \gg 1.$$

These modes do not display phase reversals with increasing r, no matter how large $|\alpha_{n0}|$. When the small-α form is used in (11.81) and set equal to the stiffnesslike reactance, equation (2.63), one can solve for the eigenvalue:

$$|\alpha_{00}|^2 = 2k^2a^3\rho c/h\rho c_p.$$

Substituted in (11.86), this does indeed match the Korteweg-Lamb result, equation (2.58a).

11.12.2 ELASTIC CYLINDRICAL HOSES AND SHELLS AS WAVEGUIDES

To start, consider a hose, defined here as a shell devoid of flexural rigidity. Combining (2.63), (2.65), and (2.66), a frequency-dependent normalized wall reactance is constructed:

$$\frac{x_w}{\rho cka} = \frac{\rho_s c_p^2 h}{\rho c^2 k^2 a^3}\left[1 - \left(\frac{kac}{c_p}\right)^2\right]$$

$$\simeq -\frac{\rho_s h}{\rho a}, \qquad k^2a^2 \gg c_p^2/c^2. \tag{11.88}$$

As anticipated, this reactance vanishes at the frequency in (7.95c) where the flexural β^2 term has been dropped since we are dealing here with a hose. It has already been pointed out that the admittances of the fluid in the pipe and of the pipe wall form, electrically speaking, a shunt circuit. When the pipe wall admittance is formulated, as in (11.88), in terms of the

parameters of a simple oscillator, the waveguide, like the bubble swarm in section 3.10, is represented by the circuit in figure 3.11, where the axisymmetric mode of the pipe wall is now equivalent to the breathing mode of the gas bubbles. One can therefore anticipate a dead zone, at least for the quasi-planar mode, when the masslike pipe wall admittance short-circuits the fluid compressibility to produce a resultant negative admittance incompatible with wave propagation. The difference with the bubble swarm situation is the existence of higher-order modes characterized by nodal circles. Propagation of these modes is compatible with a negative wall reactance in the frequency range above their respective cutoff frequencies. The existence of a dead zone therefore depends on the relative locations of the breathing mode resonance [(7.95c) with $\beta = 0$], where propagation of the I_0 mode ceases, and of the cutoff frequency of the first nonplanar mode, above which a J_0 type mode can propagate. Specifically, a dead zone exists if the normalized natural frequency $k_0 a = c_p/c$, above which the reactive impedance, equation (11.88), is negative (thus excluding the existence of a quasi-planar I_0 type mode) falls below the fundamental pressure-release eigennumber, $\alpha = 2.40$ (see figure 11.8). Consider, for example, a hose whose normalized resonance frequency is of order $ka = 0.1$, a situation typical of rubberlike materials. In the mass-controlled range $ka > 0.1$, the normalized reactance equation (11.88), may be assigned a numerical value of -0.5. The boundary condition, equation (11.77), indicates that this normalized reactance must equal $F_0(\alpha)$. Referring again to figure 11.8; the requirement $F_0 = -0.5$ is seen to call for the eigenvalue $\alpha_{01} = 1.6$. By virtue of (11.78) this is the value of ka associated with the cutoff frequency that marks the upper limit of a dead zone extending down to the breathing mode resonance $ka = 0.1$.

In the definition used here, a hose is capable of sustaining membrane stresses only. A pipe properly modeled as a cylindrical shell displays flexural rigidity embodies in the $\beta^2 k_m^4 a^4$ term of the frequency determinant in (7.95), where, in the notation of (11.76), γ_{0m} takes the place of the axial wave number k_m. An elementary form of the shell's wall reactance can be constructed by modifying (11.88) to account for flexural effects, viz., the β term in (7.95c):

$$\frac{x_w}{\rho c k a} = \frac{\rho_s c_p^2 h}{\rho c^2 k^2 a^3}\left[1 - \left(\frac{kac}{c_p}\right)^2 + \beta^2 \gamma_{0m}^4 a^4\right]. \tag{11.89}$$

It is recalled, from the discussion of the hose dispersion curve, that as

frequency rises to the natural frequency, thus causing x_w and hence $F_0(i|\alpha|)$ to tend to zero, $|\alpha_{00}|$ tends to infinity by virtue of (11.87), as does γ_{00} in (11.85). Consequently, when this resonance condition is approached, the flexural term in (11.89) grows rapidly, thus causing the pipe wall reactance to remain stiffness-controlled, and hence compatible with the existence of a quasi-planar $I_0(|\alpha_{00}|)$ mode.

Structural damping has been ignored throughout. When accounted for, wall impedances and hence the functions $F_n(\alpha)$ become complex, as do both radial and axial wave numbers. The latter represent modes that are attenuated even though capable of propagation. A similar situation arises when the waveguide is provided with an absorptive liner.

11.12.3 MODAL AMPLITUDES AND IMPEDANCE IN WAVEGUIDES

To conclude, the computation of the modal amplitudes will be illustrated. The coefficients in (11.65) are a function of the pipe wall impedance and of the axial fluid particle velocity distribution imposed by the source located in the cross section $z = 0$. This velocity is expanded in a Fourier-Bessel series:[13]

$$\dot{u}(r) = \sum_{n=0} [\dot{U}_{n0} I_n(|\alpha_{n0}|r/a) + \sum_{m=1} \dot{U}_{nm} J_n(|\alpha_{nm}|r/a)]\cos n\phi. \tag{11.90}$$

It is recalled that $\dot{U}_{n0} = 0$ when $x_w \leq 0$. Expressing the axial particle velocity in terms of $\partial p/\partial z$, as in (2.24), one can relate the modal velocity coefficients in (11.90) to the pressure coefficients in (11.76):

$$P_{nm} = \frac{\rho ck \dot{U}_{nm}}{\gamma_{nm}}. \tag{11.91}$$

The orthogonality relations of Bessel functions yield the coefficients in the Fourier-Bessel series:[13]

$$\dot{U}_{n0} = \frac{2(|\alpha_{n0}|/a)^2 \int_0^a r\dot{u}(r) I_n(|\alpha_{n0}|r/a)\, dr}{(|\alpha_{n0}|^2 + n^2) I_n^2(|\alpha_{n0}|) - |\alpha_{n0}|^2 I_n'^2(|\alpha_{n0}|)},$$

$$\dot{U}_{nm} = \frac{2(\alpha_{nm}/a)^2 \int_0^a r\dot{u}(r) J_n(\alpha_{nm}r/a)\, dr}{(\alpha_{nm}^2 - n^2) J_n^2(\alpha_{nm}) + \alpha_{nm}^2 J_n'^2(\alpha_{nm})}, \quad m > 0,$$

$$\tag{11.92}$$

where α_{nm} are eigennumbers satisfying (11.78).

Table 11.2
Wave numbers and Fourier-Bessel series coefficients for a parabolic axial velocity profile in a pressure-release lined pipe

m	α_{om}	$\mathcal{J}'_{om}$	$\dot{U}_{om}/\dot{U}$	γ_{om}/k		$P_{om}/\rho c \dot{U}$	
				$ka = 5$	$ka = 10$	$ka = 5$	$ka = 10$
1	2.40	−0.519	1.12	0.877	0.971	1.29	1.15
2	5.52	0.340	−0.140	$i0.468$	0.695	$i0.299$	−0.209
3	8.66	−0.272	0.0453	$i1.41$	0.500	$-i0.032$	0.091

This procedure for computing modal amplitudes will be illustrated for a pipe lined with a "pressure-release" material, whereby x_{w}, $F_n(\alpha)$, and $\mathcal{J}_n(\alpha_{nm}) = 0$. The sound source is a piston imposing an axisymmetric parabolic axial velocity profile:

$$\dot{u}(r) = \dot{U}(a^2 - r^2)/a^2. \tag{11.93}$$

Equation (11.92) now reduces to

$$\dot{U}_{0m} = \frac{2}{a^2 \mathcal{J}_0'^2(\alpha_{0m})} \int_0^a r\dot{u}(r)\,\mathcal{J}_0\left(\frac{\alpha_{0m}r}{a}\right) dr. \tag{11.94}$$

The three lowest eigenvalues[16] and eigenfunctions are tabulated in table 11.2. The subscript indicates the number of nodal circles, the outermost of which always coincides with the liner-fluid interface. When (11.93) is substituted in equation (11.94), the integral can be evaluated analytically:[17]

$$\dot{U}_{0m} = 8\dot{U}/\mathcal{J}_0(\alpha_{0m})\alpha_{0m}^3. \tag{11.95}$$

This is a rapidly converging series. The modal axial velocity coefficients are frequency independent because the wall reactance was selected constant. The axial wavenumbers, obtained from (11.78), and the pressure coefficients computed from (11.91), are frequency dependent even though the wall reactance is not. They are tabulated in table 11.2 for two particular values of ka.

Some practical conclusions can be drawn from these results. Even in a lined pipe that approximates a frequency-independent wall reactance, the dynamic configuration of the source must be readjusted for every test frequency if constant modal pressure amplitudes are desired. For a purely

reactive wall impedance, the modal impedances $P_{0m}/\dot{U}$ loading the source are resistive above the cutoff frequency and reactive below, never complex (provided structural damping can be ignored). Finally, at cutoff, the axial wave number vanishes [equation (11.78)], thus resulting in an infinite modal impedance [equation (11.91)]. In fact the cutoff frequency corresponds to the two-dimensional z-independent radial resonance of the fluid column. The entire semiinfinite waveguide length is thereby excited uniformly with a pressure limited only by nonlinearities or structural damping. Ignoring these effects, the infinite modal impedance predicted by (11.91) at cutoff makes the assumed mathematical model of the piston as a constant-velocity, i.e., an infinite-internal impedance source, unrealistic. In practice, therefore, a mode cannot be driven at its cutoff frequency.

References

1. E. F. F. Chladni, *Neue Beiträge zur Akustik* (Leipzig: Breitkopf und Härtel, 1817), pp. 86–87.

2. P. M. Morse, *Vibration and Sound*, 2nd ed. (New York: McGraw-Hill, 1948), pp. 383–384.

3. R. Josse and C. Lamure, "Sound Transmission by a Single Leaf Partition" (in French), *Acustica* 14:226–280 (1964).

4. U. Ingard, "On the Theory and Design of Acoustic Resonators," *J. Acoust. Soc. Am.* 25:1037–1061 (1953), section II.

5. R. Hickling, "Analysis of Echoes from a Solid Elastic Sphere in Water," *J. Acoust. Soc. Am.* 34:1582–1592 (1962), figure 2a; also K. J. Diercks and R. Hickling, "Echoes from Hollow Aluminum Spheres in Water," *J. Acoust. Soc. Am.* 41:380–393 (1967), figure 3a.

6. G. Gaunaurd and H. Überall, "Resonances in Acoustic and Elastic Wave Scattering," in V. K. Varadan and V. Varadan, eds., *Recent Developments in Classical Wave Scattering (Proceedings of a Symposium)* (New York: Pergamon Press, 1980), pp. 413–430.

7. L. M. Lyamshev, "A Method for Solving the Problem of Sound Radiation by Thin Elastic Plates and Shells," *Sov. Physics—Acoustics* 5:122–123 (1959), and L. M. Lyamshev, "Theory of Sound Radiation by Thin Elastic Plates on Shells," *Sov. Physics—Acoustics* 5:431 (1959). For a more general review, see Y. I. Belousov and A. V. Rimskii-Korsakov, "The Reciprocity Principle in Acoustics and Its Application to the Calculation of Sound Fields of Bodies," *Sov. Physics—Acoustics* 21:103–109 (1975).

8. The static response of spherical and cylindrical shells is given by P. J. Roark, *Formulas for Stress and Strain*, 4th ed. (New York: McGraw-Hill, 1965), pp. 298–299.

9. H. Lamb, *Hydrodynamics*, 6th ed. (New York: Dover, 1945), p. 155, in table where k_2 is our γ.

10. M. Hund, "Scattering of Water-Borne Sound by Flexurally Vibrating Cylindrical Shells" (in German), *Acustica* 15:88–97 (1965), figures 7 and 10.

11. P. M. Morse and K. U. Ingard, *Theoretical Acoustics* (New York; McGraw-Hill, 1968), p. 381.

12. D. Feit, "Sound Radiation from a Circular Array in an Elastic Baffle," *J. Acoust. Soc. Am.* 41:1366–1367 (1967).

13. G. N. Watson, *A Treatise on the Theory of Bessel Functions* (Cambridge: Cambridge University Press, 1966), p. 597, where $H = -F_n^{-1}$.

14. Ibid., p. 77.

15. M. Abramowitz and I. A. Stegun, eds., *Handbook of Mathematical Functions* (Washington, D.C.: NBS, Supt. of Documents, 1964), p. 375.

16. Ibid., p. 409.

17. I. N. Sneddon, *Special Functions of Mathematical Physics and Chemistry* (New York: Interscience Publishers, 1956), p. 111, equation (32.5).

High-Frequency Formulation
of Acoustic and Structural
Vibration Problems

12.1 Watson's Creeping Wave Formulation of the Diffracted Field

Wave-harmonic series, being effectively an expansion in positive powers of ka, converge slowly for large ka, particularly when a small source is located on a large baffle. Unfortunately, this is precisely the situation relevant to large spherical and cylindrical sonar arrays. Their design requires the evaluation of transfer impedances between elements, i.e., of the diffracted pressure acting on the array surface when a single element is active. The slow convergence of the series is not only inconvenient but also indicates that individual terms do not correspond to a physical reality. The approach about to be described circumvents both drawbacks. It was originated by Watson (1918) to predict the propagation loss of radio waves across the Atlantic.[1] Clearly the ray-acoustic formulation, which is commonly used to explore the diffusion of sound in a large hall, fails in this case, because it predicts that, in the absence of reflections and refraction, no signal can be detected beyond the horizon, i.e., in the "shadow zone," where the diffracted field prevails. Watson's solution expressed the diffracted field in terms of a rapidly converging series of exponentially decaying "creeping waves." This formulation has been used by workers in the field of electromagnetic wave propagation not only for point sources on spheres[2] but for line sources on cylinders.[3] In the area of acoustics, creeping waves are experimentally verifiable[4] physical phenomena. Watson's technique has been applied to various high-frequency acoustic problems, e.g., the scattering action of elastic bodies[5] and layered shells,[6] the response of point-excited submerged shells,[7] and the diffracted field generated by point sources on rigid cylindrical and spherical baffles.[8] It is this latter application that will be used to introduce Watson's technique in the next section. We then proceed to the high-frequency formulation of the response of point-excited spherical shells both in vacuo and submerged (sections 12.3–12.5). These techniques are extended to point and line sources on a rigid cylindrical baffle (section 12.6) and to point-excited cylindrical shells (section 12.7). The final two sections present the high-frequency formulation of the far-field radiated by point-excited spherical and cylindrical shells.

12.2 Point Source on a Rigid Spherical Baffle

The surface pressure generated by a small piston centered at $\theta = 0$, on a sphere of radius a is obtained by setting $R = a$ in equation (6.19), the modal acceleration amplitudes being given by equation (6.25b):

$$p(a,\theta) = \frac{-\rho \ddot{Q}}{2\pi ka^2} \sum_{n=0}^{\infty} (n + \tfrac{1}{2}) P_n(\cos \theta) \frac{h_n(ka)}{h'_n(ka)}. \tag{12.1}$$

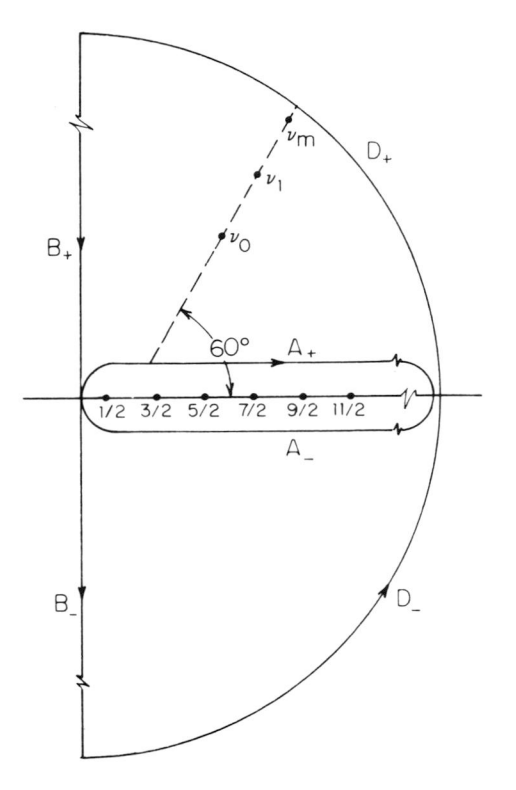

Figure 12.1
Contour integrals used in the Watson transformation of the spherical wave-harmonic series in the complex v plane.

This series can be expressed as a contour integral around the positive real axis in the plane defined by a complex variable v whose real axis contains the index numbers n in the above series. The equality to be proved is

$$\sum_{n=0}^{\infty} (n + \tfrac{1}{2}) P_n(\cos \theta) \frac{h_n(x)}{h_n'(x)} = -\frac{i}{2} \oint_C \frac{v P_{v-1/2}(-\cos \theta)}{\cos v\pi} \frac{h_{v-1/2}(x)}{h_{v-1/2}'(x)} \, dv, \qquad (12.2)$$

where C is a clockwise contour $(C = A_- + A_+)$ that encloses the real axis (figure 12.1). Here P_v is a hypergeometric function,[9] which becomes the Legendre polynomial for real integers $v = n$. The value $\theta = 0$ must be excluded, because when v is not an integer, the integral involving $P_{v-1/2}(-1)$ does not converge. The exclusion of $\theta = 0$ makes Watson's solution inappli-

cable on the piston source, that is, in the illuminated region. The contour integral can be evaluated in terms of the residues at the poles of the integrand:[10]

$$\oint_C \frac{\nu P_{\nu-1/2}(-\cos\theta)}{\cos\nu\pi} \frac{h_{\nu-1/2}(x)}{h'_{\nu-1/2}(x)}\, d\nu = -2\pi i \sum_q K_q. \tag{12.3}$$

The negative sign is used because the integration is performed clockwise. The K_q are the residues at the poles of the integrand. The functions in the numerator do not display singularities. For x real and nonzero, $h'_{\nu-1/2}$ has roots only for complex ν. Residues on the real ν-axis are therefore associated exclusively with the roots of $\cos\nu\pi$:

$$\nu_q = q + \tfrac{1}{2}, \qquad q = 0, 1, \ldots, \infty.$$

The corresponding residues are

$$K_q = \frac{\nu P_{\nu-1/2}(-\cos\theta)}{d(\cos\nu\pi)/d\nu} \frac{h_{\nu-1/2}(x)}{h'_{\nu-1/2}(x)}, \qquad \nu = \nu_q. \tag{12.4}$$

The factors in this expression are now evaluated:

$$\left.\begin{aligned} \frac{d(\cos\nu\pi)}{d\nu} &= -\pi\sin\nu\pi = -\pi(-1)^q \\[1em] P_{\nu-1/2}(-\cos\theta) &= P_q(-\cos\theta) = (-1)^q P_q(\cos\theta) \\[1em] h_{\nu-1/2}(x) &= h_q(x) \end{aligned}\right\} \quad \nu = \nu_q. \tag{12.5}$$

The residues in equation (12.4) can now be evaluated and substituted in the residue series in equation (12.3), which then becomes identical to equation (12.2).

The crucial step in Watson's analysis is to show that contour C can be deformed into an equivalent contour C' that excludes the real axis. This new contour does not therefore enclose any of the roots of $\cos\nu\pi$, but does include the complex roots of $h'_{\nu-1/2}(x)$. To construct this new contour it is noted that the original contour C is equivalent to line integrals just above and below the real axis, $\mathscr{I}_{A_+}$ and $\mathscr{I}_{A_-}$ (figure 12.1). The integrals along the two semicircles of vanishing radius ε that close the contour C at $\nu = 0$ and $\nu = \infty$ do not contribute. Hence,

Table 12.1
Numerical coefficients in equation (12.7)

m	0	1	2	3
a_m	0.7003	2.2326	3.3131	4.2363
c_m	0.4043	1.2890	1.9129	2.4459

$$\oint_C = \mathscr{I}_{A_+} + \mathscr{I}_{A_-}.$$ (12.6)

We now form a new contour C'_+ confined to the first quadrant by adding to the original line integral, the line integral $\mathscr{I}_{B_+}$ along the positive imaginary axis, and $\mathscr{I}_{D_+}$ along the quarter circle of infinite radius ($C'_+ \equiv A_+ + D_+ + B_+$). This contour now contains an infinity of poles $v = v_m$ of the integrand of equation (12.2), which correspond to the complex roots of $h'_{v-1/2}$. The location of these roots is given by[11]

$$v_m(x) = x + c_m x^{1/3} + ia_m x^{1/3}, \qquad x^{4/3} \gg 1.$$ (12.7)

The amplitude of these roots is seen to be somewhat larger than x and hence for the case $x \gg 1$ much larger than unity.

Noting that

$$h_{v-1/2}(x) = e^{-iv\pi} h_{-v-1/2}(x),$$ (12.8)

we see that another infinite series of roots lies in the third quadrant at $v = -v_m$. The coefficients a_m and c_m in equation (12.7) are given in table 12.1 for the lowest roots. The residues at these poles are

$$K_m = \frac{v P_{v-1/2}(-\cos\theta)}{\cos v\pi} \frac{h_{v-1/2}(x)}{\partial[h'_{v-1/2}(x)]/\partial v}, \qquad v = v_m.$$ (12.9)

In terms of these residues the integral around C'_+ is

$$\oint_{C'_+} = \mathscr{I}_{A_+} + \mathscr{I}_{B_+} + \mathscr{I}_{D_+} = 2\pi i \sum_{m=0}^{\infty} K_m.$$ (12.10)

Another contour, C'_- is constructed by adding to the line integral along path A_- the integrals along B_- and D_- ($C'_- \equiv D_- + B_- + A_-$). The integrand

in equation (12.2) does not have any poles in this quadrant. Hence,

$$\oint_{C'_-} = \mathscr{I}_{A_-} + \mathscr{I}_{B_-} + \mathscr{I}_{D_-} = 0. \tag{12.11}$$

Adding equations (12.10) and (12.11), we obtain

$$\oint_{C'_+} + \oint_{C'_-} = \mathscr{I}_{A_+} + \mathscr{I}_{A_-} + \mathscr{I}_{B_+} + \mathscr{I}_{B_-} + \mathscr{I}_{D_+} + \mathscr{I}_{D_-} = 2\pi i \sum_{m=0}^{\infty} K_m. \tag{12.12}$$

Sommerfeld[2] proves that the integrals along the two arcs D vanish as their radius tends to infinity. To evaluate the line integrals along B_+ and B_- we must explore whether the integrand in equation (12.2) is even or odd in v. Examining the terms one by one, we find that

$$P_{v-1/2}(x) = P_{-v-1/2}(x),$$

$$\cos v\pi = \cos(-v\pi),$$

$$v = -(-v).$$

When these relations and equations (12.8) are used in the integrand of equation (12.2), it is found that

$$\mathscr{I}_{B_+} = -\mathscr{I}_{B_-}.$$

This leaves the A integrals as the only line integrals in equation (12.2). Now, by virtue of equation (12.6), these integrals equal the contour integral around C and can therefore be used to express the summation in equation (12.12). We thus finally obtain the relation

$$\sum_{n=0}^{\infty} (n + \tfrac{1}{2}) P_n(\cos\theta) \frac{h_n(x)}{h'_n(x)} = \pi \sum_{m=0}^{\infty} K_m$$

$$= \pi \sum_{m=0}^{\infty} \frac{v P_{v-1/2}(-\cos\theta) h_{v-1/2}(x)}{\cos v\pi \, \partial[h_{v-1/2}(x)]/\partial v} \bigg|_{v=v_m} \tag{12.13}$$

To evaluate the factors in the expression for the residues, equation (12.9), we make use of large-v asymptotic forms of the three functions. First we express the θ dependence in such a way as to display the traveling-wave

nature of the representation. For this purpose, we use the Laplace approximation to the Legendre polynomial of large order:[12]

$$P_n(\cos\theta) \approx \frac{2^{1/2}}{(\pi n \sin\theta)^{1/2}} \cos\left[(n+\tfrac{1}{2})\theta - \frac{\pi}{4}\right], \qquad n\sin\theta \gg 1.$$

This result is applicable to complex orders v as long as $\mathrm{Re}(v)\sin\theta \gg 1$. The hypergeometric functions in equation (12.13) can now be written as

$$
\begin{aligned}
P_{v-1/2}(-\cos\theta) &= P_{v-1/2}[\cos(\pi-\theta)] \\
&\approx \frac{\exp[iv(\pi-\theta)-(i\pi/4)] + \exp[-iv(\pi-\theta)+(i\pi/4)]}{(2\pi v \sin\theta)^{1/2}} \\
&\simeq \frac{\exp[-i\pi(v-\tfrac{1}{4})]}{(2\pi v \sin\theta)^{1/2}}\{e^{iv\theta} + e^{iv(2\pi-\theta)}e^{-i\pi/2}\}.
\end{aligned}
\tag{12.14a}
$$

Noting that

$$\frac{e^{-i\pi/2}}{(\sin\theta)^{1/2}} = \frac{1}{(-\sin\theta)^{1/2}} = \frac{1}{[\sin(2\pi-\theta)]^{1/2}},$$

the hypergeometric function finally becomes

$$P_{v-1/2}(-\cos\theta) = \frac{\exp[-i\pi(v-\tfrac{1}{4})]}{(2\pi v)^{1/2}}\left[\frac{e^{iv\theta}}{(\sin\theta)^{1/2}} + \frac{e^{iv(2\pi-\theta)}}{[\sin(2\pi-\theta)]^{1/2}}\right].$$
$$\tag{12.14b}$$

Like the original contour integral, this expression is not valid at $\theta = 0$. Neither is it valid at $\theta = \pi$, because the condition $\mathrm{Re}(v)\sin\theta \gg 1$ is not satisfied in this region. Instead, the actual hypergeometric function $P_{v-1/2} = 1$ must be used near $\theta \approx \pi$. The cosine term in the denominator of equation (12.9) can be expanded in a power series:

$$
\begin{aligned}
\frac{1}{\cos v\pi} &= \frac{2}{\exp(-iv\pi)[1+\exp(2iv\pi)]} \\
&= 2\exp(i\pi v)\sum_{s=0,1}^{\infty}(-1)^s\exp(i2\pi vs), \qquad \mathrm{Im}\,v > 0 \\
&\approx 2\exp(i\pi v), \qquad \exp(-\mathrm{Im}\,v) \ll 1.
\end{aligned}
\tag{12.15a}
$$

The ratio of the Hankel function and its derivatives required in equation (12.9) can be approximated in terms of the asymptotic roots, equation (12.7):[11]

$$\frac{h_{\nu-1/2}(x)}{\partial h'_{\nu-1/2}(x)/\partial \nu}\bigg|_{\nu=\nu_m} = \frac{(xe^{i\pi})^{2/3}}{4c_m}, \qquad x^{4/3} \gg 1. \tag{12.15b}$$

We are now finally in a position to evaluate the summation in equation (12.13). Using the above expression and equations (12.14) and (12.15) gives

$$\sum_n = \frac{(2\pi)^{1/2}\exp(i11\pi/12)x^{2/3}}{4}$$

$$\cdot \sum_m \frac{v_m^{1/2}}{c_m}\left\{\frac{\exp(iv_m\theta)}{(\sin\theta)^{1/2}} + \frac{\exp[iv_m(2\pi-\theta)]}{[\sin(2\pi-\theta)]^{1/2}}\right\}, \qquad \theta \neq 0, \pi. \tag{12.16}$$

The first exponential term when combined with the time dependence of the form $e^{-i\omega t}$ can be interpreted as a wave propagating in the θ direction. Because of the positive imaginary component of v_m, this wave is attenuated as it propagates. The second exponential can be associated with a wave that has propagated more than halfway around the sphere before reaching this same latitude θ. It has therefore suffered a larger attenuation than the wave that has traveled the smaller distance. For $x \gg 1$, the former term predominates, except when $\theta \approx \pi$, since equation (12.16) is not valid in this region. The first term predominates whenever the traveling-wave formulation applies. If integers larger than zero had been retained in equation (12.15), these integers $s = 1, 2, \ldots$ would be associated with $1, 2, \ldots$ complete circumnavigations of the sphere. This fact is noted by Watson at the end of his paper, even though he does not elsewhere display his solution in terms of traveling waves.

The $v_m^{1/2}$ term in the numerator can be set equal to $x^{1/2}$ for $ka \gg 1$. With this approximation, an explicit expression for the pressure is finally obtained:

$$p(a,\theta) = -\frac{\rho\ddot{Q}(ka)^{1/6}}{a\sqrt{32\pi\sin\theta}}\sum_{m=0,1}^{\infty}\frac{\exp[-a_m(ka)^{1/3}\theta + i\phi_m(\theta)]}{c_m}, \qquad \theta \neq 0, \pi, \tag{12.17}$$

where

$$\phi_m(\theta) = \frac{11\pi}{12} + ka[1 + c_m(ka)^{-2/3}]\theta - \omega t. \tag{12.18}$$

For $\theta \approx 0$, the pressure on a piston in a plane baffle can be used for $ka \gg 1$. For $\theta = \pi$, where equation (12.16) is not valid, the pressure is computed directly in terms of $P_{v_m-1/2}(-\cos\theta)$ which is unity. Only the $m = 0$ term need be retained, because the higher modes are attenuated to negligible levels at this distance from the source. For the region in the vicinity of but not including $\theta = \pi$, the second term or equation (12.17) must be retained:

$$p(a, \theta) = \frac{\rho\ddot{Q}(ka)^{1/6}e^{-i\pi/12}}{a\sqrt{32\pi\sin\theta}\,c_0}(e^{iv_0\theta} - ie^{iv_0(2\pi-\theta)}). \tag{12.19}$$

The phase velocity of the traveling waves in equation (12.18) is determined by the real component of v_m. The phase velocity V_m satisfies the requirement that the Lagrangian time derivative of the phase angle be zero:

$$\frac{\partial\phi_m}{\partial t} + V_m\frac{\partial\phi_m}{\partial s} = 0, \tag{12.20a}$$

where s is the geodetic distance, $s = a\theta$. When equation (12.18) is substituted and solving for V_m, one obtains the creeping wave velocity:

$$V_m = c[1 + c_m(ka)^{-2/3}]^{-1}. \tag{12.20b}$$

This phase velocity is less than the free-space sound velocity c to which it tends as the radius of curvature becomes infinite, that is, as $ka \to \infty$. Exponential attenuation increases with both m and ka. In the region $0 < \theta < \pi/2$ the attenuation is enhanced by the factor $(\sin\theta)^{-1/2}$. For $\pi/2 < \theta < \pi$ this same factor helps to counteract the exponential decay. Physically, this can be associated with the fact that the trace length of a wave front increases with θ from 0 to $\pi/2$, thus introducing a geometrical spreading loss. In the region $\pi/2$ to π, the trace length decreases, thus producing a geometrical focusing factor.

The results of the present analysis will now be compared with those of the usual wave-harmonic series. For the spherical array a comparison of the results is shown in figure 12.2. The results show excellent agreement except near the region in the vicinity of $\theta = \pi$, where the pressures are extremely

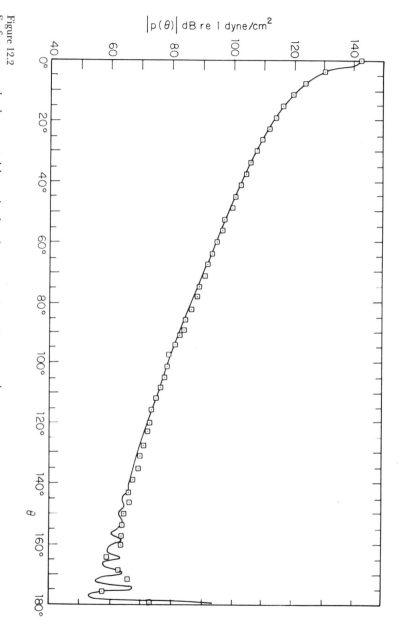

Figure 12.2
Surface-pressure levels generated by a circular piston on a sphere ($ka = 32$, $\rho c \dot{W} = 1.5 \times 10^7$ dyne/cm^2, polar angle subtended (12.1) by the active element $0 \leq \theta \leq \theta_0 = 1.875°$): □, wave-harmonic results. ———, results of present technique, (12.17) and (12.19). The Kirchhoff approximation was used for $0 \leq \theta \leq 11.25°$, and a single ($m = 0$) creeping wave mode for $\theta \gtrsim 15°$. [Reproduced from Junger[8]]

Table 12.2
Transition from the creeping wave series to the Kirchhoff approximation for the surface pressure on and near an active element on a spherical baffle[a]

Location on sphere	Number of creeping wave modes				Kirchhoff approxi- mation	Wave- harmonic results
	1	2	3	4		
0°	—	—	—	—	1.282	—
(Average over element)					−57.6°	
0°	—	—	—	—	1.501	1.340
					−60.0°	−58.7°
3.75°	0.227	0.281	0.309	0.326	0.336	0.336
	21.4°	24.1°	26.1°	27.5°	25.8°	25.4°
7.50°	0.138	0.160	0.169	0.170	0.171	0.156
	146.4°	149.9°	151.9°	153.2°	147.8°	149.1°
11.25°	0.097	0.107	0.110	0.111	0.114	0.104
	271.3°	274.9°	276.6°	277.6°	268.6°	271.0°

a. Values of the parameters are the same as for figure 12.2. The upper number in each square gives the surface pressure amplitude in 10^7 dyne/cm,[2] the lower number gives the pressure phase angle. The pressures are point pressures except as indicated for 0° (see Junger[8]).

small. At $\theta = \pi$, where $P_{\nu-1/2} = 1$, the pressure is enhanced by a focusing of the creeping wave. The convergence of the creeping wave series to the Kirchhoff pressure (plane baffle) is shown in table 12.2.

12.3 High-Frequency Response of a Spherical Shell

In the preceding section, we directed our attention to the problem of the surface pressure generated on a sphere by a high-frequency velocity source. We shall now employ the same methods to study the problem of the response of a spherical elastic shell caused by a high-frequency excitation force, for both the submerged shell and the shell in vacuo. The response of the submerged spherical shell excited by a concentrated force is stated in terms of modal impedance, in equation (9.13). The sound field is constructed by multiplying each term by $i\rho c h_n(kR)/h'_n(ka)$, equation (6.19):

$$p(R, \theta) = \frac{i\rho c F}{2\pi a^2} \sum_{n=0}^{\infty} \frac{(n + \frac{1}{2}) h_n(kR) P_n(\cos \theta)}{(Z_n + z_n) h'_n(ka)}, \tag{12.21}$$

where z_n is given by equation (6.29) and $\mathcal{Z}_n$ by equation (7.121), except that in the present chapter we shall denote Poisson's ratio by μ instead of ν.

Equations (9.13) and (12.21) are examples of modal series similar to that given by equation (12.1), either one of which can be written in the form

$$S = \sum_{n=0}^{\infty} \frac{(n + \tfrac{1}{2}) P_n(\cos\theta) g(n, kR)}{f(n, ka)}. \tag{12.22}$$

An integral representation of this series is

$$S = -\frac{i}{2} \oint_C \frac{\nu P_{\nu-1/2}(-\cos\theta) g(\nu - \tfrac{1}{2}, kR)}{\cos\nu\pi f(\nu - \tfrac{1}{2}, ka)} \, d\nu, \tag{12.23a}$$

which is analogous to the expression in equation (12.2). Using the same methods as before, the sum S can be written as

$$S = 2\pi i \left[-\frac{i}{2} \sum_j \frac{\nu P_{\nu-1/2}(-\cos\theta) g(\nu - \tfrac{1}{2}, kR)}{\cos\nu\pi \, \partial f/\partial\nu} \bigg|_{\nu=\nu_j} \right], \tag{12.23b}$$

where ν_j are the roots of $f(\nu - \tfrac{1}{2}, ka)$ included within the contour C, since $P_{\nu-1/2}(-\cos\theta) g(\nu - \tfrac{1}{2}, kR)$ has no poles in the complex ν plane.

The crucial problem now is to evaluate the roots of f. One could proceed to evaluate the roots of f, a transcendental equation, by numerical techniques. However, the real reason for formulating the problem in the foregoing manner was to provide a useful solution in the high-frequency range. Fortunately, in this range various simplifying assumptions as to the form of f can be made so that useful asymptotic approximations for the roots can be obtained. In the discussion up to this point, we have considered the shell to be submerged in an acoustic fluid. It is also of interest to consider the high-frequency response of structures in vacuo. Accordingly we shall now proceed to find the roots of f for two different cases: (1) spherical shell in vacuo, and (2) spherical shell submerged in an acoustic fluid.

12.4 The Point-Excited Spherical Shell in Vacuo

The normal-mode series for the shell response is given by equation (9.13), where $z_n = 0$ in vacuo. The structural impedances, equation (7.121), can be rewritten in the form

$$\mathcal{Z}_n \equiv \mathcal{Z}(\Omega, \lambda_n)$$

$$= \frac{i\rho_s c_p \beta^2}{\Omega(1 + \beta^2)} \frac{h}{a} \frac{\lambda_n^3 + c_2 \lambda_n^2 + c_1 \lambda_n + c_0}{\lambda_n + c_3}, \tag{12.24}$$

where $\lambda_n = n(n+1)$ and

$$c_0 = (1 - \mu)[\Omega^2 - 2(1 + \mu)]\left[1 + \frac{1}{\beta^2}\left(\frac{\Omega^2}{1 - \mu} + 1\right)\right] \sim \frac{\Omega^4}{\beta^2} + O(\Omega^2, \beta^{-2}),$$

$$c_1 = \left(5 - \mu^2 + \frac{1 - \mu^2}{\beta^2} - \mu\Omega^2 - \frac{\Omega^2}{\beta^2}\right) \sim -\frac{\Omega^2}{\beta^2} + O(\Omega^2, \beta^{-2}),$$

$$c_2 = -(4 + \Omega^2) \sim -\Omega^2 + O(1), \tag{12.25}$$

$$c_3 = -\left(1 - \mu + \frac{\Omega^2}{1 + \beta^2}\right) \sim -\Omega^2 + O(1).$$

The approximations in equation (12.25) are made on the basis of $\Omega^2 \gg 1$ (high frequencies) and $\beta^2 \ll 1$ (thin shells). In the derivation of the shell equations (chapter 7), the effects of rotatory inertia and transverse shear were neglected, so that in applying the high-frequency approximation we must keep in mind the requirement that the shell thickness must be such that in the frequency range of interest the structural wavelength is much larger than the thickness. In contrast to the modal impedances in equation (9.11), which are intended for the low- and middle-frequency range, where the structural wavelengths are relatively long, the short-wavelength regime considered here requires that the flexural, β^2 terms be retained, even though the shell is thin. Using the approximations in equation (12.25), the structural impedance becomes

$$\mathcal{Z}(\Omega, \lambda_n) \approx \frac{i\rho_s c_p \beta^2}{\Omega} \frac{h}{a} \frac{\lambda_n^3 - \lambda_n^2 \Omega^2 - \lambda_n \Omega^2 / \beta^2 + \Omega^4 / \beta^2}{\lambda_n - \Omega^2}$$

$$= \frac{i\rho_s c_p \beta^2}{\Omega} \frac{h}{a}(\lambda_n^2 - \Omega^2 / \beta^2). \tag{12.26}$$

Therefore for the case at hand, f can be written as

$$f(v - \tfrac{1}{2}, ka) = \frac{i}{F} \frac{\pi}{4} \frac{\mathcal{Z}_p}{\alpha_f^2}[(v^2 - 1/4)^2 - \alpha_f^4], \tag{12.27}$$

where $\alpha_f = \sqrt{\Omega/\beta} = k_f a$. The flexural wavenumber k_f is given in equation (7.62), and Z_p, the drive-point impedance of the infinite plate, in equation (7.68). The four roots of f are

$$v_1 = (\alpha_f^2 + \tfrac{1}{4})^{1/2} \simeq \alpha_f, \tag{12.28a}$$

$$v_2 = -(\alpha_f^2 + \tfrac{1}{4})^{1/2} \simeq -\alpha_f, \tag{12.28b}$$

$$v_3 = i(\alpha_f^2 - \tfrac{1}{4})^{1/2} \simeq i\alpha_f, \tag{12.28c}$$

$$v_4 = -i(\alpha_f^2 - \tfrac{1}{4})^{1/2} \simeq -i\alpha_f. \tag{12.28d}$$

Introducing the structural loss factor, the dimensionless frequency becomes

$$\Omega = \frac{\omega a}{c_p (1 - i\eta_s)^{1/2}} \sim \frac{\omega a}{c_p}\left(1 + \frac{i\eta_s}{2}\right), \qquad \eta_s \ll 1. \tag{12.29}$$

The only roots that are therefore included within the proper contour for this case are v_1 and v_3.

Combining equations (9.13), (12.23), (12.27), and (12.28), we obtain an alternative representation for the radial response of a shell in vacuo:

$$\dot{w}(\theta) = -\frac{iF}{Z_p}\left\{\frac{P_{v_1-1/2}(-\cos\theta)}{\cos v_1 \pi} - \frac{P_{v_3-1/2}(-\cos\theta)}{\cos v_3 \pi}\right\}. \tag{12.30}$$

From equation (12.28) we see that for high frequencies and thin shells the order of the Legendre functions appearing in equation (12.30) is large in magnitude; therefore, using the asymptotic representations given in equations (12.14a) and (12.15),

$$\dot{w}(\theta) = -\frac{iFe^{i\pi/4}}{Z_p}\sqrt{\frac{2}{\pi}}\left\{\frac{e^{iv_1\theta}}{\sqrt{v_1 \sin\theta}} + \frac{e^{iv_1(2\pi-\theta)}}{\sqrt{v_1 \sin(2\pi-\theta)}}\right.$$
$$\left. - \frac{e^{iv_3\theta}}{\sqrt{v_3 \sin\theta}} - \frac{e^{iv_3(2\pi-\theta)}}{\sqrt{v_3 \sin(2\pi-\theta)}}\right\}, \qquad 0 < \theta < \pi. \tag{12.31}$$

The solution is displayed in terms of waves propagating on the sphere in opposite directions. When combined with the $e^{-i\omega t}$ time dependence, the phase angle of the two terms associated with v_1 yields a velocity,

$$V = \frac{\omega a}{\mathrm{Re}(v_1)} = \frac{c_f}{[1 + (4\alpha_f^2)^{-1}]^{1/2}},$$

that approaches the phase velocity of straight crested waves in an infinite elastic plate of thickness h, as α_f goes to infinity.

Of course, equation (12.31) as it stands cannot be used at either the drive point, $\theta = 0$, or at the antipode $\theta = \pi$. In order to calculate the response at $\theta = \pi$ from (12.30), we note that $P_{v-1/2}(1) = 1$. Equation (12.30) now yields

$$\dot{w}(\pi) = -\frac{iF}{Z_p}\left[\frac{1}{\cos v_1 \pi} - \frac{1}{\cos v_3 \pi}\right]. \tag{12.32}$$

For high frequencies, the second term becomes negligibly small since

$$|\cos v_3 \pi| = \left|\cosh\left[\alpha_f \pi\left(1 + \frac{i\eta_s}{4}\right)\right]\right| \gg 1,$$

so that $\dot{w}(\pi)$ can be approximated by

$$\dot{w}(\pi) = -\frac{iF}{Z_p}\frac{\cos(\alpha_f \pi)\cosh[\alpha_f(\pi\eta_s/4)] + \sin(\alpha_f \pi)\sinh[\alpha_f(\pi\eta_s/4)]}{\cos^2(\alpha_f \pi) + \sinh^2(\alpha_f \pi\eta_s/4)}. \tag{12.33}$$

The above formula yields a maximum response whenever $\cos(\alpha_f \pi) = 0$. This allows us to define resonance frequencies in the form

$$\omega_n = \frac{hc_p}{\sqrt{12a^2}}\left(\frac{2n + 1}{2}\right)^2, \qquad n \gg 1. \tag{12.34}$$

Recalling the expression for the flexural wave number, equation (7.62), we see that a resonance occurs when a great circle of the sphere measures an odd number of half-flexural wavelengths. At these frequencies, the response can be put in the convenient form

$$|\dot{w}(\pi)| = \frac{F}{Z_p}\frac{1}{\sinh[(\pi\eta_s/8)(2n + 1)]}, \tag{12.35}$$

which shows that the antipodal response decreases with increasing mode order n, that is, frequency, or an increased amount of damping.

The situation at the drive point, $\theta = 0$, is more difficult to resolve. Returning to equation (9.13) with $z_n = 0$, $\theta = 0$, $P_n(1) = 1$, we find that

$$\dot{w}(0) = \frac{F}{2\pi a^2} \sum_{n=0}^{\infty} \frac{(n + \frac{1}{2})}{Z_n}. \tag{12.36}$$

The above summation is equivalent to the integral

$$\mathscr{I} = -\frac{i}{2} \oint_C v \frac{\sin v\pi}{\cos v\pi} \frac{dv}{Z(\Omega, v^2 - \frac{1}{4})},$$

where $C = A_- + A_+$ is the contour shown in figure 12.1 and Z is given by equation (12.26). The contour C is again closed by the paths D_+, B_+, B_-, and D_-. The contribution to the integral from the paths D_+ and D_- vanishes, while the integral along the paths B_+ and B_- becomes

$$\mathscr{I}_{B_+B_-} = \int_0^{\infty} \frac{\rho \tanh \rho\pi \, d\rho}{Z(\Omega, -\rho^2 - \frac{1}{4})}.$$

Using Cauchy's integral theorem we then find that

$$\mathscr{I} = 2\pi i \sum_j K_j - \mathscr{I}_{B_+B_-}, \tag{12.37}$$

where K_j denotes the jth residue of the integrand located within the contour C. Using the results of equation (12.37), equation (12.36) can be evaluated in the form

$$\dot{w}(0) = -\frac{iF}{Z_p} \left[\tan v_1 \pi - \tan v_4 \pi - \frac{4\alpha_f^2}{\pi} \mathscr{I}_{B_+B_-} \right].$$

$\mathscr{I}_{B_+B_-}$ can be evaluated in closed form, but for our purposes we are only interested in the large-α_f approximation,[13] in which case

$$\mathscr{I}_{B_+B_-} = \frac{i\pi}{4\alpha_f^2} + O\left(\frac{1}{\alpha_f^4}\right).$$

Using the definition of v_4 given by equation (12.28d), we see that

$$\lim_{\alpha_f \to \infty} \tan v_4 \pi = -i,$$

so that the leading term of the integral $\mathscr{I}_{B_+B_-}$ cancels the residue contribution of v_4 allowing us to arrive at the high-frequency approximation of the drive-point velocity:

$$\dot{w}(0) = -\frac{iF}{Z_p} \tan v_1 \pi. \tag{12.38}$$

The high-frequency limit of this expression reduces to the flat plate solution, since $\lim_{\alpha_f \to \infty} \tan v_1 \pi = i$.

Using the above results, we can interpret the response of the spherical shell in terms of the infinite plate solution. Near the drive point the two solutions agree exactly, while away from the excitation point the curvature of the sphere and the fact that it is a closed surface cause the waves that travel around the sphere to form standing waves whose wavelength is characteristic of straight-crested waves in an infinite plate.

One might argue that with the current speed of computers the slow convergence of the modal series for high-frequency excitation should not be bothersome and therefore that the foregoing analysis would not be necessary. This would be true if one were interested only in the numerical value of the final results, but no amount of computation with the modal series solution would have been able to produce the interpretation of the results that has been given in the preceding paragraph. The numerical accuracy of this alternative formulation can be best illustrated by comparing the results using the two methods.

We have considered an aluminum shell with $h/a = 0.01$, being excited at a frequency corresponding to $ka = 50$ in water, giving $\Omega = 13.8$, and having a loss factor of $\eta_s = 0.04$. In summing the modal series we have used 250 terms. Table 12.3 shows the very close agreement of the two solutions for most of the range of θ except for two regions approximately 5° wide in the vicinity of the drive point and the antipode $\theta = \pi$. Now that we have considered the spherical shell in vacuo, we shall move on to the case of a submerged spherical shell.

Table 12.3
Radial velocity of point-driven spherical shell[a]

θ (degrees)	Normal mode series [equation (9.13)]		Traveling wave representation [equation (12.31)]	
	$\dfrac{Z_p|\dot{w}|}{F} \times 100$	Phase[b] (degrees)	$\dfrac{Z_p|\dot{w}|}{F} \times 100$	Phase[b] (degrees)
0	98.05	3.1	97.85	0.1
1	77.36	33.8	81.66	37.8
2	53.94	91.1	54.83	92.4
3	40.09	158.4	40.04	159.4
4	33.68	230.8	33.90	231.2
5	30.66	299.5	30.54	299.3
10	20.49	284.9	20.51	285.1
20	12.92	258.0	12.92	256.6
30	9.22	229.5	9.27	227.8
50	5.79	168.1	5.76	165.7
70	4.49	111.6	4.49	107.8
90	3.12	60.4	3.16	56.7
110	2.27	348.0	2.25	343.7
130	2.89	293.1	2.90	287.3
150	2.36	266.3	2.38	260.6
170	1.51	131.2	1.54	124.1
175	3.57	114.1	3.74	106.8
176	5.23	283.7	5.16	277.5
177	8.87	287.9	9.03	281.6
178	.27	350.9	.70	304.7
179	14.96	107.3	14.99	100.6
180	22.38	107.8	22.60	101.2

a. See Feit and Junger.[13]
b. With respect to driving force.

12.5 The Submerged Spherical Shell

The case of a submerged spherical shell is more complicated in that we must include z_v in the denominator of equation (12.23). Using equation (12.37) and the relation between spherical and cylindrical Hankel functions,

$$h_{v-1/2}(x) = \left(\frac{\pi}{2x}\right)^{1/2} H_v(x),$$

$$h'_{v-1/2}(x) = \left(\frac{\pi}{2x}\right)^{1/2} \left[H'_v(x) - \frac{1}{2x} H_v(x) \right] \tag{12.39}$$

$$\simeq \left(\frac{\pi}{2x}\right)^{1/2} H'_v(x), \qquad x \gg 1,$$

the function f in equation (12.23) becomes

$$f(v - \tfrac{1}{2}, ka) = \frac{i}{F} \left\{ \frac{\pi Z_p}{(4\alpha_f^2)} [(v^2 - \tfrac{1}{4})^2 - \alpha_f^4] + 2\pi a^2 \rho c \frac{H_v(ka)}{H'_v(ka)} \right\}. \tag{12.40}$$

This function has two sets of roots, one group predominantly associated with structure-borne waves and the other associated primarily with fluid-borne waves. The roots associated with the former satisfy the condition $O(|v - ka|) = O(ka)$. Substituting the approximation[14]

$$\frac{H_v(\alpha)}{H'_v(\alpha)} \simeq -\alpha(v^2 - \alpha^2)^{-1/2}, \qquad \text{Re}(v^2 - \alpha^2)^{1/2} > 0 \tag{12.41}$$

in the above expression, we obtain an asymptotic form of f:

$$f(v - \tfrac{1}{2}, ka) \simeq \frac{i\pi Z_p}{4F\alpha_f^2} \left\{ [(v^2 - \tfrac{1}{4})^2 - \alpha_f^4] - \frac{8a^2 \rho c \alpha_f^2 ka}{Z_p(v^2 - k^2 a^2)^{1/2}} \right\}. \tag{12.42}$$

To find the roots, we make the substitution

$$\zeta = (v^2 - k^2 a^2)^{1/2}. \tag{12.43}$$

The roots satisfy the equation

$$\zeta^5 + 2(ka)^2 \zeta^3 + (k^4 a^4 - \alpha_f^4)\zeta - \left(\frac{\rho}{\rho_s}\right)\left(\frac{a}{h}\right)\alpha_f^4 = 0, \qquad v \gg 1. \tag{12.44}$$

Table 12.4
Roots of the Airy integrals $A(q)$ and of its derivative $A'(q')$

m	0	1	2	3
q_m	3.3721	5.8958	7.9620	9.7881
q'_m	1.4694	4.6847	6.9518	8.8890

If the substitution $\zeta = ka\tau$ is made, equation (12.44) becomes precisely the characteristic equation for the wave numbers of a thin infinite submerged plate [equation (8.8)]. This quintic has five roots, which for a very thin shell, $h/a \ll 1$, can be approximated by

$$\zeta_q \simeq \left(\frac{\rho}{\rho_s}\right)^{1/5} \left(\frac{a}{h}\right)^{1/5} \alpha_f^{4/5} e^{i2\pi q/5}, \qquad q = 0, 1, 2, 3, 4. \tag{12.45}$$

The corresponding values of v that satisfy the condition $\mathrm{Im}(v) \geq 0$ are

$$\begin{aligned} v_q &= (k^2 a^2 + \zeta_q^2)^{1/2}, \qquad q = 0, 1, \\ v_q &= -(k^2 a^2 + \zeta_q^2)^{1/2}, \qquad q = 4. \end{aligned} \tag{12.46}$$

The set of roots of equation (12.40) associated primarily with the water-borne waves that diffract around the shell is obtained in a different manner. The roots in this case will be found to satisfy the condition that $O(|v - ka|) = O[(ka)^{1/3}]$, $ka \gg 1$. In this range $H_v/H'_v = O(ka)$. When this is substituted in equation (12.40), we find that for a thin shell in water, the $\mathcal{Z}_p$ term is negligible compared with the ρc term. Consequently, the roots of equation (12.40) can be approximated by the zeros of $H_v(ka)$:

$$v_m = ka + e^{i\pi/3} \left(\frac{ka}{6}\right)^{1/3} q_m, \tag{12.47}$$

where the q_m are given in table 12.4. The coefficients of the roots of H'_v, (12.7), can be similarly computed from the roots q'_m of the derivative A' of the Airy function

$$c_m + i a_m = \frac{q'_m}{6^{1/3}} e^{i\pi/3}.$$

The relation between Airy integrals and Hankel functions was developed in the first edition of this book (mathematical appendix to chapter 13).

Now that we have derived values for the roots of equation (12.40), in order to obtain the residue series, equation (12.23), we must calculate expressions for $\partial f/\partial v|_{v=v_j}$. For the structure-borne waves v_q, using equation (12.40), this is

$$\left.\frac{\partial f}{\partial v}\right|_{v=v_q} = \frac{i\pi v_q}{F}\left[Z_p\left(\frac{v_q}{\alpha_f}\right)^2 + \frac{2\rho ca^2 ka}{(v_q^2 - k^2 a^2)^{3/2}}\right]. \tag{12.48}$$

To complete the calculation for the other set of roots v_m we find that

$$\left.\frac{df}{\partial v}\right|_{v=v_m} = \frac{i}{F}\left\{\frac{\pi Z_p v^3}{\alpha_f^2} + 2\pi a^2 \rho c \frac{\partial}{\partial v}\left[\frac{H_v(ka)}{H_v'(ka)}\right]\right\}\Bigg|_{v=v_m}. \tag{12.49}$$

With the same degree of approximation that has already been introduced,

$$\frac{\partial}{\partial v}\left[\frac{H_v(ka)}{H_v'(ka)}\right]_{v=v_m} \simeq -1,$$

equation (12.49) now reduces to

$$\left.\frac{\partial f}{\partial v}\right|_{v=v_m} = \frac{i\pi}{F}\left\{Z_p\frac{v_m^3}{\alpha_f^2} - 2\rho ca^2\right\}. \tag{12.50}$$

We are now in a position to give an alternative expression for the radial velocity response of a submerged spherical shell. Combining equations (12.23), (12.48), and (12.50), the response expressed as a residue series becomes

$$\dot{w}(\theta) = -iF\sqrt{\frac{2}{\pi \sin\theta}}\left\{\sum_q \frac{\cos[v_q(\pi - \theta) - \pi/4]}{v_q^{1/2}\cos v_q\pi\,[Z_p(v_q/\alpha_f)^2 + 2\rho ca^2 ka\,(v_q^2 - k^2 a^2)^{-3/2}]}\right.$$
$$\left. + \sum_{m=0}^{\infty} \frac{v_m^{1/2}\cos[v_m(\pi - \theta) - \pi/4]}{\cos v_m\pi\,[Z_p(v_m^3/\alpha_f^2) - 2\rho ca^2]}\right\}, \tag{12.51}$$

where the v_q are the structure-borne wave roots in equation (12.46), and the

v_m are the diffracted wave roots in equation (12.47). The summation over q includes only a finite number of terms corresponding to $q = 0, 1, 4$, while the summation over m includes an infinite number of terms, which, however, converge rapidly except for small θ, because of the large imaginary component of v_m.

12.6 The Point Source on a Cylindrical Surface

The methods used in the earlier sections can also be extended to cases in which the radiating surface is an infinite cylinder. As an example we consider a point source located on a cylindrical surface. Using cylindrical coordinates (r, ϕ, z), we consider the source as centered at $z = 0$, $\phi = 0$, and on the surface $r = a$. The acceleration distribution of the cylinder is given by

$$\ddot{w}(\phi, z) = \ddot{W}, \qquad -z_0 < z < z_0, \qquad -\phi_0 < \phi < \phi_0$$
$$= 0, \qquad |z| > z_0, \qquad |\phi| > \phi_0. \tag{12.52}$$

When this acceleration distribution is expanded in a Fourier series in ϕ and Fourier transformed with respect to z, one obtains the transform series

$$\tilde{\ddot{w}}(\phi; \gamma) = \tilde{f}(\gamma) \sum_{n=0,1}^{\infty} \ddot{W}_n \cos n\phi$$

$$= \frac{2\phi_0 z_0 \ddot{W}}{\pi} \frac{\sin \gamma z_0}{\gamma z_0} \sum_{n=0,1}^{\infty} \varepsilon_n \frac{\sin n\phi_0}{n\phi_0} \cos n\phi \tag{12.53}$$

$$\approx \frac{2\phi_0 z_0 \ddot{W}}{\pi} \sum_{n=0}^{\infty} \varepsilon_n \cos n\phi, \qquad \phi_0, \frac{z_0}{a} \ll 1.$$

This can be expressed conveniently in terms of the volume acceleration amplitude $4a\phi_0 z_0 \ddot{W} = \ddot{Q}$:

$$\tilde{\ddot{w}}(\phi; \gamma) = \frac{\ddot{Q}}{2a\pi} \sum_{n=0}^{\infty} \varepsilon_n \cos n\phi. \tag{12.54}$$

This result is now substituted in the wave-harmonic series for the pressure equation (6.49). Setting r equal to a, this equation finally yields the surface pressure

$$p(a, \phi, z) = -\frac{\rho \ddot{Q} \exp(-i\omega t)}{4\pi^2 a} \int_{-\infty}^{\infty} \frac{e^{i\gamma z}}{(k^2 - \gamma^2)^{1/2}}$$

$$\cdot \sum_{n=0}^{\infty} \varepsilon_n \frac{H_n[(k^2 - \gamma^2)^{1/2} a]}{H_n'[(k^2 - \gamma^2)^{1/2} a]} \cos n\phi \, d\gamma. \tag{12.55}$$

Not only is this series impractical because of its slow convergence for large ka, but the integrals cannot be evaluated analytically. We shall see that both drawbacks are circumvented in the creeping wave formulation. Our first task is to transform the series in n by writing

$$\sum_{n=0,1}^{\infty} \frac{\varepsilon_n H_n(x)}{H_n'(x)} \cos n\phi = \sum_{n=-\infty}^{\infty} \frac{H_n(x)}{H_n'(x)} e^{in\phi}. \tag{12.56}$$

Proceeding as for the sphere, we express the summation in terms of a contour integral, which now however includes the entire real axis (see figure 12.3):

$$\sum_{n=-\infty}^{\infty} \frac{H_n(x)}{H_n'(x)} e^{in\phi} = \frac{i}{2} \oint_C \frac{H_\nu(x) \exp i\nu(\phi - \pi)}{H_\nu'(x) \sin \nu\pi} \, d\nu, \tag{12.57}$$

where the contour $C = A_- + A_+$ encloses the real axis in a clockwise fashion. Noting that the only poles on the real axis are those associated with $\sin \nu\pi$, where ν is a real positive or negative integer, the contour integral can be written as the residue series

$$\oint_C = -2\pi i \sum_{\nu=n} \frac{H_\nu(x) \exp[i\nu(\phi - \pi)]}{H_\nu'(x) \, d(\sin \nu\pi)/d\nu}, \qquad n = 0, \pm 1, \pm 2, \dots. \tag{12.58}$$

It can be verified that this equation is correct by noting that

$$\left. \frac{d(\sin \nu\pi)}{d\nu} \right|_{\nu=n} = \pi \cos n\pi = (-1)^n \pi, \tag{12.59}$$

$$e^{i\nu\pi}|_{\nu=n} = (-1)^n. \tag{12.60}$$

We now proceed to transform the contour C into a contour C', which contains only the poles of H_ν'. For this purpose a contour C_+' is formed in the upper half-plane, and a similar contour C_-' in the lower plane. Each contour

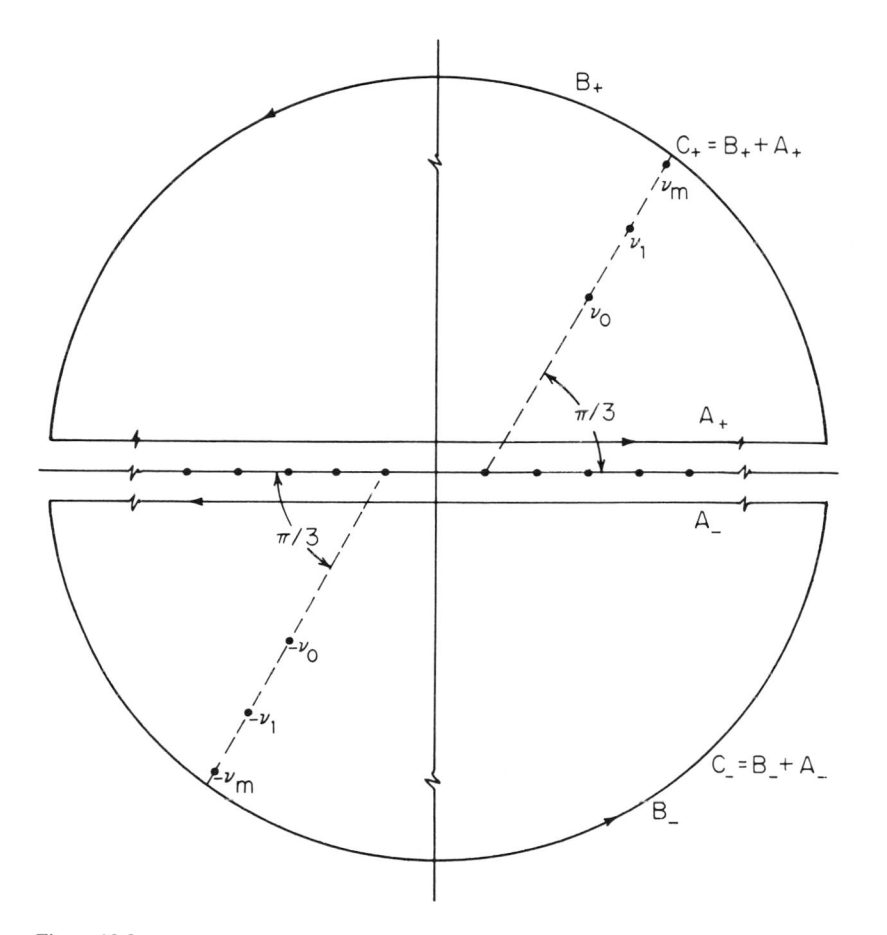

Figure 12.3
Contour integrals used in the Watson transformation of the cylindrical wave-harmonic series in the complex ν plane.

consists of a semicircle of infinite radius, and of a parallel to the real axis (figure 12.3). The poles in the upper half-plane are again given by equation (12.7) since, for large argument, $H'_\nu(x)$ and $h'_{\nu-1/2}(x)$ have nearly the same roots, as verified from the relations between spherical and cylindrical functions [equation (12.39)]. As before, $\nu = -\nu_m$ are roots of H'_ν located in the lower half-plane. Both the C_+ and C_- contour integrals equal a sum of residues:

$$\oint_{C_+} = \mathscr{I}_{A_+} + \mathscr{I}_{B_+} = 2\pi i \sum_{\nu=+\nu_m} K_m,$$
$$\oint_{C_-} = \mathscr{I}_{A_-} + \mathscr{I}_{B_-} = 2\pi i \sum_{\nu=-\nu_m} K_m, \tag{12.61}$$

where the residues are given by

$$K_m = \frac{\exp[i\nu(\phi - \pi)]}{\sin \nu\pi} \left. \frac{H_\nu(x)}{\partial H'_\nu(x)/\partial \nu} \right|_{\nu=\pm\nu_m}. \tag{12.62}$$

The integrals along the semicircles B_+ and B_- vanish as the radius $|\nu|$ tends to infinity. Thus, adding the two equations (12.6), we obtain

$$\oint_{C_+} + \oint_{C_-} = 2\pi i \sum_{\nu=\pm\nu_m} K_m$$
$$= \mathscr{I}_{A_+} + \mathscr{I}_{A_-} = \oint_C; \tag{12.63}$$

By virtue of equation (12.57), this implies the equality

$$\sum_n = \frac{i}{2} \oint_C$$
$$= -\pi \sum_{\nu=\pm\nu_m} K_m. \tag{12.64}$$

Substituting equation (12.62), we obtain

$$\sum_n = -\pi \sum \left\{ \frac{\exp[iv_m(\phi - \pi)]}{\sin v_m \pi} \left(\frac{H_v(x)}{\partial H_v'(x)/\partial v} \right)_{v=v_m} \right.$$

$$\left. + \frac{\exp[-iv_m(\phi - \pi)]}{-\sin v_m \pi} \left(\frac{H_v(x)}{H_v'(x)/\partial v} \right)_{v=-v_m} \right\}. \tag{12.65}$$

For large x, where the approximation in the latter of equations (12.39) applies, the ratio $H_v/(\partial H_v'/\partial v)$ at $v = v_m$ is again given by equation (12.15b). At $v = -v_m$, where the coefficients in table 12.1, specifically the coefficients c_m, must be multiplied by (-1), the ratio takes on the opposite sign:

$$\left(\frac{H_v}{\partial H_v'/\partial v} \right)_{v=v_m} = -\left(\frac{H_v}{dH_v'/\partial v} \right)_{v=-v_m}. \tag{12.66}$$

The two terms in equation (12.65) can then be combined:

$$\sum_n = -2\pi \sum_m \frac{\cos v_m(\phi - \pi)}{\sin v_m \pi} \left(\frac{H_v(x)}{\partial H_v'(x)/\partial v} \right)_{v=+v_m}. \tag{12.67}$$

To express the ϕ dependence in terms of traveling waves, we make use of the expansion

$$\frac{1}{\sin v\pi} = \frac{-2i}{\exp(-iv\pi)[1 - \exp(i2v\pi)]},$$

$$= -2i \sum_{s=0,1}^{\infty} \exp[iv\pi(1 + 2s)], \qquad \text{Im } v > 0 \tag{12.68}$$

where, as in the spherical case, we need only retain the $s = 0$ term. The traveling-wave representation of the trigonometric functions thus becomes

$$\frac{\cos v(\phi - \pi)}{\sin v\pi} = \frac{-2i \exp(iv\pi)\{\exp[iv(\phi - \pi)] + \exp[-iv(\phi - \pi)]\}}{2}$$

$$= -i\{\exp(iv\phi) + \exp[iv(2\pi - \phi)]\}. \tag{12.69}$$

The first term represents a wave that has propagated counterclockwise to a point ϕ, the second term a wave that has reached this point by traveling clockwise. Once again, as in the spherical case, the second term can be

neglected except when $\phi \approx \pi$. The comments on exponential attenuation and phase velocity made in connection with equation (12.16) apply without modification. By virtue of equations (12.39), the Hankel functions in equation (12.67) can again be computed from equation (12.56). We thus finally obtain the traveling-wave equivalent of the wave-harmonic series in equation (12.56):

$$\sum_n = \frac{i\pi e^{i2\pi/3} x^{2/3}}{2} \sum_m \frac{\exp[i(x + c_m x^{1/3})\phi - a_m x^{1/3}\phi]}{c_m}. \tag{12.70}$$

When we set $x = (k^2 - \gamma^2)^{1/2} a$, and substitute the result in the expression for the pressure, equation (12.55), we obtain a series of integrals that are of the form of equation (5.18) and can therefore be evaluated by stationary-phase integration:

$$p(a, \phi, z) = \frac{\rho \ddot{Q}}{8\pi a^{1/3}} \exp\left(-i\omega t + i\frac{\pi}{6}\right)$$

$$\cdot \sum_m \int_{-\infty}^{\infty} \Phi_m(\phi, \gamma) \exp[i\Psi_m(\phi, z, \gamma)] \, d\gamma, \tag{12.71}$$

where

$$\Phi_m(\phi, \gamma) = \frac{\exp[-a_m(k^2 - \gamma^2)^{1/6} a^{1/3}\phi]}{(k^2 - \gamma^2)^{1/6} c_m}, \tag{12.72}$$

$$\Psi_m(\phi, z, \gamma) = \gamma z + (k^2 - \gamma^2)^{1/2} a\phi + c_m(k^2 - \gamma^2)^{1/6} a^{1/3}\phi.$$

The point of stationary phase occurs when the derivative of the phase angle with respect to γ vanishes. Dropping higher-order terms in negative powers of ka,

$$\frac{\partial \Psi_m}{\partial \gamma} = z - \frac{\gamma a\phi}{(k^2 - \gamma^2)^{1/2}}.$$

It is convenient to express the axial cylindrical coordinate z in terms of the helical angle:

$$\alpha = \tan^{-1}\left(\frac{a\phi}{z}\right). \tag{12.73}$$

The condition of a stationary phase now becomes

$$\cot \alpha - \gamma (k^2 - \gamma^2)^{-1/2} = 0,$$

which admits the solution

$$\bar{\gamma} - k \cos \alpha,$$
$$(k^2 - \bar{\gamma}^2)^{1/2} a = ka \sin \alpha. \tag{12.74}$$

The stationary-phase values of Φ_m, Ψ_m, and of the second derivative of Ψ_m with respect to γ, which are required in the stationary-phase approximation, equation (5.21b), can now be computed:

$$\Phi_m (\bar{\gamma}, \alpha) = \frac{a^{1/3}}{(ka \sin \alpha)^{1/3} c_m} \exp[-a_m (ka \sin \alpha)^{1/3} \phi], \tag{12.75}$$

$$\Psi_m (\bar{\gamma}, \alpha, \phi) = \frac{ka \phi}{\sin \alpha} + c_m (ka \sin \alpha)^{1/3} \phi, \tag{12.76}$$

$$\left. \frac{\partial^2 \Psi_m}{\partial \gamma^2} \right|_{\gamma = \bar{\gamma}} = -\frac{k^2 a \phi}{(k^2 - \bar{\gamma}^2)^{3/2}}$$

$$= -\frac{a^2 \phi}{ka \sin^3 \alpha}, \qquad (ka \sin \alpha)^{-2/3} \ll 1. \tag{12.77}$$

Terms of order $(ka \sin \alpha)^{-2/3}$ are neglected. This approximation breaks down as $\alpha \to 0$, that is, in the illuminated zone, where the Watson formulation does not apply anyway. When equations (12.75), (12.76), and (12.77) are used in the stationary-phase approximation, the integrals in Equation (12.71) become

$$\int_{-\infty}^{\infty} = \frac{(2\pi)^{1/2} a^{1/3} (ka)^{1/6} (\sin \alpha)^{7/6}}{c_m a \phi^{1/2}} \exp[-a_m (ka \sin \alpha)^{1/3} \phi]$$

$$\cdot \exp\left[-\frac{i\pi}{4} + \frac{ika\phi}{\sin \alpha} + ic_m (ka \sin \alpha)^{1/3} \phi\right]. \tag{12.78}$$

When this is in turn substituted in the original expression for the pressure, one obtains an explicit rapidly converging series in terms of creeping waves:

$$p(\alpha, \phi) = \frac{\rho \, (ka)^{1/6} (\sin \alpha)^{7/6} \ddot{Q}}{4a \, (2\pi\phi)^{1/2}}$$

$$\cdot \sum_{m=0}^{\infty} \frac{\exp[-a_m (ka \sin \alpha)^{1/3} \phi + i\phi_m]}{c_m}, \qquad \phi \neq 0, \pi, \qquad (12.79)$$

$$\phi_m = -\frac{\pi}{12} + \frac{ka\phi}{\sin \alpha} + c_m (ka \sin \alpha)^{1/3} \phi - \omega t.$$

In the interference region $\phi \approx \pi$, waves propagating in the positive and negative ϕ directions are of comparable magnitude, but only the lowest creeping wave ($m = 0$) is of significant amplitude. In this region, the ($m = 0$) integral in equation (12.71) must be supplemented with a second ($m = 0$) integral of argument

$$\alpha = \tan^{-1} \left[(2\pi - \phi) \frac{a}{z} \right]. \qquad (12.80)$$

Its point of stationary phase is obtained by substituting the foregoing expression in equation (12.74).

The phase velocity again satisfies equation (12.20a), where the path length is now given by

$$s = (a^2 \phi^2 + z^2)^{1/2} = \frac{a\phi}{\sin \alpha}. \qquad (12.81)$$

The phase velocity is therefore of the form

$$V_m = c \left[1 + c_m \left(\frac{\sin^2 \alpha}{ka} \right)^{2/3} \right]^{-1}. \qquad (12.82)$$

As the helical angle α increases, there is a gradual decrease in phase velocity from $V_m = c$ for axially spaced points ($\alpha = 0$) to circumferentially spaced points ($\alpha = \pi/2$, $\sin \alpha = 1$), where V_m takes on the value found for a sphere of equal ka, equation (12.20b). The exponential attenuation becomes more rapid as $\sin \alpha$ increases, that is, as the radius of curvature of the geodetic path between the source and the field point at (ϕ, α) decreases from infinity at $\alpha = 0$ to a at $\alpha = \pi/2$. One thus arrives at a picture of wave fronts spreading circumferentially with the creeping wave velocity and axially with

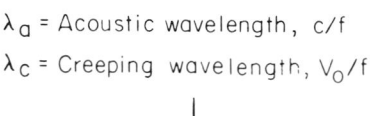

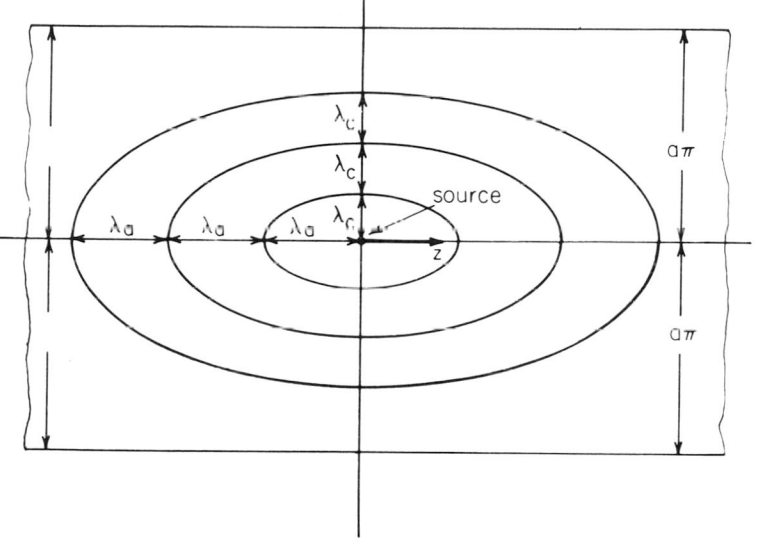

Figure 12.4
Traces of constant phase surfaces on the developed cylinder.

the free-space sound velocity. Developing the cylinder into an infinite strip of width $2\pi a$, the trace of a constant phase surface on this plane forms an ellipse whose major axis is parallel to the cylindrical axis, that is, to the infinite dimension of the strip (figure 12.4). The eccentricity remains constant as the wave front spreads from the source. The pressure decays more rapidly with circumferential than with axial spacing. An interference pattern between clockwise and counterclockwise traveling waves is formed in the regions diametrically opposite the source.

For small values of ϕ, the ϕ dependence of the creeping waves resembles the θ dependence of the sphere in the region where $\sin \theta \approx \theta$. However, because the length of the trace keeps increasing with ϕ, the focussing effect observed for the sphere in the region $\pi > \theta > \pi/2$ does not occur. The exclusive dependence of the attenuation and of the phase velocity of the creeping waves on the curvature of the geodetic path between the source and a field is consistent with a theory of diffraction developed by Keller[15] using a generalization of Fermat's principle.

A similar degree of agreement is obtained for a line source forming the generator of a circular cylindrical baffle. This source configuration was not dealt with earlier, because it is of lesser practical importance than a point source on a cylinder and can be obtained by replacing the appropriate cylindrical wave-harmonic series with the Watson residue series derived in equation (12.70). The wave-harmonic series for the surface pressure generated on a cylinder by a line source located at $\phi = 0$ is given by equation (6.41) with $r = a$ and $m = 0$. The coefficients of the Fourier series representing the displacement distribution over the cylinder are obtained from equation (6.36) with $k_m = 0$:

$$\ddot{W}_{0n} = \frac{\ddot{q}}{2\pi a} \varepsilon_n, \tag{12.83}$$

where $\ddot{q}$ is the volume acceleration per unit length of line source. When this result is substituted in equation (6.41), on setting $r = a$ the surface-pressure wave-harmonic series becomes

$$p(a, \phi) = -\frac{\rho \ddot{q}}{2\pi ka} \sum_{n=0} \frac{\varepsilon_n H_n(ka)}{H_n'(ka)} \cos n\phi. \tag{12.84}$$

This series can be transformed into the Watson residue series by applying equation (12.70). The creeping wave formulation of the surface pressure thus becomes:

$$p(a, \phi) = -\frac{i^{1/2} \rho (ka)^{-1/3} \ddot{q}}{4} \sum \frac{1}{c_m} \{\exp[i\phi_m(\phi) - a_m(ka)^{1/3}\phi]$$
$$+ \exp[i\phi_m(2\pi - \phi) - a_m(ka)^{1/3}(2\pi - \phi)]\}, \tag{12.85}$$

where the ϕ-dependent phase angle ϕ_m is defined in equation (12.18) with θ replaced by ϕ. Creeping wave velocity and exponential attenuation are seen to be the same as for a sphere of similar ka. There is, of course, no geometric spreading loss or focusing at the center of a shadow zone, because a diffracted wave spreading from the line source maintains a constant trace length as its wave front envelops the cylinder. The results in table 12.5 show that the agreement is good except near the center of the shadow zone. This region, where the interference pattern between clockwise and counterclockwise traveling diffracted waves occurs, is depicted in figure 12.5. The

Table 12.5
Surface pressure generated by an infinite strip set in a cylinder[a]

Angular spacing from center line of strip (degrees)	Normalized pressure amplitude $|p(\theta)|/\rho c \dot{W}$		Phase angle of pressure (degrees)	
	Wave-harmonic series	Present technique	Wave-harmonic series	Present technique
0^b	1.3504	1.364	20.8	-21.1
2.5^c	0.6715	0.690	9.1	10.4
5^d	0.4365	0.409	124.4	123.4
10	0.2771	0.282	-47.4	-51.1
40	0.0689	0.0730	-12.3	-17.1
80	0.0134	0.0144	39.8	41.9
120	0.0026	0.0027	92.5	92.9
160	5.6×10^{-4}	6.4×10^{-4}	152.8	145.4

a. See also figure 12.5 and Junger.[8]
b. Angle subtended by strip $\equiv 2\phi_0 = 5°$; $ka\phi_0 = \pi/2$; $ka = 36$.
c. Kirchhoff approximation.
d. Creeping wave series [equation (12.85)] with $m = 0,1$ and 2 for $\phi \leq 7.5°$, $m = 0$ and 1 for $\phi \leq 12.5°$, and $m = 0$ elsewhere.

wave-harmonic series predicts approximately the same mean pressure as the creeping wave series; but, presumably because of a truncation error, it predicts seemingly random fluctuations instead of the regular interference pattern associated with the phase velocity of the $(m = 0)$ creeping wave mode.

The creeping wave formulation for a rectangular piston on a cylindrical baffle was checked against the wave-harmonic results for circumferentially and axially spaced field points, in tables 12.6 and 12.7, respectively. The agreement, especially for the latter, is generally less satisfactory than for the other geometries. It is not clear whether this is caused by a failure of the Kirchhoff approximation in the illuminated strip subtended by the active piston, with errors resulting from the truncation of the wave-harmonic series, or errors resulting from the numerical evaluation of the inverse Fourier transform in equation (12.55).

12.7 Cylindrical Shells

We shall now go on to consider the high-frequency vibrational response of a point-excited infinite cylindrical shell by the same methods as those used in the previous sections of this chapter.

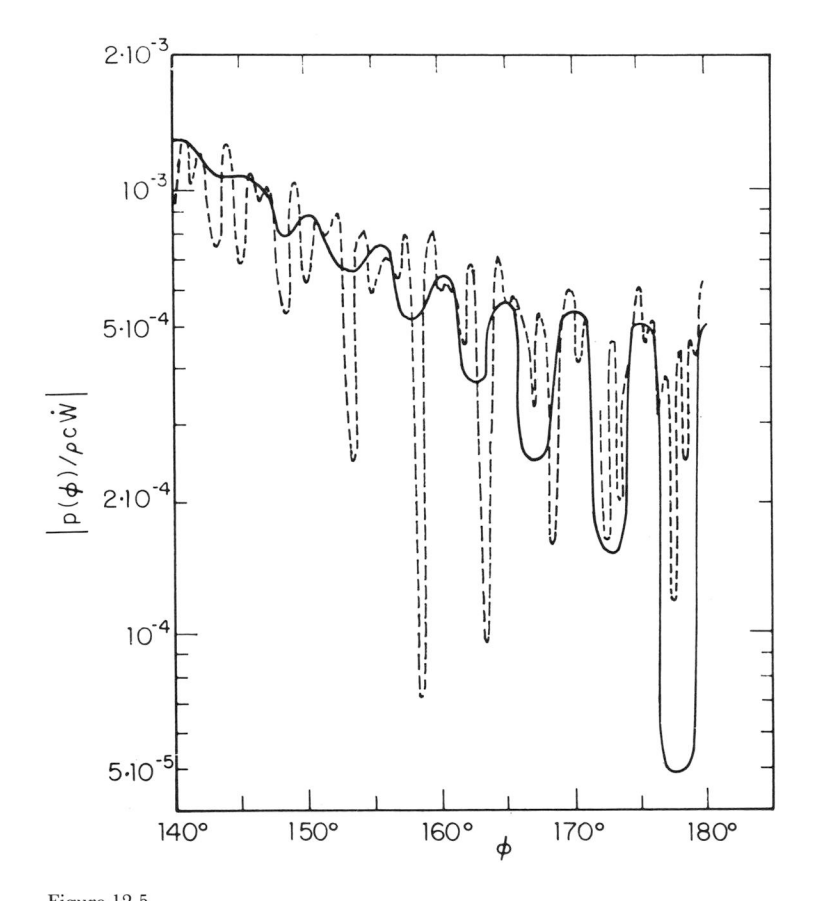

Figure 12.5
Normalized surface-pressure amplitudes generated on a cylinder near the center of the shadow zone by an infinite strip: $-\phi_0 \leq \phi \leq \phi_0 = 2.5°$; $ka\phi_0 = \pi/2$; - - -, wave-harmonic results; ———, results of present technique. [Reproduced from Junger[8]]

Table 12.6
Surface pressure generated by a square element at points spaced circumferentially around a cylinder [a]

Angular spacing from piston center (degrees)	Normalized pressure amplitude $|p(\theta,0)|/\rho c \dot{W}$		Phase angle of pressure (degrees)	
	Wave-harmonic series	Present technique	Wave-harmonic series	Present technique
0	1.534	1.513[b]	−40.9	−39.0[b]
1.61	1.172	—	−33.1	—
2.5	—	0.829[b]	—	−21.1[b]
3.23	0.4109	0.410[c]	11.9	23.0[c]
4.84	0.3590	0.286[c]	74.8	78.4[c]
6.45	0.2426	0.255[c]	133.9	139.2[c]
7.50	—	0.196[c]	—	172.9[c]
8.07	0.1960	0.180[d]	−175.8	−170.6[d]
9.68	0.1478	0.150[d]	−106.1	−103.9[d]
11.29	0.1078	0.122[d]	−43.3	−44.8[d]
12.79	0.1029	0.107[d]	7.9	12.1[d]
14.52	0.0737	0.0848[e]	69.0	76.8[e]
16.14	0.0822	0.0758	131.0	136.8
17.75	0.0694	0.0677	−168.0	−162.8
19.36	0.0734	0.0606	−105.0	−102.9
20.98	0.0538	0.0545	−48.2	−42.2
22.59	0.522	0.0494	20.2	17.8

a. $ka = 36$, piston dimensions $(2a\phi_0) \times (2z_0)$, $\phi_0 = 2.5°$, $kz_0 = ka\phi_0 = \pi/2$ (see Junger[8]).
b. Kirchhoff approximation.
c. Creeping wave series equation (12.79) with $m = 0, 1,$ and 2.
d. Creeping wave series ($m = 0$ and 1).
e. Creeping wave series ($m = 0$ for $\phi \geq 14.5°$).

Table 12.7
Surface pressures generated by a square element at points spaced axially along a cylinder[a]

| Axial spacing from piston center[b] | Normalized pressure amplitude $|p(o, z)|/\rho c \dot{W}$ | | Phase angle of pressure (degrees) | |
	Wave-harmonic series	Present technique	Wave-harmonic series	Present technique
0	1.534	1.513[c]	−40.9	−39.0[c]
z_0	0.8449	0.829[c]	−23.0	−21.1[c]
$2z_0$	0.2841	0.337	80.3	82.2
$3z_0$	0.2255	0.227	176.2	174.7
$4z_0$	0.1996	0.168	−90.9	−93.0

a. See Junger.[8]
b. Piston dimensions $(2z_0) \times (2a\phi_0)$, $\phi_0 = 2.5°$, $kz_0 = ka\phi_0 = \pi/2$.
c. Kirchhoff approximation for the square piston. All other results obtained from the present technique use the Kirchhoff approximation with a circular piston of equal area.

Consider an elastic cylindrical shell of radius a and uniform thickness h submerged in an acoustic medium of density ρ and sound speed c. A concentrated time harmonic force of magnitude F is applied to the cylindrical surface $r = a$ at the angle $\phi = 0$ and at the longitudinal location $z = 0$.

Using equation (6.49), we see that the z Fourier transform of the radiated pressure for a distribution of acceleration $\ddot{w}(z, \phi)$ on the surface of a cylinder defined by $r = a$ is given by

$$\tilde{p}(r, \phi; \gamma) = -\rho \sum_{n=0}^{\infty} \frac{\ddot{W}_n \cos n\phi \tilde{f}_n(\gamma) H_n[(k^2 - \gamma^2)^{1/2} r]}{(k^2 - \gamma^2)^{1/2} H_n'[(k^2 - \gamma^2)^{1/2} a]}, \tag{12.86}$$

where

$$\ddot{W}_n \tilde{f}_n(\gamma) = \frac{\varepsilon_n}{2\pi} \int_0^{2\pi} d\phi \int_{-\infty}^{\infty} \ddot{w}(z, \phi) \cos n\phi \, e^{-i\gamma z} \, dz \tag{12.87}$$

and we have assumed that $\ddot{w}(z, \phi)$ is an even function of ϕ. If we use the definition of $\tilde{z}_n$ given by equation (9.27) and the relation $\ddot{W}_n = -i\omega \dot{W}_n$, we can write the transform of the acoustic pressure on the surface $r = a$ as

$$\tilde{p}(a, \phi; \gamma) = \sum_{n=0}^{\infty} \tilde{z}_n \dot{W}_n \tilde{f}_n(\gamma) \cos n\phi. \tag{12.88}$$

We use the equations of motion in equation (7.80) with harmonic time dependence and assume a load distribution of the form

$$p_a(\phi, z) = \frac{F \delta(z) \delta(\phi)}{a} - p(a, \phi, z), \tag{12.89}$$

where p is the pressure exerted on the cylinder by the acoustic fluid. If in the above-mentioned set of equations we make the substitution

$$u = \frac{1}{2\pi} \sum_{n=0}^{\infty} \cos n\phi \int_{\infty}^{\infty} U_n e^{i\gamma z} \, d\gamma,$$

$$v = \frac{1}{2\pi} \sum_{n=1}^{\infty} \sin n\phi \int_{-\infty}^{\infty} V_n e^{i\gamma z} \, d\gamma, \tag{12.90}$$

$$w = \frac{1}{2\pi} \sum_{n=0}^{\infty} \cos n\phi \int_{-\infty}^{\infty} W_n e^{i\gamma z} \, d\gamma,$$

and then solve for $\dot{W}_n$, we can arrive at an expression for $\dot{w}$:

$$\dot{w}(z, \phi) = \frac{F}{(2\pi)^2 a} \sum_{n=0}^{\infty} \varepsilon_n \cos n\phi \int_{-\infty}^{\infty} \frac{e^{i\gamma z} \, d\gamma}{\tilde{Z}_n + \tilde{z}_n}, \tag{12.91}$$

where $\tilde{Z}_n$ is a complicated expression, which, for our current solution can be approximated in a manner analogous to Equation (12.26):

$$\tilde{Z}_n(\gamma^2, \Omega) \cong -i \frac{\rho_s c_p}{\Omega} \frac{h}{a} [\Omega^2 - 1 - \beta^2 (\gamma^2 a^2 + n^2)^2] + O\left(\frac{1}{\Omega^3}\right). \tag{12.92}$$

The problem we are now faced with is finding an alternative expression to equation (12.91) that will prove useful at high frequencies. As for the spherical shell, we proceed to find an integral representation for the sum. For this purpose, we rewrite the series:

$$\dot{w}(z, \phi) = \frac{F}{(2\pi)^2 a} \sum_{n=0}^{\infty} \varepsilon_n \cos n\phi \int_{-\infty}^{\infty} \frac{e^{i\gamma z} \, d\gamma}{\tilde{Z}_n + \tilde{z}_n}$$

$$= \frac{F}{(2\pi)^2 a} \sum_{n=-\infty}^{\infty} e^{in\phi} \int_{-\infty}^{\infty} \frac{e^{i\gamma z} \, d\gamma}{\tilde{Z}_n + \tilde{z}_n}. \tag{12.93}$$

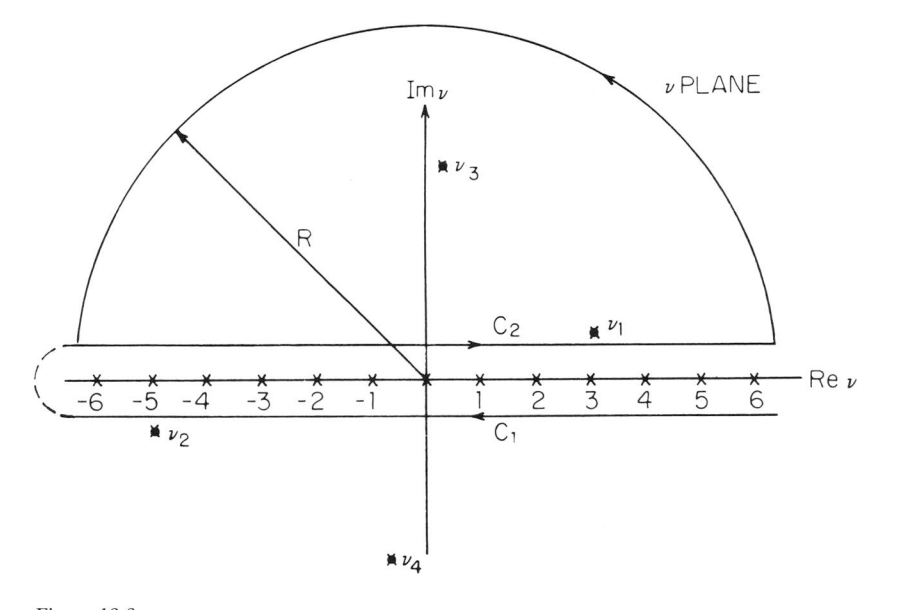

Figure 12.6
Contour of integration [equation (12.94)].

The latter summation can be written as a contour integral:

$$\mathscr{I} = \frac{i}{2} \frac{F}{(2\pi)^2 a} \oint_{C_1+C_2} \frac{dv\, e^{iv(\phi-\pi)}}{\sin v\pi} \int_{-\infty}^{\infty} \frac{dy\, e^{iyz}}{\tilde{\tilde{Z}}_v + \tilde{z}_v}, \tag{12.94}$$

where the contour $C_1 + C_2$ in the v plane encloses the entire real v axis in a clockwise manner (figure 12.6). Using residue theory, it is then shown that $\mathscr{I}$ is equivalent to the sum in equation (12.93). If we consider the integral along C_1,

$$\mathscr{I}_{C_1} = \frac{i}{2} \frac{F}{(2\pi)^2 a} \int_{C_1} \frac{dv\, e^{iv(\phi-\pi)}}{\sin v\pi} \int_{-\infty}^{\infty} \frac{dy\, e^{iyz}}{\tilde{\tilde{Z}}_v + \tilde{z}_v}, \tag{12.95}$$

and reverse the direction of integration,

$$\mathscr{I}_{C_1} = \frac{i}{2} \frac{F}{(2\pi)^2 a} \int_{C_2} \frac{dv\, e^{-iv(\phi-\pi)}}{\sin v\pi} \int_{-\infty}^{\infty} \frac{dy\, e^{iyz}}{\tilde{\tilde{Z}}_v + \tilde{z}_v}, \tag{12.96}$$

so that $\mathscr{I}$ can be written as

$$\mathscr{I} = \frac{F}{(2\pi)^2 a} \int_{C_2} \frac{dv \cos v(\phi - \pi)}{\sin v\pi} \int_{-\infty}^{\infty} \frac{dy\, e^{iyz}}{\tilde{\tilde{z}}_v + \tilde{z}_v}, \tag{12.97}$$

For $0 \leq \phi < 2\pi$ the integrand in equation (12.97) vanishes exponentially on the semicircle $v = Re^{i\theta}$; $0 < \theta < \pi$. Using this fact and residue theory, we find that

$$\mathscr{I} = -\frac{F}{2\pi a} \sum_j \int_{-\infty}^{\infty} \frac{\cos v(\phi - \pi)}{\sin v\pi\, \partial/\partial v(\tilde{\tilde{z}}_v + \tilde{z}_v)} \Bigg|_{v=v_j}^{e^{iyz}} dy, \tag{12.98}$$

where v_j are the complex zeros of the denominator that occur in the upper half plane, $\mathrm{Im}(v) > 0$. As was the case earlier, we now face the task of evaluating the zeros of $\Delta(v)$. We shall do this only for the case of the in vacuo response of the cylindrical shell.

12.7.1 CYLINDRICAL SHELL IN VACUO
In this case the acoustic impedance $\tilde{z}_v = 0$. Using equation (12.92), we find that, for high frequency,

$$\tilde{\tilde{z}}_v = -\frac{i\rho_s c_p}{\Omega} \frac{h}{a} [\Omega^2 - 1 - \beta^2 (\gamma^2 a^2 + v^2)^2]. \tag{12.99}$$

The zeros in the upper half-plane are

$$v_1 = \left(\frac{\Omega}{\beta} - \gamma^2 a^2\right)^{1/2} = a(k_f^2 - \gamma^2)^{1/2} \left.\begin{array}{c} \\ \\ \end{array}\right\} \tag{12.100}$$

$$\Omega^2 \gg 1.$$

$$v_3 = i\left(\frac{\Omega}{\beta} + \gamma^2 a^2\right) = ia(k_f^2 + \gamma^2)^{1/2} \tag{12.101}$$

We can now solve equation (12.98) explicitly:

$$\dot{w}(\phi, z) = \frac{iFa}{\pi \tilde{z}_p} \left[\int_{-\infty}^{\infty} \frac{\cos v_1(\phi - \pi) e^{iyz}}{v_1 \sin v_1 \pi} dy - \int_{-\infty}^{\infty} \frac{\cos v_3(\phi - \pi) e^{iyz}}{v_3 \sin v_3 \pi} dy\right]. \tag{12.102}$$

This expression is analogous to equation (12.31) for the spherical shell. The task that still remains however, in this case is the inversion of the Fourier transform over the z coordinate. It is convenient to use the identity equation (12.68) and the relationship

$$\cos v(\phi - \pi) = \tfrac{1}{2}[e^{iv(\phi-\pi)} + e^{-iv(\phi-\pi)}].$$

Using these identities, equation (12.102) can be put in the form

$$\dot{w}(\phi, z) = \frac{F}{\pi Z_p} \sum_{m=0}^{\infty} [A_m(\phi) + A_m(2\pi - \phi) + iB_m(\phi) + iB_m(2\pi - \phi)], \tag{12.103}$$

where

$$A_m(\phi) = \int_{-\infty}^{\infty} \frac{e^{i[\gamma z + v_1(\phi + 2m\pi)]}}{(k_f^2 - \gamma^2)^{1/2}} \, d\gamma, \tag{12.104}$$

$$B_m(\phi) = \int_{-\infty}^{\infty} \frac{e^{i[\gamma z + v_3(\phi + 2m\pi)]}}{(k_f^2 + \gamma^2)^{1/2}} \, d\gamma. \tag{12.105}$$

The integrals appearing in equation (12.104) and equation (12.105) can be evaluated in closed form. Making the transformation $\gamma = k_f \sin w$ and letting

$$z = R_m \sin \psi,$$
$$a(\phi + 2m\pi) = R_m \cos \psi, \tag{12.106}$$

where $R_m = [z^2 + a^2(\phi + 2m\pi)^2]^{1/2}$ is the helical distance measured on the surface of the cylinder from the drive point to the observation point in the direction of increasing ϕ, equation (12.104) can be put in the form

$$A_m(\phi) = \int_{-\pi/2+i\infty}^{\pi/2-i\infty} \exp\{ik_f R_m \cos(w - \psi)\} \, dw, \tag{12.107}$$

where the path of integration W_1 is shown in figure 12.7. This representation can be shown to be equivalent to that of the Hankel function first derived

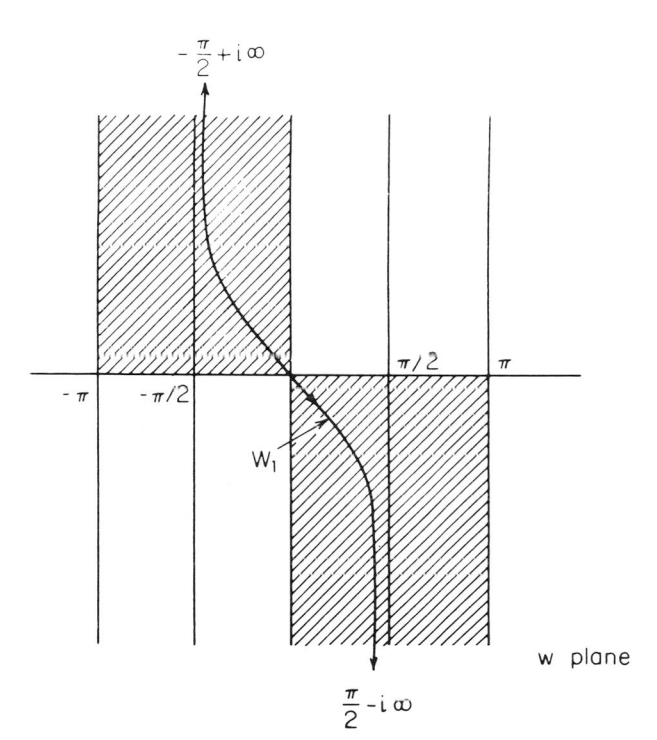

Figure 12.7
Contour of integration [equation (12.107)].

by Sommerfeld.[16] We finally obtain

$$A_m(\phi) = \pi H_0^{(1)}(k_f R_m).$$ (12.108)

In figure 12.7, the cross-hatched region denotes the region in which the path of integration must be taken so that the integral converges.

The B_m defined by equation (12.105) can be evaluated as an inverse Fourier cosine transform directly;[17]

$$B_m(\phi) = 2K_0(k_f R_m).$$ (12.109)

Using Equations (12.103), (12.108), and (12.109), we finally arrive at an expression for the radial velocity response:

$$\dot{w}(z,\phi) = \frac{F}{Z_p} \sum_{m=0}^{\infty} \left\{ H_0(k_f R_m) \right.$$

$$\left. + H_0(k_f R'_m) + \frac{2i}{\pi} [K_0(k_f R_m) + K_0(k_f R'_m)] \right\},$$

(12.110)

where $R'_m = [z^2 + a^2(2\pi - \phi + 2m\pi)^2]^{1/2}$ again represents the helical distance from the drive point to the observation point, but this time measured in the direction of decreasing ϕ.

Comparing equation (7.66) and equation (12.110), we see that the response of the point-excited shell can be interpreted in terms of the solution to the point-excited plate. The shell response is given by a superposition of terms, each of which is exactly in the form of equation (7.66) but whose spatial dependence corresponds to distinct circumnavigations of the cylinder along helical paths. For high frequencies and field points not too near the drive point, $k_f R_m, k_f R'_m \gg 1$, equation (12.110) is well approximated by

$$\dot{w}(z,\phi) = \frac{F}{Z_p} \sum_{m=0}^{\infty} \left[\sqrt{\frac{2\pi}{k_f R_m}} e^{i(k_f R_m - \pi/4)} + \sqrt{\frac{2\pi}{k_f R'_m}} e^{i(k_f R'_m - \pi/4)} \right].$$

(12.111)

In equation (12.111) we have neglected the contribution of $K_0(k_f R_m)$ and $K_0(k_f R'_m)$ because these give rise only to exponentially decaying near fields and are only of significance near the drive point.

The radial response calculated from equation (12.111) for $h/a = 0.01$, $\eta_s = 0.04$, and $\Omega = 13.8$ is shown in figure 12.8. Here we have plotted the envelope of the maxima and minima of the radial velocity along two specified directions, one along the generator and the other around the circumference of the shell. An interesting feature is that the responses are the same up to a certain point, but then diverge, with the envelope of the response along the circumference always including the response along the generator. This is a result of well-defined standing waves in the circumferential direction formed by the interference between waves that travel around the cylinder in two opposite directions. Near the drive point the standing wave is less apparent because the response is primarily caused by the disturbance that has traveled through the least distance. The wave that has traversed the cylinder in the opposite direction must travel through a substantially larger distance given by $s = a(2\pi - \phi)$. In the vicinity of $\phi = \pi$, the path lengths of the two $m = 0$ waves, for example, are much closer, and hence their relative contributions to the total response are comparable.

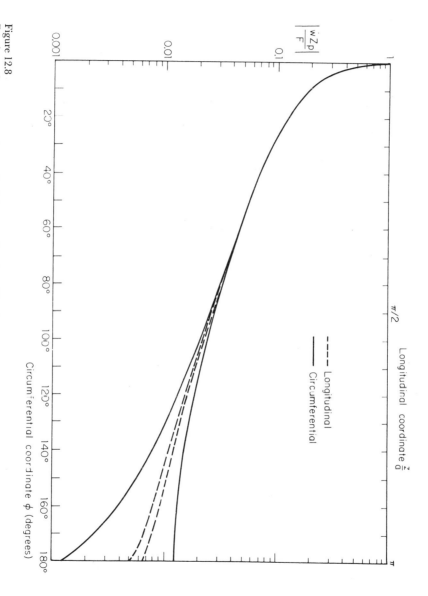

Figure 12.8
Envelope of response maxima and minima as a function of longitudinal and circumferential coordinates. The mobility of the shell has been normalized by multiplying it by the point impedance ζ_p of the infinite plate, equation (7.68).

In figure 12.9 is a comparison of the envelopes of maximum and minimum responses of the cylindrical and spherical shells. The sphere displays larger buildup at the antipode ($\theta = \pi$). This is a result of the fact that in the sphere all the rays' paths converge at the antipode, producing a focusing of energy. This is, of course, not the case for the cylinder, since waves that are launched at an angle that is not perpendicular to the longitudinal axis do not converge on the back side of the cylinder.

12.8 Pressure Radiated by a Point-Excited Cylindrical Shell

To complete the analysis of the high-frequency structural, acoustic interaction problem we shall now direct our attention to the radiated pressures. The far-field radiated pressure of a cylindrical radiator, equation (6.53), is a modal series that for high frequencies, $ka \gg 1$, presents convergence difficulties. If we combine equation (6.53) with equation (12.91), we obtain the expression

$$p(R, \theta, \phi) = \frac{\rho \omega F}{2\pi^2} \frac{e^{ikR}}{R} \sum_{n=0}^{\infty} \frac{\varepsilon_n e^{-in\pi/2} \cos n\phi}{(\tilde{\tilde{Z}}_n + \tilde{z}_n) ka \sin \theta \, H_n'(ka \sin \theta)}, \qquad (12.112)$$

where $\tilde{\tilde{Z}}_n$ and $\tilde{z}_n$ are evaluated at $\bar{\gamma} = k \cos \theta$, using equations (9.27) and (12.92).

Now consider the sum in equation (12.112) and manipulate it into a form that can be readily calculated. If we use the relation

$$H_{-n}(x) = e^{in\pi} H_n(x) \qquad (12.113)$$

and the fact that $\tilde{\tilde{Z}}_n$ and $\tilde{z}_n$ are even functions of n, the summation in equation (12.112) can be written as

$$S = \sum_{n=-\infty}^{\infty} \frac{e^{-in\pi/2} e^{-in\phi}}{(\tilde{\tilde{Z}}_n + \tilde{z}_n) x H_n'(x)}, \qquad (12.114)$$

where $x = ka \sin \theta$. If we then consider the integral

$$\mathscr{I} = \frac{i}{2x} \oint_{C_1 + C_2} \frac{e^{-iv(\phi - \pi/2)}}{(\tilde{\tilde{Z}}_v + \tilde{z}_v) H_v'(x) \sin v\pi} \, dv, \qquad (12.115)$$

where $C_1 + C_2$ is again a contour that encloses the entire real axis as shown

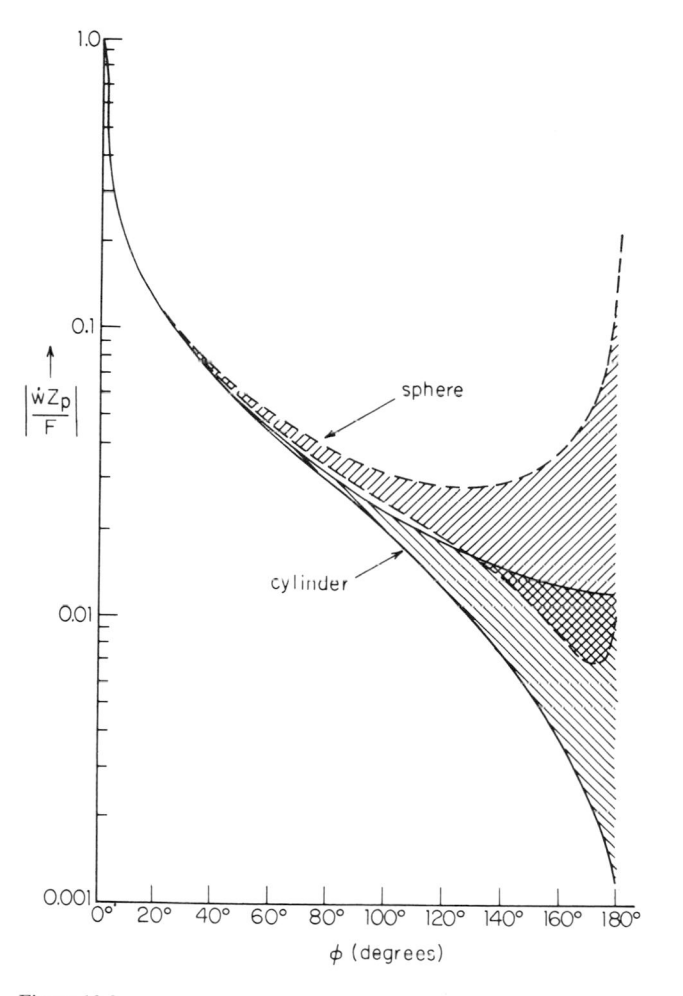

Figure 12.9
Comparison of the response envelopes of spherical and cylindrical shells. The shell mobility has been normalized as in the preceding figure.

in figure 12.6. Using residue theory, it can be shown that $\mathscr{I}$ is identical to the summation S. If on C_1 we reverse the direction of integration,

$$\mathscr{I} = \frac{i}{2x} \int_{C_2} \frac{e^{-iv\pi} \left[e^{iv(3\pi/2 - \phi)} + e^{iv(\phi - \pi/2)} \right]}{(\tilde{Z}_v + \tilde{z}_v) H'_v(x) \sin v\pi} \, dv. \tag{12.116}$$

On the contour C_2, $\operatorname{Im} v > 0$, so that we can again make use of the identity given by equation (12.68). Equation (12.116) now becomes

$$\mathscr{I} = \sum_{m=0}^{\infty} \left\{ C_m \left(\frac{3\pi}{2} - \phi \right) + C_m \left(\phi - \frac{\pi}{2} \right) \right\}, \tag{12.117}$$

where

$$C_m(\psi) = \frac{1}{x} \int_{C_1} \frac{e^{iv(\psi + 2m\pi)}}{(\tilde{Z}_v + \tilde{z}_v) H'_v(x)} \, dv. \tag{12.118}$$

In our previous use of integral expressions such as the foregoing, we would have located the poles of the integrand in the upper half of the v plane and then used residue theory to represent the integral. The residue series thus calculated would as a result of its convergence properties represent the solution only in the shadow zone. For our present purposes we are interested in the illuminated region and we therefore proceed in a different manner in our consideration of $C_m(\psi)$. We first examine the behavior of the integrand along the contour C. In the region $v \gg ka \sin \theta$, the magnitude of the integral behaves like

$$\exp\left(-v \log \frac{2v}{ka \sin \theta} \right).$$

For values of $ka \sin \theta$ not too small, this portion of the contour contributes little to the value of the integral. It is more convenient to rewrite equation (12.118) as

$$C_m(\psi) = \frac{-i}{\rho \omega a} \int_{C_1} \frac{b(v) e^{iv(\psi + 2m\pi)}}{H_v(ka \sin \theta)} \, dv, \tag{12.119}$$

where

$$b(v) = \left(\frac{\tilde{z}_v}{\tilde{z}_v} + 1\right)^{-1}.$$

This form of the integral allows us to go to the limit of a very thin shell or to the case of a force applied directly to the acoustic fluid without first reformulating the solution. In the range $v \ll ka \sin \theta$, we can make use of Sommerfeld's second-order saddle point approximation:[18]

$$H_v(x) \simeq \sqrt{\frac{2}{\pi x \sin \alpha}} \exp\left[ix(\sin \alpha - \alpha \cos \alpha) - \frac{\pi}{4} i \right], \tag{12.120}$$

where $\cos \alpha = v/x$ and $\alpha > 0$. Equation (12.118) now becomes

$$C_m(\psi) = \frac{e^{-i\pi/4}}{\rho \omega a} \sqrt{\frac{\pi x}{2}} \int_{C_1} b(v) \sqrt{\sin \alpha} \; e^{if_m(v)} \, dv, \tag{12.121}$$

where $x = ka \sin \theta$ and

$$f_m(v) = v(\psi + 2m\pi) + x(\alpha \cos \alpha - \sin \alpha). \tag{12.122}$$

As given by equation (12.121), $C_m(\psi)$ is now in a form suitable for evaluation by the method of stationary phase. The point of stationary phase v is given as a solution of $f'_m(v) = 0$, viz.:

$$\alpha + \psi + 2m\pi = 0. \tag{12.123}$$

The condition for the validity of the approximation given in equation (12.120) is that $\alpha > 0$, so that the integrand has a stationary phase only if $\psi + 2m\pi < 0$.

If we consider that $0 \le \phi < 2\pi$ for its principal value, the condition for the existence of a stationary phase point limits the summation in equation (12.117) to the case $m = 0$ and $0 < \phi < \pi/2$, $3\pi/2 < \phi < 2\pi$. If $m \ne 0$, the integral cannot be evaluated by the method of stationary phase, because there is no stationary phase point for the integral. The stationary phase integration in equation (5.11) yields

$$\mathscr{I} = -\frac{i\pi \sin \theta \cos \phi \, e^{-ika \sin \theta \cos \phi}}{\rho c} b(|\bar{v}|), \qquad \bar{v} = ka \sin \theta \sin \phi. \tag{12.124}$$

Combining the high-frequency approximation for the shell impedance as given by equation (12.92) and the high-frequency asymptotic form of the acoustic impedance equation (9.27),

$$\tilde{z}_v(k\cos\theta) = \frac{i\rho c}{\sin\theta}\frac{H_v(ka\sin\theta)}{H_v'(ka\sin\theta)} \simeq \rho c (k^2 a^2 \sin^2\theta - v^2)^{-1/2}, \tag{12.125}$$

we arrive at an expression for the radiated pressure:

$$p(R,\theta,\phi)$$

$$= -\frac{ikF}{2\pi R}e^{ik(R-a\sin\theta\cos\phi)} \tag{12.126}$$

$$\cdot \frac{\sin\theta\cos\phi}{1 - [i\omega\rho_s h/(\rho c)]\cos\phi\sin\theta\{1 - (\omega^2/\omega_c^2)[\cos^2\theta + \sin^2\theta\sin^2\phi]^2\}}.$$

The interesting feature of this result is that it is analogous to the pressure radiated by a point-excited infinite plate. This is exactly what one would expect when the radius of curvature becomes infinite compared with the wavelength.

12.9 Spherical Shell Radiated Field
Procedures quite similar to those used in calculating the radiated field of a point-excited cylindrical shell can also be used for a spherical shell. The far-field pressure is given in equation (9.14), where, for our purposes, $(-i)^n$ will be expressed as $\exp(-in\pi/2)$. Again, in order to obtain a useful approximation for high frequency, $ka \gg 1$, the wave-harmonic series for the far-field is formulated as an integral. The series being in the form of equation (12.22), we immediately obtain the integral representation

$$S = \frac{-i}{2}\oint_C \frac{vP_{v-1/2}(-\cos\theta)e^{-i(v-1/2)\pi/2}\,dv}{\cos v\pi\,(1 + \tilde{Z}_{v-1/2}/z_{v-1/2})h_{v-1/2}(ka)}, \tag{12.127}$$

where C is again the contour shown in figure 12.1. Utilizing the fact that the integrand is an odd function, we can replace the lower half of the contour by its reflection in the origin. This latter contour in conjunction with the

upper half of C is denoted by the contour C'. If we replace $(\cos v\pi)^{-1}$ by the identity equation (12.15), which is valid for $\mathrm{Im}\, v > 0$ and is satisfied on the contour C', equation (12.127) becomes

$$S = e^{-i\pi/4} \sum_{m=0}^{\infty} e^{im\pi} \oint_C \frac{v P_{v-1/2}(-\cos\theta)\, e^{iv\pi(2m+1/2)}}{(1 + \mathcal{Z}_{v-1/2}/z_{v-1/2})\, h_{v-1/2}(ka)}\, dv. \tag{12.128}$$

We now represent the Legendre function in the above integrand in the form[19]

$$P_{v-1/2}(-\cos\theta) = \frac{1}{2\pi} \int_{-\pi}^{\pi} \exp[(v - \tfrac{1}{2}) \\ \cdot \{\log[\cos(\theta - \pi) + i\sin(\theta - \pi)\cos\eta]\}]\, d\eta \tag{12.129}$$

and use the relation

$$h_{v-1/2}(ka) = \sqrt{\frac{2}{\pi k a}}\, H_v(ka) \simeq \frac{e^{i(ka\sin\alpha - v\alpha - \pi/4)}}{ka\sqrt{\sin\alpha}},$$

where $v = ka\cos\alpha$. This then leads to an expression for S in the form

$$S = \frac{-ika}{2\pi} \sum_{m=0}^{\infty} e^{im\pi} \iint A(v)\, e^{if(v,\eta)}\, dv\, d\eta, \tag{12.130}$$

where

$$A(v) = v\sqrt{\sin\alpha} \left(1 + \frac{\mathcal{Z}_{v-1/2}}{z_{v-1/2}}\right)^{-1} \\ \cdot \exp(v - \tfrac{1}{2}) \log|\cos(\theta - \pi) + i\cos\eta\sin(\theta - \pi)|,$$

$$f(v,\eta) = v\left(2m\pi + \frac{\pi}{2} + \alpha\right) - ka\sin\alpha + (v - \tfrac{1}{2})\tan^{-1}[\tan(\theta - \pi)\cos\eta].$$

We can now evaluate the double integral, equation (12.130), by the method of stationary phase as applied to double integrals [equation (5.40)]. We find that the only term of the m series that contributes is the $m = 0$ term:

$$S = \frac{(ka)^2 \cos \theta \, e^{-ika \cos \theta}}{1 + \mathcal{Z}_{v-1/2}/z_{v-1/2}} \bigg|_{v=ka \sin \theta}, \tag{12.131}$$

which is valid for $\theta < \pi/2$. If we substitute this result into equation (9.14), we arrive at the expression

$$p(R, \theta) = \frac{iFke^{ikR}}{2\pi R} \frac{\cos \theta \, e^{-ika \cos \theta}}{1 + \mathcal{Z}_{v-1/2}/z_{v-1/2}} \bigg|_{v=ka \sin \theta},$$

which again leads to an expression analogous to that given by radiation from a point-excited plate.

In summary, we see that for high frequencies the far-field pressures radiated by point-excited cylindrical and spherical shells are exactly analogous to the pressure radiated by a point-excited infinite plate.

References

1. G. N. Watson, "The Diffraction of Electric Waves by the Earth; the Transmission of Electric Waves round the Earth, "*Proc. Royal Soc. London* A95:83–99, 546–563 (1918).

2. See A. Sommerfeld, *Partial Differential Equations in Physics* (New York: Academic Press, 1949), pp. 279–289.

3. J. R. Wait, *Electromagnetic Radiation from Cylindrical Structures* (London: Pergamon Press, 1959), chapter 8.

4. For a review of the acoustics literature on creeping waves as well as their experimental study, see M. L. Harbold and B. N. Steinberg, "Direct Experimental Verification of Creeping Waves," *J. Acoust. Soc. Am.* 45:592–603 (1969).

5. An extensive body of literature was developed by Überall and his coworkers: e.g., for cylinders, R. D. Doolittle, H. Überall, and P. Ugincius, "Sound Scattering by Elastic Cylinders," *J. Acoust. Soc. Am.* 43:1–14 (1968); more generally, H. Überall, L. R. Dragonette, and L. Flax, "Relation between Creeping Waves and Normal Modes of Vibration of a Curved Body," *J. Acoust. Soc. Am.* 61:711–718 (1977).

6. G. C. Gaunaurd, "High-Frequency Acoustic Scattering from Submerged Cylindrical Shells Coated with Viscoelastic Layers," *J. Acoust. Soc. Am.* 62:503–512 (1977).

7. M. C. Junger and D. Feit, "High-Frequency Response of Point-Excited Submerged Spherical Shells," *J. Acoust. Soc. Am.* 45:630–636 (1969).

8. M. C. Junger, "Surface Pressures Generated by Pistons on Large Spherical and Cylindrical Baffles," *J. Acoust. Soc. Am.* 41:1336–1346 (1967).

9. M. Abramowitz and I. A. Stegun, *Handbook of Mathematical Functions* (Washington, D.C.: NBS Supt. of Doc., 1969) p. 332, formula (8.12).

10. P. M. Morse and H. Feshbach, *Methods of Theoretical Physics* (New York: McGraw-Hill, 1953), pp. 363ff.

11. Sommerfeld, *Partial Differential Equations*, p. 288, equation (21). A detailed derivation of the roots in (12.7) can be found in the mathematical appendix of chapter 13 in the first edition of Sound, Structures, and Their Interaction.

12. Abramowitz and Stegun, *Handbook*, p. 336, formula (8.10.7).

13. D. Feit and M. C. Junger, "High Frequency Response of an Elastic Spherical Shell," *J. Appl. Mech.* 36:859–864 (1969).

14. W. Streifer and R. S. Kodis, "On the Solution of a Transcendental Equation Arising in the Theory of Scattering by a Dielectric Cylinder," *Quart. Appl. Math.* 21:285–298 (1964).

15. J. B. Keller, "Geometric Theory of Diffraction," *J. Opt. Soc. Am.* 52:116–130 (1962); also, "A Geometrical Theory of Diffraction," in *Calculus of Variations and Its Application* (*Proceedings Symposium Appl. Math.*) (New York: McGraw-Hill, 1958), Vol. 8, pp. 27–52.

16. Sommerfeld, *Partial Differential Equations*, p. 89.

17. H. Bateman, *Tables of Integral Transforms*, Vol. 1, A. Erdélyi, ed. (New York: McGraw-Hill, 1954), p. 17, equation (27).

18. Sommerfeld, *Partial Differential Equations*, p. 119.

19. H. Bateman, *Higher Transcendental Functions*, Vol. 1, A. Erdélyi, ed. (New York: McGraw-Hill, 1953), p. 158, equation (23).

Glossary

Note: Vectors are shown in boldface. A tilde indicates a transform. Other conventions are indicated in equations (2.13), (2.14), and (2.15). An overbar on a natural frequency identifies the resonance of a radiation-loaded structure; an overbar on a spatial transform parameter identifies the point of stationary phase.

a	radius
B	bulk modulus
c	sound velocity
c_b	low-frequency phase velocity of compressional waves in a beam, $= (E/\rho_s)^{1/2}$
c_f	velocity of flexural waves in an elastic plate, in vacuo in chapter 7 and when fluid-loaded in figure 8.1
c, g, l	subscripts identifying bubble swarm parameters pertaining to, respectively, the equivalent homogeneous medium, the gaseous cavities, and the liquid matrix
c_0	phase velocity in liquid-filled, elastic pipe or hose
c_p	low-frequency phase velocity of compressional waves in an elastic plate, (2.53)
c_R	Rayleigh wave velocity
c_s	phase velocity of structure-borne sound waves; in chapter 7 shear velocity, $= (G/\rho_s)^{1/2}$
D	flexural rigidity, $= Eh^3/12(1 - v^2)$
e	half-distance between two point sources
E	Young's modulus
F	concentrated force acting normally to a plate or shell; also shear force in a beam
F_{mn}	modal generalized force
f_{mn}	mode shape of rectangular plate

f_n modal generalized force per unit area of a spherical or cylindrical shell, positive outward

f_0 natural frequency of a bubble, (3.34a)

G shear modulus, $= E/2(1 + v)$

h plate or shell thickness

I acoustic intensity $[(3.21), (3.22),$ and $(3.26)]$; also, cross-sectional moment of inertia of a beam; also integral

$\mathscr{I}$ integral

k wave number, $= \omega/c$

k_f flexural wave number, in vacuo, (7.16) and (7.62)

k_n, k_m wave numbers identifying standing wave configuration on planar or cylindrical radiators (see figure 5.9)

k_0 spring stiffness per unit area

k_s structural wave number (see figure 5.10); helical wave number of cylindrical shell, (7.100)

$\mathbf{k}_i$ vector wave number of amplitude k pointing in the direction of propagation of a plane incident wave

L half-length of array or of cylindrical or other slender sound radiator or scatterer; also, length of simply supported beam

L_x, L_y half-length of x and y dimensions of rectangular radiators, respectively, in chapters 5 and 8; in chapter 7, dimensions of simply supported plates

M bending moment in beam; also total mass of beam

M_{mn} modal mass of plate, (11.17)

M_n modal mass per unit area of cylindrical or spherical shell

M_p, R_p resultant accession to inertia and radiation resistance of a circular piston in a plane baffle, respectively (see figure 5.6)

m_n accession to inertia per unit area, $= -\operatorname{Im}(z_n)/\omega$

m_s	accession to inertia per unit area on a plane boundary, for a standing wave field of infinite extent, of structural wave number k_s (see figure 5.16)
P_s	hydrostatic pressure
p	acoustic pressure
p_a	transverse force per unit area applied to a plate or shell
p_i	pressure in a plane wave incident upon a scatterer or reflector
p_r	pressure radiated by an elastic scatterer excited by an incident wave $(p_{s\infty} - p_{se})$; in section 3.3, reflected pressure
p_s	pressure field of point source [see equation (3.2)]
p_{se}	pressure scattered by an elastic boundary
p_{s0}	pressure scattered by a "pressure release" boundary
$p_{s\infty}$	pressure scattered by a boundary of infinite impedance
p_t	pressure transmitted through an elastic plate, sections 11.3 and 11.4
q	distribution of forces per unit length acting on a beam
$\ddot{Q}$	amplitude of volume acceleration of a sound source
R, θ, ϕ	spherical coordinates
$\mathbf{R}$	position vector locating a field point
$\mathbf{R}_0$	position vector locating a source point
r_n	specific acoustic resistance, $= \mathrm{Re}(z_n)$
r_s	specific acoustic resistance associated with m_s (see figure 5.16); also, structural resistance of shell associated with η_s [see equations (8.95)]
r, ϕ, z	cylindrical coordinates
S	surface; also, cross-sectional area of beam
S_0	radiating surface

S_F area of the first Fresnel zone

s Laplace transform parameter

T period of vibration, $= 2\pi/\omega$; also, kinetic energy

$T(R, \theta)$ pressure radiated by a plate excited by a unit force, (8.30) or (8.31) with $F = 1$

t time

U, V group and phase velocities, respectively

$\dot{U}_s, \dot{V}_s, \dot{W}_s$ shell velocity components

u, v displacement components tangential to a boundary

V potential energy; in chapters 10 and 11, scatterer volume

V_f velocity of flexural waves in beams, (7.40)

V_m phase velocity of creeping waves

w displacement component normal to a boundary; also, transverse displacement of beam

X_{mn} reactance, $= \mathrm{Im}(Z_{mn})$, in units of pressure over velocity

x_w specific acoustic reactance of a locally reacting wave guide boundary

x, y, z rectangular coordinates, with z normal to a planar boundary

Y mechanical mobility, in units of velocity over force

$\tilde{Z}_p, \tilde{Z}_a$ transform of the structural impedance of a plate, (8.15) and (8.17), and of the acoustic medium in contact with a planar boundary, (8.19), in units of pressure over velocity

Z_n, Z_{mn} modal impedance of a shell vibrating in vacuo in units of pressure over velocity

Z_p drive-point impedance of point excited infinite plate, in units of force over velocity, (7.68)

z_n	specific acoustic impedance, in units of pressure over velocity
α	half-angle subtended by a circular piston on a spherical baffle (see figure 6.2); also, in chapters 3 and 8, fractional air volume in a bubble swarm
α_{nm}	radial wave number of a waveguide mode
α_f	$= k_f a$
β	$\dfrac{1}{12^{1/2}} \dfrac{h}{a}$
$\gamma, \gamma_x, \gamma_y$	spatial transform parameters
γ_{nm}	axial wave number of a waveguide mode
γ	ratio of specific heats of a gas at respectively constant pressure and constant temperature $= 1.41$ for air; in chapters 6 and 9, ratio of entrained mass to the mass of the structure; in chapter 8, (8.10) and (8.12), structural wavenumber of fluid-loaded plate
ε	fluid loading parameter, $= \rho c / \omega_c \rho_s h$
ε_n	Neumann function, $= 1$ for $n = 0$, and 2 for $n > 0$
η	$\cos \theta$
η_s	structural loss factor
$\langle \eta_n \rangle$	radiation loss factor associated with radiation of acoustic power by the nth structural mode, space averaged over the radiating surface
$\theta_0, \theta_m, \theta_c$	critical or coincidence angle for which the trace of the acoustic wavelength matches the structural wavelength, $= \sin^{-1} k_s / k$ when measured from normal to radiating surface
λ	acoustic wavelength, $= 2\pi / k$
λ_n	$= n(n + 1)$ in chapters 7 and 12
v	Poisson's ratio; also, in chapter 12, complex wave-harmonic order number

$\hat{\xi}_0$ unit vector pointing into the acoustic medium (see figure 4.2)

Π acoustic power

ρ density of acoustic medium

ρ_s density of structural material

σ, ε stress and strain, respectively

Φ, Ψ respectively, modulus and phase angle of integrand suitable for stationary phase integration

Ω dimensionless frequency of a vibrating spherical or cylindrical shell, $= \omega a/c_p = (c/c_p)ka$, in chapters 7 and 9; also, ω/ω_c in chapter 8

ω circular frequency

ω_c circular coincidence or critical frequency at which flexural wavelength in an elastic plate equals the acoustic wavelength in adjoining fluid, $= \sqrt{12c^2/hc_p}$

ω_m characteristic frequency above which the inertial reactance of a plate exceeds the ambient fluid's characteristic impedance, $= \rho c/\rho_s h$

ω_n, ω_{mn} circular natural frequency of a shell vibrating in vacuo

ω_0 circular natural frequency of a bubble, (3.34)

τ $[(c/c_f)^2 - 1]^{1/2}$

ζ $(\gamma^2 - k^2)^{1/2}$ in chapter 8; $(\nu^2 - k^2a^2)^{1/2}$ in chapter 12

Functions

$F_n(\alpha)$ normalized modal impedance of the boundary of a fluid cylinder, (11.77)

$G(\mathbf{R}|\mathbf{R}_0)$ Green's function satisfying Neumann conditions on the boundary [see equation (4.22)]

$g(|\mathbf{R} - \mathbf{R}_0|)$ free-space Green's function [see equation (4.8)]

$H(x)$ Heaviside function [see equation (5.2)]

$H_n(x)$	cylindrical Hankel function of the first kind [see equation (6.40)]
$h_n(x)$	spherical Hankel function of the first kind [see equation (6.11)]
$\mathcal{J}_n(x)$	cylindrical Bessel function of the first kind
$j_n(x)$	spherical Bessel function of the first kind
$j_0(x)$	$\equiv \sin x / x$
$K_n(x)$	modified Hankel function [see equation (6.43)]
$P_n(\eta)$	Legendre polynomial
$P_n^m(\eta)$	associated Legendre function [see equation (6.5)]
$Y_n(x)$	cylindrical Bessel function of the second kind
$y_n(x)$	spherical Bessel function of the second kind
$\delta(\mathbf{R} - \mathbf{R}_0)$	Dirac delta function [see equations (4.2)]
∇^2	Laplace operator
∇_σ^2	surface Laplace operator

Index

Accession to inertia. *See* Radiation loading; Specific acoustic reactance
Adiabatic gas law, 18
Airy integrals, 406
Arrays
  end fire, 59
  line, 54
  product theorem, 112
  rectangular, 109
Assumptions of linear acoustics, 2, 3
  of Bernoulli-Euler beams, 198
  of thin shell theory, 216

Bessel functions
  cylindrical
    addition theorems, 145–147, 322
    asymptotic expressions, 103, 109, 143, 176
    integral of, 106, 146, 147
    integral representation, 96, 145, 331
  spherical
    addition theorems, 319
    asymptotic expressions, 35, 158
Bessel's differential equation, 167
Boundary conditions
  beams, 207
  cylindrical radiator, 166
  cylindrical shells, 224
  elastic scatterer, 342
  Neumann, 86
  plane wave, 26
  plates, 213
  rigid baffle, 316
  spherical source, 30
Boyle's law, 18
Breathing mode
  cylindrical shell, submerged, 292
  cylindrical shell in a vacuum, 223
  gas bubble, 67
  spherical shell, submerged, 288
Bulk modulus, 17

Cauchy's theorem. *See* Residue theory
Characteristic impedance, 21
Coincidence angle, 247
  for cylindrical radiators, 179
  effect on elastic baffle performance, 375
  effect on transmission loss, 349
  for planar radiators, 121, 270
Coincidence frequency, 236
Condensation
  related to bulk modulus, 17
  restricted by small-signal assumption, 2
Corner mode, 123, 144, 127

Coupling by mutual modal impedances, 266
Creeping waves
  on cylindrical surface, 409
  definition, 389
  phase velocity of, 395
  on submerged spherical shells, 401
Critical angle. *See* Coincidence angle
Critical frequency. *See* Coincidence frequency
Cylindrical radiator
  finite, 176
  high-frequency formulation, 408
  infinite, 168
Cylindrical shell, sound radiated
  free-free, 305
  high-frequency, 429
  simply supported, 301
Cylindrical shell vibrations
  equations of motion, 217
  extensional and inextensional type modes, 219
  high-frequency, 427
  nonplanar vibrations, 222
    natural frequencies, 226
    submerged, 301
  planar vibrations
    mode shapes, 218
    natural frequencies, 219

Damping. *See also* Structural damping
  by sound radiation, 357
Decibels. *See* Sound pressure level
Dirac delta function, 76
Directivity factor
  circular piston, 97
  line array, 56
  two point sources, 51
Doublet, 54

Edge mode
  of cylindrical source, 182
  of rectangular source, 127, 141
Entrained mass. *See* Specific acoustic reactance
Equations of motion of structures coupled to a fluid
  fluid-filled spherical shell, 287
  submerged cylindrical shell, 290
  submerged elastic plate, 237
  submerged spherical shell, 281

Far-field conditions, 60
Finite beams

Finite beams (cont.)
  characteristic equation, 207
  natural frequencies, 208
  orthogonality of characteristic functions,
    209, 304
Flexural vibrations of thick plates, Timoshenko-Mindlin plate theory, 214
Flexural vibrations of thin plates
  boundary conditions, 213
  classical theory, 210
  drive point impedance, infinite plate, 212
  finite plate mobility, 214
  flexural wave number, 211
  natural frequencies, 214
  propagating wave, 212
  scattering form discontinuities, 255
Force applied to fluid, 165
  related to free-floating shell, 305
Fourier series
  in space, 119, 152, 208
  in time, 22
Fourier transform
  beam vibration, use in, 205
  solution of acoustic problems
    finite cylinders, 173
    infinite strip, 135
    rectangular sources, 112
Fraunhofer zone, definition, 61
Fredholm integral equation, 9, 83
Free-floating cylinder, 304
Fresnel near-field, 61
Fresnel zone, 332
Fresnel's theory of diffraction, 84

Gas bubble, pulsations of, 67
Gauss's integral theorem, 78
Green's function, acoustical, 8
  free-space, 79
  obeying Neumann boundary conditions, 86
  for spherical radiators, 154
Green's integral theorem, 78
Group velocity, 201

Hankel function, cylindrical
  asymptotic form, 176, 432
  definition of, 167
  modified, 168
Hankel function, spherical
  asymptotic form, 158
  definition of, 155
Hankel transform
  axisymmetric sources, 94, 100

in plate vibrations, in vacuo, 211
in sound field of plate, 245
in submerged plate vibrations, 242
Harmonic time dependence, 22
Heaviside step function, 93
Heckl's damping criterion, 272
Helmholtz equation
  one-dimensional, 16
  separability, 87
  in spherical coordinates, 77
Helmholtz integral equation, 80, 82
High-frequency shell vibrations
  cylindrical, 424
  spherical, 398
High-frequency sound diffraction
  cylindrical baffle, 408
  spherical baffle, 388
High-frequency sound radiation
  by cylindrical shells, 429
  by spherical shells, 433
Hypergeometric function, 389

Image source, 48
Infinite beam
  impedance, 206
  nonpropagating near-field, 206
  phase velocity, 206
Influence coefficient, 5
Intensity, acoustic, 65, 187
Interaction, structure-fluid, 8

Keller, J., theory of diffraction, 12, 416

Laplace operator
  cylindrical coordinates, 41
  rectangular coordinates, 28
  spherical coordinates, 29
Legendre functions
  addition theorem, 155
  definition, 152
  integral representation, 434
  integrals of, 159
  large-order Laplace approximation, 393
Line source on cylindrical baffle, 417
Loss factor, radiation, 282, 69, 307
  structural, 250, 271, 272
Low-impedance layer
  insertion loss, 275
  thickness resonance, 276

Mechanical impedance
  drive point impedance of beam, 206

modal impedances of cylindrical shell, 222
modal impedances of spherical shell, 233
of plate, 212
Mobility, definition, 6
of cylindrical shell, 227

Natural frequencies
of beams, 208
gas bubble, 68
of plates, 214
shells in vacuo
cylindrical, 219, 226, 223
spherical, 232
spherical shell, fluid-filled, 287
submerged shells
cylindrical, 291, 292, 301
spherical, 282
Near-field
of circular piston, 99
transition from, 62
Newton's isothermal sound velocity, 21
Normal modes
coupled by fluid loading, 267, 297, 310
Rayleigh's criterion, 309

Orthogonality conditions
Legendre functions, 157
plate normal modes, 214
submerged structure, 294, 309
trigonometric functions, 209, 157

Peak pressures
radiated by efficiently radiating modes,
268
radiated by resonant modes, 270
Phase velocity
of creeping waves, 395, 415
flexural waves, 200
high-frequency limit, 401
in submerged plate, 239
Piston radiator
circular in plate baffle, 95
circular in spherical baffle, 159
rectangular in cylindrical baffle, 177
rectangular in plane buffle, 111
Plane waves
asymptotic short-wavelength solution, 85
differential equation, 20
impedance, 21
steady-state solution, 25
Plates, submerged
with line discontinuity, 255

phase velocity of flexural wave, 239
point-excited, 240
power radiated by, 262
pressure field radiated by, 251, 258
transition to infinite plate
of cylindrical shells, 433
of finite plates, 270
of spherical shells, 435
Point source
on elastic baffles
curved, 376
planar, 373
near planar boundary
pressure-release, 52
rigid, 48
on rigid cylinder, 177, 408
on rigid sphere, 160, 388
in space, 45
Power, acoustic
affected by structural damping, 307, 357
by arbitrary source, 66
by cylindrical source, 182
by elastic plate, 262
radiated by plate, 262
by spherical source, 65
Pressure field radiated by elastic structures
finite cylindrical shell, 302, 305
finite plate, 270
high-frequency formulation, 429
infinite elastic plate, 240, 258
by resonant modes
of cylindrical shells, 301, 305
of plates, 271
of spherical shells, 286, 357
spherical shell, 285, 435

Radiation loading. See also Specific acoustic
reactance; Specific acoustic resistance
of circular piston in plane baffle, 109
of circular piston on spherical baffle, 163
of cylindrical shell, 299
effect on mode shapes, 282
of finite cylinder, 298
of rectangular plates, 142
spherical radiator, 162
of spherical shell modes, 282
Rayleigh scattering, 320, 262, 366
Rayleigh wave, 205
Rayleigh's formula
applied to circular piston, 96
applied to rectangular sources, 112
definition, 89

Rayleigh's formula (cont.)
  equivalent to stationary-phase integration,
    117
Reciprocity principle
  Helmholtz, 338
  Rayleigh, 352, 360, 377
Reference sound pressure, 17
Reflected field
  elastic plate, 344
  "pressure-release" boundary, 52
  rigid boundary, 48
Residue theory, 205, 212, 390
Resonant modes. See Pressure field radiated
    by elastic structures

Scattered pressure
  bodies of revolution, 326
  cylindrical approximation, 376
  cylindrical shell, 369
  finite cylinder, 323, 329
  infinite cylinder, 321
  Kirchhoff scattering, 329
    geometric acoustics, 332
    physical acoustics, 331
    rectangular baffle, 334
    pressure-release sphere, 360
    Rayleigh scatterers, 362, 366, 320
    resonant modes, 355, 357, 371
    rigid sphere, 318
Scattering of flexural waves, 255
Sommerfeld radiation condition, 83
Sound pressure level, 16
  generated by concentrated force, 165
Sound velocity
  in fluids, 20
  in gases, 21
Specific acoustic reactance
  of bodies of revolution, 192
  of circular piston, 106
  definition, 27
  of infinite cylindrical radiator, 169, 171
  of infinite planar radiator, 134
  of pulsating sphere, 33
  of spherical radiator, 162
Specific acoustic resistance
  of circular piston, 106
  definition, 27
  of finite cylindrical radiator, 182, 301
  of infinite cylindrical radiator, 172
  of infinite plane radiators, 132
  of pulsating sphere, 32
  of rectangular radiator, 142

of spherical radiator, 162
Spherical shell vibrations
  fluid-filled, 287
  high-frequency response, 397
  submerged, 284
  in vacuo, 230
Spreading loss
  anomalies, 64
  spherical, 31
Stationary-phase integration
  equivalent to Rayleigh's formula, 117
  Fourier transform, 115, 414
  Hankel transform, 102
Stiffening frames, effect on submerged
    plates, 255
Structural damping, 249, 307, 250
  effect on radiated pressure, 249, 308, 357,
    375
Struve function, 106
Surface mode, 123, 268

Timoshenko-Mindlin plate theory, 215
  pressure radiated by point-excited plate,
    246, 248

Vibrations of submerged structure. See also
    Plates, submerged
  cylindrical shells, 289
  spherical shells, 280

Watson's transformation, 12, 388
Wave equation
  one-dimensional, 20
  three-dimensional, 29
Wave harmonics, 87
Wave number, acoustic, 24
  flexural, 200, 211
Waveguides, elastic, 40, 378
Wavelength, acoustic, 25
Weber function, 331
Wronskian
  cylindrical harmonics, 169
  spherical harmonics, 156

thickness does one talk

...ce waves rather than plate

...etc are plate waves received

...a function of plate thickness?

3) What is the flexural wave speed for
thick plates.